Electrochemical Science: Advanced Principles and Applications

Electrochemical Science: Advanced Principles and Applications

Edited by
Harper Newman

Willford Press
www.willfordpress.com

Published by Willford Press,
118-35 Queens Blvd., Suite 400,
Forest Hills, NY 11375, USA

ISBN: 978-1-68285-622-2

Cataloging-in-Publication Data

Electrochemical science : advanced principles and applications / edited by Harper Newman.
p. cm.
Includes bibliographical references and index.
ISBN 978-1-68285-622-2
1. Electrochemistry. 2. Electrochemistry, Industrial. I. Newman, Harper.
QD553 .E44 2019
541.37--dc23

For information on all Willford Press publications
visit our website at www.willfordpress.com

Willford Press

Printed in China.

Contents

Preface

This book has been a concerted effort by a group of academicians, researchers and scientists, who have contributed their research works for the realization of the book. This book has materialized in the wake of emerging advancements and innovations in this field. Therefore, the need of the hour was to compile all the required researches and disseminate the knowledge to a broad spectrum of people comprising of students, researchers and specialists of the field.

Electrochemistry is a branch of physical chemistry that deals with the study of the interaction of electrical energy and chemical change. It studies electric current either as a cause or effect of a chemical change. It also studies reactions like electrolysis, electrochemical reactions and redox reactions. The most important practical application of electrochemical science is in the construction of battery. Other applications include coating of materials and various quality tests in the food industry. Emerging technologies like fuel cells, large format lithium-ion batteries and super-capacitors are being used commercially. The ever-growing need for advanced technology is the reason that has fueled the research in the field of electrochemical energy in recent times. This book is a compilation of chapters that discuss the most vital concepts and emerging trends in this field. It attempts to assist those with a goal of delving into the domain of electrochemical science.

At the end of the preface, I would like to thank the authors for their brilliant chapters and the publisher for guiding us all-through the making of the book till its final stage. Also, I would like to thank my family for providing the support and encouragement throughout my academic career and research projects.

Editor

1

Square wave and differential pulse voltammetric methods for the analysis of olivetol at gold electrode

Jyoti T. Bagalkoti, Vijay P. Pattar and Sharanappa T. Nandibewoor✉

P.G. Department of Studies in Chemistry, Karnatak University, Dharwad 580 003, India

✉Corresponding author: stnandibewoor@yahoo.com

Abstract

Few electrochemical techniques have been employed for the determination of the powerful antitumor agent olivetol (OLV) in real samples at gold electrode. An intense and well pronounced oxidation peak was obtained at 1.04 V in the phosphate buffer pH 5.0 as the supporting electrolyte. The effects of pH and scan rate on the oxidation peak were studied. The electrochemical behavior of OLV was investigated using cyclic voltammetric (CV), square wave voltammetric (SWV) and differential pulse voltammetric (DPV) techniques. In DPV and SWV, the gold electrode showed a good sensitivity for OLV in a linear range of 0.1-1.5 μM and 0.1-1.3 μM and detection limits of 1.936×10^{-9} M and 4.754×10^{-9} M, respectively. A plausible mechanism involving an adsorption controlled oxidation reaction was deduced. The effect of various excipients was also studied and the method was successfully applied for the determination of OLV in human biological samples.

Keywords

Olivetol, electrooxidation, voltammetric techniques, gold electrode

Introduction

Olivetol (OLV) is one of the naturally occurring alkyl resorcinol and is also known as 5-pentylresorcinol or 5-pentyl-1,3-benzenediol. OLV is benzendiol with an odd numbered carbon chain attached to 5th carbon of the benzene ring, as shown in **Scheme 1.**

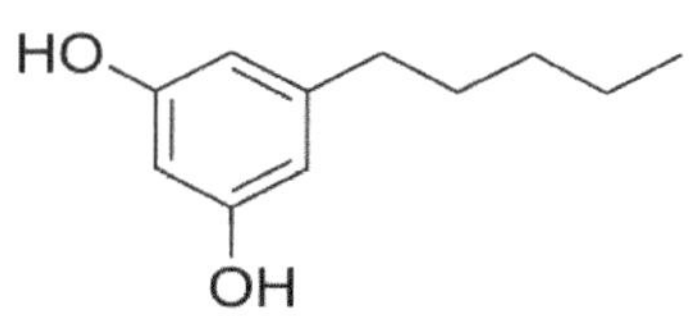

Scheme 1. *Structure of Olivetol*

OLV shows an antibiotic behavior [1] and is known to reduce risks of cancer, cardiovascular disease, diabetes and other diseases associated with ageing. OLV acts as an anti-tumor against KB cells that are reported to contain human papilloma viruses closely related to human genital, cervical and oropharyngeal cancers. The cytotoxic potency of OLV is measured, as it produced a CD_{50} (concentration of a compound required for 50 % cytotoxic dose of cell growth) of 105.10 μM in 3-(4,5-dimethylthiazol-2-yl)-2,5-diphenyl-2H-tetrazolium bromide (MTT) assay and 116.69 μM in sulforhodamine B protein (SRB) assay [2]. Several techniques like HPLC [3], gas chromatography (GC) [4] and colorimetry [4,5] have already been used for the determination of alkyl resorcinols, but have proved, however, to be quite expensive and time consuming. Therefore, a sensitive, accurate, cost effective and less time consuming technique is required. Previously, we have studied the electrochemical property of OLV using cyclic voltammetry and differential pulse voltammetry (DPV) with the help of pencil graphite (PGE), glassy carbon (GCE) and carbon paste electrodes (CPE) [6]. In continuation of our previous work on the determination of OLV, another analytical method for the determination of OLV is proposed in the present study. The objective of the study is to find suitable experimental conditions and to investigate the oxidation mechanism of OLV on the gold electrode using square wave voltammetry (SWV), differential pulse voltammetry (DPV) and cyclic voltammetric (CV) techniques. The gold electrode has been chosen because of its stability, wide potential window and fast rate of electron transfer. The method has been proved to be effective even at lower concentrations and also applied for the determination OLV in human fluid samples and in investigation of pharmacokinetics.

Materials and methods

Reagents and chemicals

95 % pure OLV was purchased from Sigma-Aldrich. 1.0 mM stock solution of OLV was prepared using doubly distilled water and stored in the dark and cool place. 0.2 M phosphate buffer stock solutions were prepared within the pH range of 3.0–10.0 using 98.5 % pure H_3PO_4, Na_2HPO_4, Na_3PO_4 and KH_2PO_4 [7,8] and double distilled water. Authentic and analytical grade reagents and doubly distilled water were used in preparation of all solutions.

Instrumentation

CHI630 electrochemical analyzer (CH Instruments Inc., USA) was used in the present electrochemical study. The voltammetric experiments were carried out in a conventional three-electrode system containing the working gold electrode, the platinum wire auxiliary electrode and the Ag/AgCl/KCl reference electrode. pH measurements were made using an Elico LI 120 pH meter (Elico Ltd., India).

Working electrode

The working electrode used here is the Au disk of 2 mm in diameter that was placed at the bottom of the three-electrode cell. Since some diminishing of sensitivity of gold electrode due to poisoning of the electrode surface has frequently been observed with successive usage, a pretreatment of the electrode prior to every use is needed to acquire reproducible results. With this aim and prior to each experiment, the Au working electrode was rinsed with double distilled water and polished to get a mirror-like finish using 0.3 μm alumina powder on a smooth polishing cloth or polishing pads and rinsed again with double distilled water and dried. Experiments were conducted at room temperature, 25 ± 0.2 °C.

Area of the working electrode

The area of the gold working electrode was determined by variation of the scan rate in cyclic voltammetric experiments in 0.1 M KCl containing 0.001 M potassium ferricyanide. The effect of the scan rate on the peak current of the ($Fe(CN)_6^{3-/4-}$) redox reaction is given by the well-known Randles–Sevcik equation [9]:

$$I_{pa} = 0.4463\ (F^3/RT)^{1/2} n^{3/2} A D_O{}^{1/2} C_O\ \upsilon^{1/2} \quad (1)$$

In eq. (1), I_{pa} is the anodic peak current, n represents the number electrons in the reaction, ν is is the scan rate (V/s), A is the electrode area (cm^2), C_o is the concentration of the probe molecule in the bulk solution (mol/cm^3), while D_o is its diffusion coefficient (7.6×10^{-6} cm^2 s^{-1} [9]). Using eq. (1) and by knowing the constants, the surface area of gold electrode was calculated to be 0.399 cm^2.

Experimental procedure

Before the beginning of the experiment, all three electrodes were rinsed using double distilled water and then connected to the electrochemical work station. pH was determined within a pH range of 3–10. The working parameters for CV experiments were chosen as: initial potential –0.4 V; final potential 1.6 V; scan rate 0.1 V/s; sensitivity 1×10^{-4} A/V. Effect of the scan rate change was studied at pH 5.0 at the same working parameter values. The working parameters for SWV experiments were chosen as: initial potential 0.4 V; final potential 0.9 V; sensitivity 1×10^{-5} A/V, while for DPV experiments the working parameters were chosen as: initial potential 0.4 V; final potential 1.4 V and sensitivity 1×10^{-5} A/V. All the working parameters were set and reviewed thoroughly before commencing the experiment.

Results and discussion

Cyclic voltammetric behavior of olivetol

Cyclic voltammetric behavior of olivetol (OLV) at the gold electrode was studied in 0.2 M phosphate buffer, pH 5.0 solution as the supporting electrolyte. The cyclic voltammograms were measured for the bare gold electrode in either presence (Fig. 1(a)) or absence (Fig. 1(b)) of OLV. Fig. 1(a) shows one anodic peak at 1.041 V. On scanning in the negative direction, a reduction peak at about 0.5 V was observed as previously [10]. This peak can be ascribed to the reduction of gold oxides and will not be considered further.

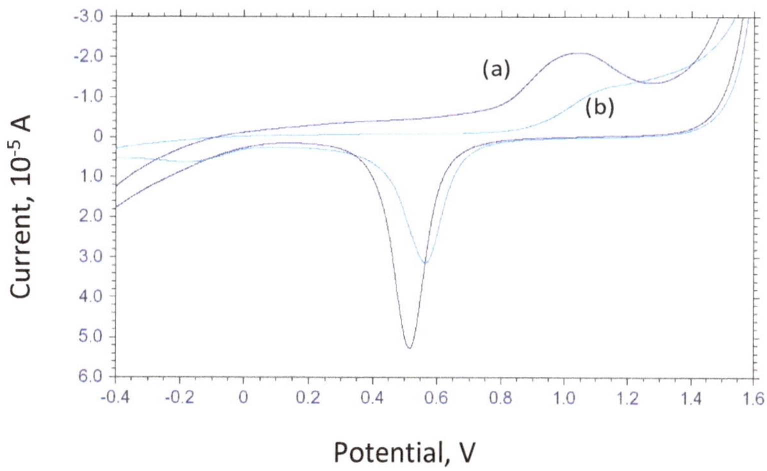

Fig. 1. *Cyclic voltammogram of gold electrode in phosphate buffer pH 5.0 solution, recorded at scan rate of 0.1 V s^{-1} in: (a) presence of 0.1 mM OLV; (b) absence of OLV*

In CV experiments, the voltammograms obtained from the first run were always considered. This was due to apparent decrease in oxidative peak current values with successive sweeps which is usually ascribed to adsorption of oxidative products over the electrode surface [11].

Effect of variation of pH

Effect of pH change on the electrooxidation of OLV at gold electrode is shown in Fig. 2. It can be observed in Fig. 2(a) that the peak potential values are displaced towards more negative values as the pH was increased from 3.0 to 10.0 (curves (i-viii)). The highest peak current was observed for pH 5.0 and so all further experiments were carried at this pH value. The graph of E_{pa} versus pH drawn in Fig. 2(b) showed a linear relationship with a slope of 58 $mVpH^{-1}$ and its linear regression equation is given by:

$E_{pa} = 1.299 - 0.058$ pH ($r^2 = 0.874$).

The slope value of 58 mV pH^{-1} is close to the theoretical value of 59 mV pH^{-1} [12,13], indicating transfer of equal number of protons and electrons involved in the electrochemical reaction.

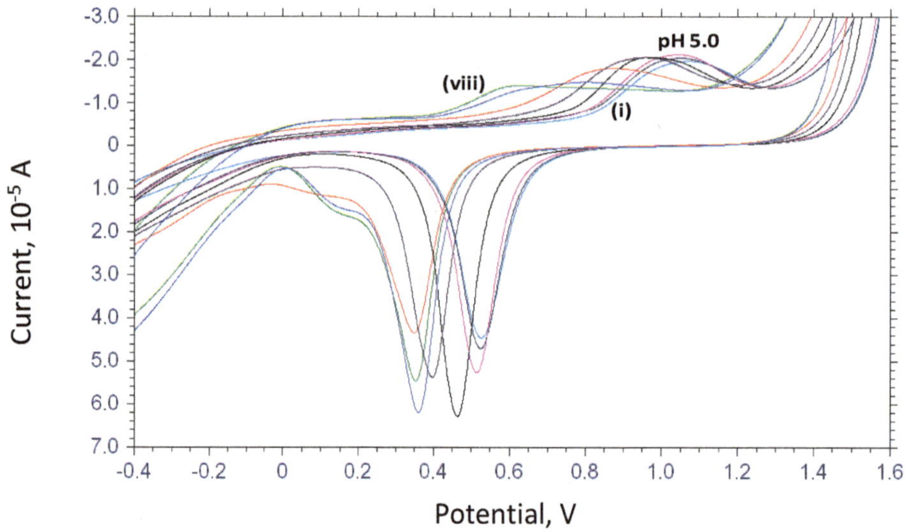

Fig. 2(a). *Cyclic voltammograms of gold electrode for 0.1 mM OLV in the phosphate buffer solution recorded at scan rate of 0.1 Vs^{-1} and pH 3.0-10.0 (i-viii)*

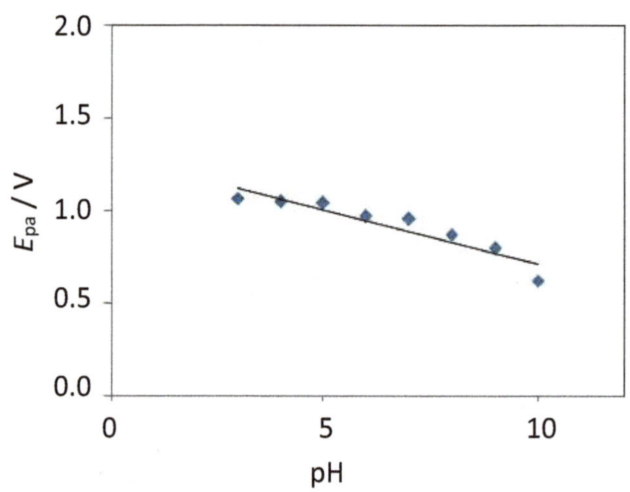

Fig. 2(b). *Peak potential* E_{pa} vs. *pH dependence*

Influence of scan rate

Effects of variation of the scan rate over domain of 0.1 – 0.5 V/s on CVs of gold electrode in presence of 1.0 mM OLV are shown in Fig. 3. The shapes of the voltammograms are obviously dependent on the applied scan rates. When the scan rates were varied in the ascending order (from 0.1-0.5 V/s), there was a steady increase in peak current values, showing a sensitivity of 1.0×10^{-4} A/V (Fig. 3(a)).

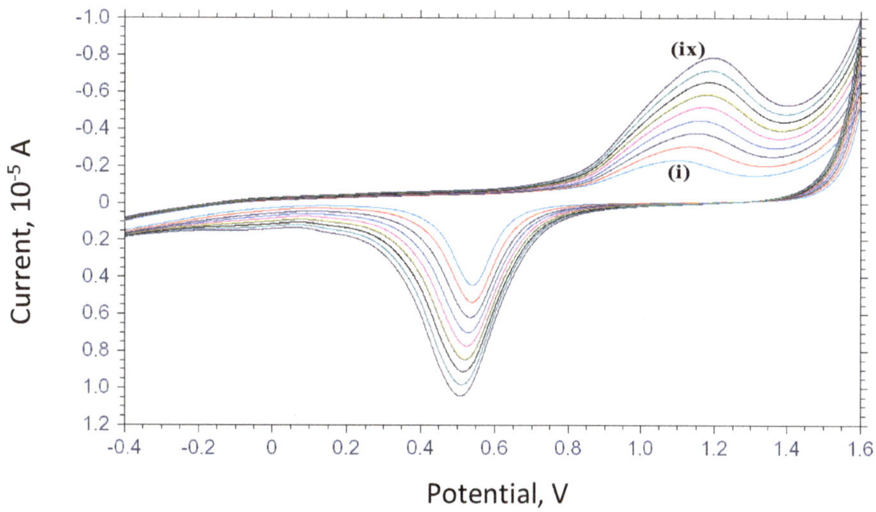

Fig. 3. (a). *Cyclic voltammograms of gold electrode in phosphate buffer, pH 5.0 solution containing 0.1 mM OLV at different scan rates (0.1-0.5 Vs^{-1})*

The peak current versus scan rate drawn in Fig. 3(b) showed a linear dependence, indicating adsorption regime. The straight line equation is given by:

$I_{pa} = 12.73\ \nu + 0.858\ (r^2=0.999)$.

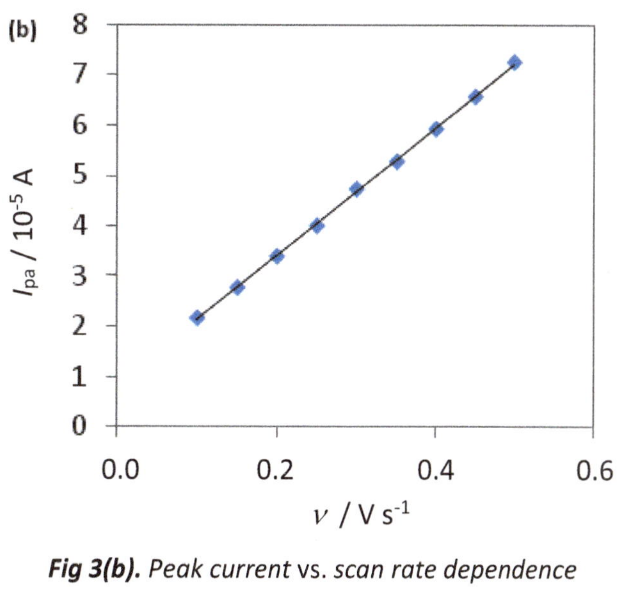

Fig 3(b). *Peak current* vs. *scan rate dependence*

From the graph of log of the I_{pa} *vs.* log of the scan rate given in Fig. 3(c), a straight line was obtained with a slope value of 0.760. This value is relatively close to 1.0 that is the theoretical value for the adsorption controlled electrode process. The obtained equation is given by:

$\log I_{pa} = 0.760 \log \nu + 1.075\ (r^2=0.995)$.

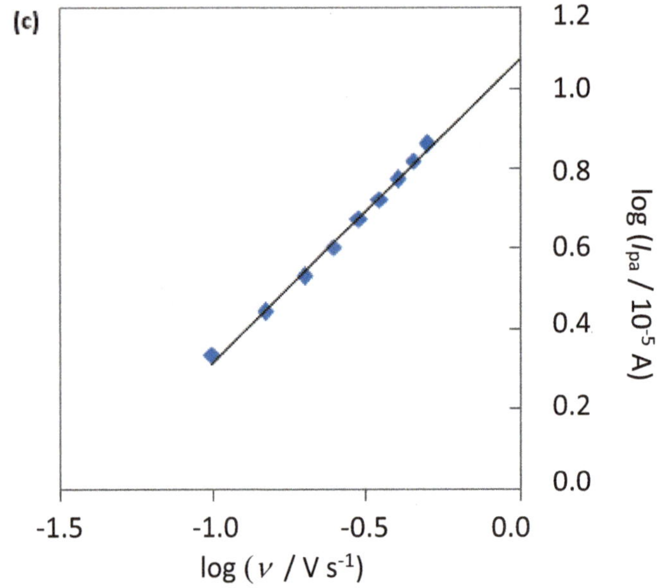

Fig. 3(c) *log of peak current vs. log of scan rate dependence*

Two more graphs showing E_{pa} *vs.* ν (Fig. 3(d)) and E_{pa} vs. log ν (Fig. 3(e)) were also drawn, which gave slope values of 0.225 and 0.137, respectively. With the reference to these graphs, it was observed that with the increase of the scan rate, the oxidation potential shifted towards more positive values. The dependence of E_p on log log ν is expressed as:

E_{pa} = 0.137 log ν + 1.242.

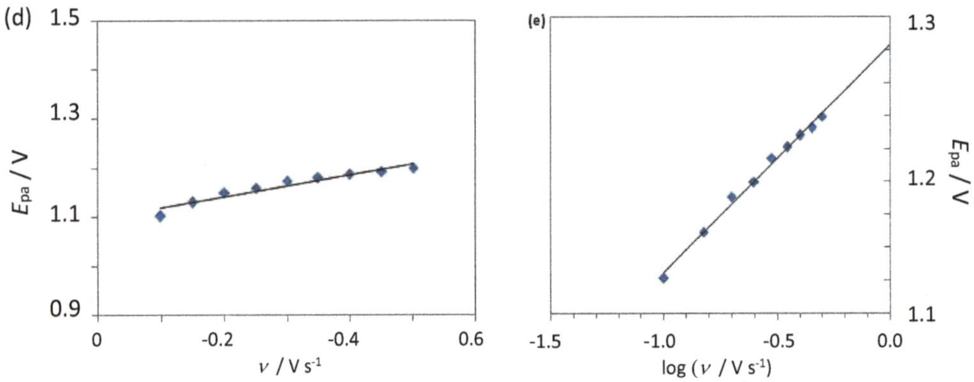

Fig 3(d). *Peak potential* (E_{pa}) *vs. scan rate* (ν). ***(e)*** *Peak potential* (E_{pa}) vs. *log of scan rate (log* ν*).*

In order to study the adsorption-controlled irreversible process at the electrode, Laviron [14] suggested the following equation

$$E_{pa} = E^0 + \left(\frac{2.303RT}{\alpha nF}\right)\log\left(\frac{RT\,k^0}{\alpha nF}\right) + \left(\frac{2.303RT}{\alpha nF}\right)\log\nu \tag{2}$$

In eq. (2), E^o is the formal redox potential, *R* is the universal gas constant (8.314 J mol^{-1} K^{-1}), T is room temperature (298.15 K), α is electron transfer coefficient, *n* is the number of transferred electrons, *F* is the Faraday constant (96 480 C mol^{-1}), k^o is the standard rate constant of the surface reaction and ν is the scan rate. From the experimental results obtained for the linear graph shown in Fig 3(e), the value of $\alpha \times n$ was determined from its slope (m = 0.137). From the eq. (2) we have:

$$m = \frac{2.303RT}{\alpha nF} \tag{3}$$

α was calculated from the Bard and Faulkner's [15] equation:

$$\alpha/\text{mV} = \frac{47.7}{E_{\text{pa}} - E_{\text{pa/2}}} \tag{4}$$

In eq. (4), $E_{\text{pa/2}}$ is the potential measured at the half of the peak current value and hence α = 0.206. The calculated number of electrons was 2.095 ≈ 2. By extending the line in the graph of E_{pa} *vs.* ν to the intercept where υ =0 [16], we get E^0= 1.096 V which was used to calculate the k° value using eq. (2):

$$C = E^0 + \left(\frac{2.303RT}{\alpha nF}\right)\log\left(\frac{RTk^0}{\alpha nF}\right) \tag{5}$$

In eq. (5), C is the intercept value obtained from the graph of E_{pa} *vs.* log v. From the above equation, it was possible to determine the value of k° as 1.95×10^2 s^{-1}.

Mechanism

From the previous results, the number of electrons involved in the reaction was found to be equal to the number of protons and calculated to be equal to 2. Based on the earlier literature [6], a probable mechanism was proposed according to the Scheme 2.

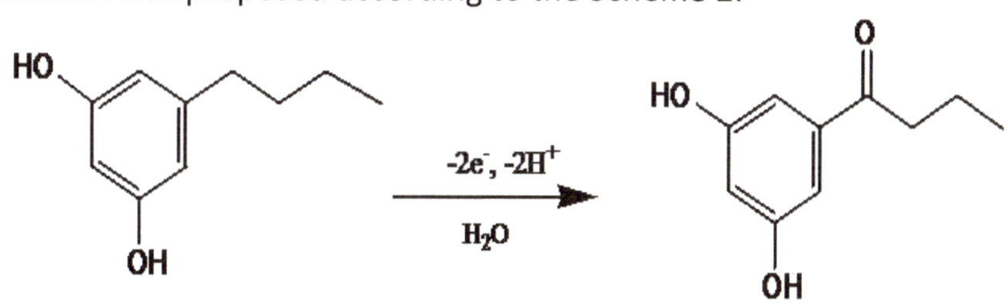

***Scheme 2**. Probable mechanism for the electrooxidation of OLV*

Calibration curve and reproducibility

SWV and DPV voltammograms recorded between 400–850 mV at various concentrations of OLV are presented in Fig. 4.

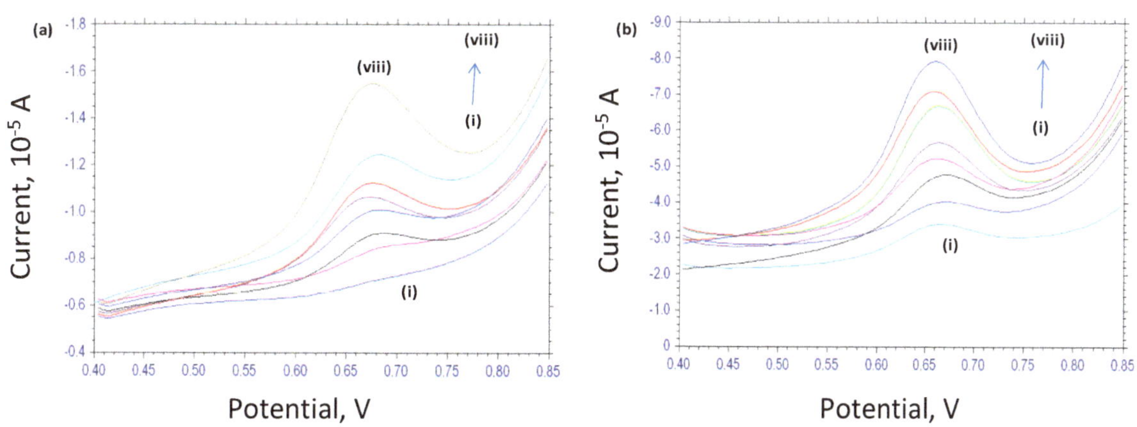

***Fig 4. (a)** Square wave voltammograms of gold electrode recorded at various concentrations of OLV (i-viii): 0.1, 0.3, 0.5, 0.7, 0.9, 1.1, 1.3, 1.5 μM; **(b)** Differential pulse voltammograms of gold electrode recorded at various concentrations of OLV (i-viii): 0.1, 0.3, 0.5, 0.7, 1.0, 1.3, 1.5, 1.7 μM*

To check the precision of the method, five experiments were carried out on the same day at the same standard conditions which were authenticated by plotting the graphs of peak current values (I_{pa}) versus concentration of OLV, as is shown in Fig. 5. The graphs gave linear calibration curves in the concentration range of 0.1–1.3 μM (Fig 5(a)) and 0.1-1.5 μM (Fig 5(b)), respectively. A slight deviation in the calibration curve was observed as the [OLV] was increased, which is due to adsorption behaviour of OLV or its oxidation products on the electrode surface. Characteristics of the calibration plot are given in Table 1. The Limit of Detection (LOD) and the Limit of Quantification (LOQ) values shown in the Table 1 were calculated using the following two equations:

$$\text{LOD} = 3\, s/m \quad \text{and} \quad \text{LOQ} = 10\, s/m. \tag{6}$$

In eq. (6), s is standard deviation of the intercept and m is the slope of calibration curve, respectively.

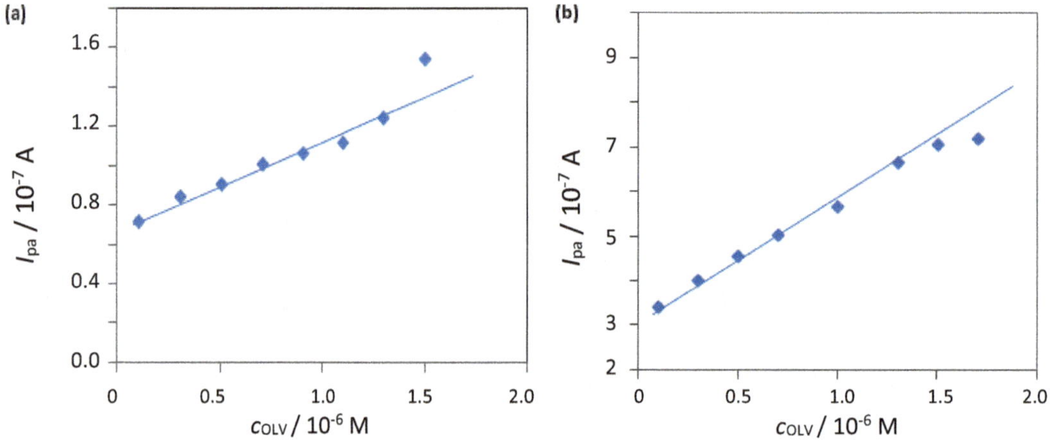

Fig. 5. ***(a)*** *Calibration plot (I_{pa} vs. concentration of OLV) obtained using SWV;* ***(b)*** *Calibration plot (I_{pa} vs. concentration of OLV) obtained using DPV*

Table 1*. Characteristics of OLV calibration plot obtained using SWV and DPV techniques at gold electrode.*

	SWV	DPV
Linearity, μM	0.1-1.3	0.1-1.5
Slope	0.307	0.408
Intercept	3.105	0.701
Correlation coefficient	0.990	0.986
RSD, % (slope)	0.569	0.856
RSD, % (intercept)	0.357	0.599
LOD, M	4.754×10^{-9}	1.936×10^{-9}
LOQ, M	1.585×10^{-8}	6.453×10^{-9}

Effect of excipients

For probable analytical applications, the effect of certain common excipients used in pharmaceutical preparations on OLV determination has to be been checked. Here, effects of different and potential excipients were studied using the DPV technique and 1.0 μM OLV in the presence of 0.1 mM of the chosen excipient. The results are listed in Table 2. The results shown in Table 2 suggest that presence of even hundred-fold excess concentration of an excipient like glucose, citric acid etc., along with certain ions like Mg^{2+}, K^{+}, Cu^{2+} *etc.*, does not interfere much. Thus, the present technique can be used to assay OLV in the presence of various excipients.

Table 2. Influence of various potential excipients on the DPV response for 1.0 μM OLV.

Interferents, 0.1 mM	E_{pa} / V	RSD, %
Oxalic acid	0.7028	-0.26
Gum acacia	0.7001	0.01
Lactose	0.6969	0.33
Glucose	0.6938	0.64
Starch	0.6880	1.22
Citric acid	0.6854	1.48
Sucrose	0.6798	2.04
$MgSO_4$	0.6850	1.52
KCl	0.6806	1.96
$CuCl_2$	0.6519	4.83
Sodium citrate	0.6850	1.52

Effect of surfactants

It is well known that surfactants affect strongly the electrode processes even if they are present in trace quantities [17]. Surfactants prove to be effective in the electroanalysis of biological compounds and drugs [18]. It was also shown that anionic surfactants could be used to improve the accumulation of some electroactive organic molecules [19]. Cationic and anionic surfactants with varying hydrocarbon chains affect the redox behaviour of electroactive species and bring some irregularities in the voltammograms [20,21]. In present study, the DPV technique has been adapted to study the effect of surfactants. Addition of the cationic surfactant, cetyltrimethyl ammonium bromide (CTAB), to the solution did not have any effect on the peak potential value, but the peak current decreased with increase of the concentration of CTAB that is accumulated at the interface. The repulsion between the hydrophobic end of the CTAB and water phase reduces the peak current as the concentration of CTAB is increased [22]. The anionic surfactant, sodium dodecyl sulfate (SDS), influenced the voltammetric behaviour in an opposite way *i.e.* the peak potentials remained the same but the peak current increased with the increase in concentration of SDS. In both cases only the peak current of OLV showed the surfactant concentration-dependent behaviour.

Detection of OLV in urine samples

The most striking feature of the method applied to urine samples is that not any prior extraction step is needed. Urine samples from a healthy human volunteer were obtained for the study and the DPV method for the determination of OLV was applied. The urine samples were diluted 100 times using buffer solution before the analysis and not any pre-treatment was made. The recoveries were measured by spiking urine samples with known concentration of drug. The recovery results for three samples are listed in Table 3, together with standard deviations (SD) and relative standard deviations (RSD) values. It is shown in Table 3 that the percentage recovery was in the range of 95–101 %

***Table 3**: Determination of OLV in urine sample*

OLV added, μM	OLV found[(a)], μM	Recovery, %	SD ± RSD, %	Bias, %
0.2	0.19	95.00	0.0016±0.82	0.050
0.6	0.61	101.66	0.0017±0.27	0.017
0.8	0.81	101.25	0.0019±0.24	-0.012

(a) Average of five determinations

Application of pharmacokinetics studies

Clinical applications of pharmacokinetics have resulted in significant improvement in the drug therapy. Pharmacokinetic properties of a particular drug is important to determine the route of administration, dose, onset of action, peak action time, duration of action and frequency of dosing. Pharmacokinetics studies help primarily in enhancing efficacy and decreasing toxicity during a drug therapy. An investigation of effects of various concentrations of drug at the site of action has always been of importance. Direct measurement of drug concentrations at the receptor sites is hardly practicable as the receptor sites might be distributed all over the body. Concentration of drug in blood or plasma, urine, saliva and other easily sampled fluids can, however, be measured. The results suggest that the disposition of OLV was comfortable to a one compartment open model [23]. The rate constant of elimination (k_e) in a one compartment system is obtained from the plot of concentration of OLV versus time (Fig.6). Using this 'k_e' value, half life of the drug was calculated which helps in understanding the time interval between the doses and is very important in design of infusion system.

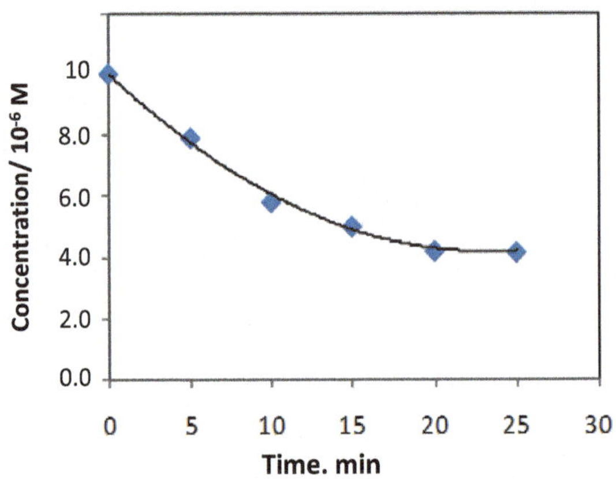

Fig 6. *Concentration of OLV* vs. *time function in spiked urine sample*

Table 4 presents the peak current values of 10 μM OLV recorded at different times (0–25 minutes), and the values of the elimination rate constant and half-life of the drug. The analytical data obtained suggest that the method is sensitive, reproducible and precise for the determination of OLV.

Table 4. *Response of peak current of 10μM OLV in urine sample at different time intervals*

Time, min	Peak current, 10^{-7} A	Concentration of OLV, 10^{-6} M
0	9.58	10.00
5	7.59	7.91
10	5.59	5.82
15	4.82	5.03
20	4.06	4.23
25	4.02	4.19
Elimination rate constant (k_e), h^{-1}	1.12	
Half-life of drug, h	0.67	

Detection of OLV in serum samples

DPV was also applied for the determination of OLV in spiked human serum plasma sample. The recoveries were measured by spiking drug free serum plasma with known amounts of OLV. A quantitative study was made by adding known concentration of standard drug solution in the system of plasma sample which was determined successfully. The results of determination obtained for three plasma samples with recovery and RSD are listed in Table 5.

Table 5: *Determination of OLV in human blood serum*

Added, µM	Found[a], µM	Recovery, %	SD± RSD, %	Bias, %
0.35	0.34	97.14	0.0019±0.72	0.028
0.55	0.56	101.18	0.0011±0.32	-0.018
0.75	0.73	97.33	0.0021±0.18	0.026

(a) Average of five determinations

Conclusion

The voltammetric oxidation of OLV at gold electrode was successfully studied using phosphate buffer, pH 5.0 solution. The oxidation reaction is adsorption controlled and irreversible process. It was found that OLV undergoes a transfer of two electrons and two protons and a plausible mechanism was proposed. A linear relation between peak current and concentration of drug was obtained within a certain range and even 4.7 nM can be used to assay the drug in human urine and blood serum samples. This work has also provided a sensitive and selective method for investigation of the pharmacokinetics of OLV and its possible utilization in the clinical practice.

Acknowledgement: *The authors thank University Grants Commission, New Delhi for the award of UGC-BSR faculty fellowship to Dr. S.T.Nandibewoor, and BSR fellowship to Jyoti T.Bagalkoti.*

References

[1] P. Lopez, G. Ferrao, *Phytotherapy Research* **25** (2011) 271–276

[2] S. H. Baek, Y. O. Kim, J. S. Kwag, K. E. Choi, W. YJung, D. S. Han, *Archives of Pharmacal Research* **21** (1998) 353-356.

[3] W. J. Mullin, J. P. H. Emery. *Journal of Agricultural and Food Chemistry* **40** (1992) 2127-2130.

[4] R. Landberg , A. A. M. Andersson, P. Åman, A. Kamal-Eldin, *Food Chemistry* **113** (2009) 1363-1369.

[5] W. Mejbaum-Katzenllenbogen, F. Tluscik, A. Kozubek, A. Sikorski, Z .Maresz, *Acta Societatis Botanicorum Poloniae* **44** (1975) 479-487.

[6] D. G. Patil, V. P. Pattar, S. T. Nandibewoor, *Physical Chemistry Communications* **4** (2017) 10-23.

[7] J. I. Gowda, S. T. Nandibewoor, *Asian Journal of Pharmaceutical Sciences* ***9*** (2014), 42-49.

[8] G. D. Christian, W. C. Purdy, *Journal of Electroanalytical Chemistry* **3** (1962) 363-367.

[9] B. Rezaei, S. Damiri, *Sensors and Actuators B* **134** (2008) 324-331.

[10] C. Barus, P. Gros, M. Comtat, S. Daunes-Marion, R. Tarroux, *Electrochimica Acta* **52** (2007) 7978-7985.

[11] S. S. Kalanur, J. Seetharamappa, G. P. Mamatha, M. D. Hadagali, P. B. Kandagal, *International Journal of Electrochemical Science* **3** (2008) 756-767.

[12] I. Martins, F.C.Cristiani, S.C. Larissa, A. Francisco, M.C. Letícia, R. Susanne, *Talanta* **85** (2011) 1-7.

[13] T. Łuczak, *Electrochimica Acta* **53** (2008) 5725-5731.
[14] E. Laviron, *Journal of Electroanalytical Chemistry* **101** (1979) 19-28.
[15] A. J. Bard, L. R. Faulkner, *Electrochemical methods; Fundamentals and Applications*, Wiley, New York, 2nd ed, (2004), 236
[16] W. Yunhua, J. Xiaobo, H. Shengshui, *Bioelectrochemistry* **64** (2004) 91-97.
[17] R. Vittal, H. Gomathi, K.-J. Kim, *Advances in Colloid and Interface Science* **119** (2006) 55-68
[18] B. Hoyer, N. Jensen, *Electrochemistry Communications* **8** (2006) 323-328
[19] L. Huang, L. Bu, F. Zhao, B. Zeng, *Journal of Solid State Electrochemistry* **8** (2004) 976-981
[20] United Kingdom Poisoning Information Database (UKPID), *ABPI Compdendium of Data Sheets and Summaries of Products*, Characteristics Datapharm Publications Ltd. (1996–1997).
[21] J. Herovsky, J. Kuta, *Principles of Polarography*; Academic Press: New York, 1966.
[22] J. Abbar, S. T. Nandibewoor, *Journal of Critical Reviews in Analytical Chemistry* **42** (2012) 272-281
[23] J. I. Gowda, A. T. Buddanavar, S. T. Nandibewoor, *Journal of Pharmaceutical Analysis* **5** (2015) 231-238.

2

An improved method of water electrolysis – effect of complexing agent

Seetharaman Swaminathan*,**, Balaji Rengarajan*, Ramya Krishnan*,✉, Kaveripatnam Samban Dhathathreyan*, Manickam Velan**

**Centre for Fuel Cell Technology, International Advanced research Centre for Powder Metallurgy and New Materials(ARCI), II Floor, IIT-M Research Park, No.6 Kanagam road, Taramani, Chennai – 600 113, India*

***Department of Chemical Engineering, A.C. College of Technology, Anna University, Chennai-600 025, India*

✉Corresponding Author: ramya.k.krishnan@gmail.com

Abstract

The present work investigates the efficiency of an alkaline water electrolysis process in the presence of a complexing agent like citric acid (CA) when added directly into the electrolyte during the electrolytic process. High surface area nickel electrodes prepared by electrodeposition technique were used as the electrode to evaluate the efficiency of the oxygen evolution reaction (OER) by the polarization measurements and cyclic voltammetry. The quantity of the complexing agent CA in the electrolyte was varied from 0-1 wt. %. An increase in the current density of about 25 % resulted at a temperature of 30 °C in the presence of 0.2 wt. % of CA at 1.0 V vs. *Hg/HgO. CA was found to improve performance by forming a complex with the alloy electrode and by formation of the high surface area catalyst for efficient OER.*

Keywords

Citric acid; Alkaline water electrolysis; Oxygen evolution reaction; Complexing agent

---

Introduction

Clean energy is considered as the solution to world's increasing energy demand and to concerns regarding pollution and contamination. Hydrogen is important energy alternatives as its combustion product is only water. Water electrolysis is the best technology for producing hydrogen and oxygen with no resulting greenhouse gas emissions [1-3]. The energy required for production of hydrogen by electrolytic method is very high (4.5-5 kWh N m^{-3} of H_2) in most industrial electrolyzers. Alkaline water electrolysis (AWE) offers the advantages of the use of non-noble metal catalysts, ease of manufacturing and scalability. Therefore, research on suitable catalyst materials and methods for modifying the electrodes to improve efficiency of water electrolysis techniques are always being carried out. Efficiency in electrolysis can be improved by reducing the cell voltage through the use of catalysts with increased surface area or by changing the nature of the overpotential of the reaction. In AWE anodic overpotential is a major factor that limits the efficiency and hence is researched frequently [4-6]. Ni based electrodes, being less corrosive in alkaline solution, offer low cost solution despite its higher overpotential in alkaline solutions when compared to costly RuO_2 and IrO_2 based catalyst used in acidic media [7]. Further, it has been found that incorporation of zinc and sulphur in the nickel electrode increases surface area and catalytic activity, lowers overvoltage and helps in increasing the current density by removal of the gas bubbles from the electrode surface [8].

Citric acid (CA) is a chelating agent often used for depositing metals at a controlled rate in electroplating and electroless plating operations. It is also used to buffer the pH of the plating bath. The effects of additives in the electroless nickel deposition process have been studied by electrochemical analysis [9,10]. It has been shown that the additives significantly improved the deposition rate and helped in forming finer grain structure during the plating process. The importance of additives such as boric acid, citric acid and ascorbic acids in influencing the electrodeposition of nickel alloys was described by Kieling [11]. It was shown that complexing agents influence the kinetics of powder electrodeposition, as well as the morphology of the Ni powders. According to the conclusion of the authors finer powders were produced in the presence of citric acid than those obtained with oxalic acid. A study on organic additive, tri sodium citrate as a stabilizing agent reveals that they can be used for structure related factors to control the nucleation, growth and alignment of crystal phases [12,13].

Nikolić *et al.* [14] have shown that alkaline electrolytic hydrogen production can be made efficient (~15 %) by the addition of activating compounds of tungsten and cobalt in the ionic and complex form into the electrolyte (6 M KOH) during electrolytic process. Additions of ionic activators were found to increase surface area, porosity and performance due to the synergistic effect of the two different metals on the surface of the electrode. The effect of ionic activators and complexes in alkaline water electrolysis has been emphasized by improved performance in the presence of molybdate, chromate and cobalt based compounds [15,16]. The ability of citric acid to form complexes with Ni is very well established. It has been used to decrease the precipitation of Ni as hydroxides thus stabilizing it in the electrolyte. It has been used in both the electroless and electrodeposition of Ni and its alloys [17,18]. It was hence decided to study the effect of this stabilizer during the electrolysis of water by adding it to the electrolyte, *i.e.* to study whether the Ni citrate complex formed will act similar to the ionic activator and improve the efficiency of the electrolytic process. In the present system the activator is formed in situ from the electrode and the complexing agent in the solution compared to those in the literature where the metal complexes are externally added. A systematic study on OER has been carried out by varying the concentration of citric acid in the electrolyte between 0 and 1 % by weight.

Experimental

1. Materials

Nickel sulphate (Merck), nickel chloride (Merck), zinc sulphate (Merck), ammonium nickel sulfate (Merck), sodium thiocyanate (Fisher), potassium hydroxide (Merck), citric acid (Merck) and other chemicals were all used as received. Ni mesh (procured from Champion Manufacturing Co, India with 0.10×0.32×0.20 mm nominal thickness, long diagonals of the diamond and strand width respectively) of area 9.0 cm^2 size were used for preparation of the electrodes.

2. Electrode fabrication

The surface of nickel mesh was electrochemically cleaned before electroplating using the bath solution containing nickel chloride (240 g L^{-1}) and conc. HCl (125 ml L^{-1}). A plating current density of 50 mA cm^{-2} was used for cleaning. The plating bath consisting of nickel (II) sulphate (70 g l^{-1}), ammonium nickel sulfate (40 g L^{-1}), zinc sulphate (35 g L^{-1}) and sodium thiocyanate (12 g L^{-1}) were used for plating. Electrodes were prepared by the pulse electrodeposition technique as mentioned in our previous paper with a pH of 5-6, a current density of 40 mA cm^{-2} and duty cycle of 40 % under the influence of ultrasonic vibration [19]. Once the electrodeposition was concluded, the electrodes were treated with 28 g L^{-1} of KOH to leach out excess zinc to give a porous electrode having a high surface area.

3. Cell frame-up

Electrochemical measurements were carried out using a Solartron analytical potentiostat/galvanostat 1400 cell test system in a small undivided beaker cell (volume 50 cm^3). As prepared nickel alloy electrode (working electrode), large area platinum gauze (counter electrode) and Hg/HgO reference electrode in 1 M NaOH were used for the study. The area of the electrode was 9 cm^2 and 30 % KOH solution to which CA between 0 and 1 wt. % has been added was used as the electrolyte. Figure 1 shows the cell set up used for testing of the electrodes of the present study.

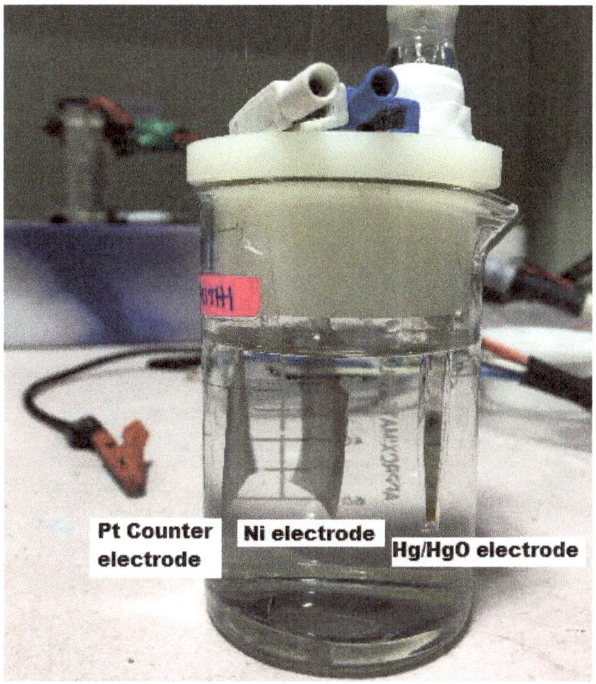

Figure 1. *Three electrode cell set up used for electrochemical characterization*

4. Electrode characterization

The surface morphology of the electrodes was studied using Hitachi SU1500 Scanning Electron Microscopy (SEM). Energy dispersive analysis of X-ray (EDAX) using Hitachi 4300 was used to study the composition of metals on the electrode. Rigakuminiflex X-Ray diffraction (XRD) with CuK_α

(λ- 0.154 nm) radiation was used for X-ray characterization of the electrodeposits. Perkin-Elmer system was used to obtain the Fourier Transform Infrared spectra (FT-IR).

Results and discussion

1. FT-IR analysis

Figure 2 shows FT-IR spectra of the nickel electrode with and without the addition of stabilizer. The bands due carboxyl groups of citric acid are present between 1800 and 1300 cm^{-1}. The band at 1721 cm^{-1} has been ascribed to free COO^- group. The band at 1630 cm^{-1} has been ascribed to the carboxylic group forming intramolecular hydrogen bonds. The band at 1425 cm^{-1} may be assigned to the bidentate carboxylate anion. For nickel alloy electrode in citric acid spectra, the band around 3450 cm^{-1} was very broad, due to hydration by water molecule indicating the electrolytes ability to wet the electrode. In spectra without citric acid due to less adsorbed water molecule this peak was small. The bands around 1150 and 650 cm^{-1} represent the stretching vibrations of the nickel sulphate used for electroplating of nickel [12]. These studies confirm the formation of nickel citrate complex on dipping the electrode in the electrolyte containing CA. Literature studies have shown that upon immersion of a nickel electrode into the solution of aqueous alkali, a film of Ni(II) oxide species (α-$Ni(OH)_2$) is formed spontaneously [20]. This on ageing dehydrates and gets converted into β–$Ni(OH)_2$. The electroprecipitation of this oxide forms Ni plaques and may lead to decrease in performance. On addition of citric acid into the electrolyte Ni forms citrate complexes that may help in cleaning the surface of the oxide layer. The complexes formed may improve the wetting ability of the electrolyte by decreasing the surface tension of the electrodes.

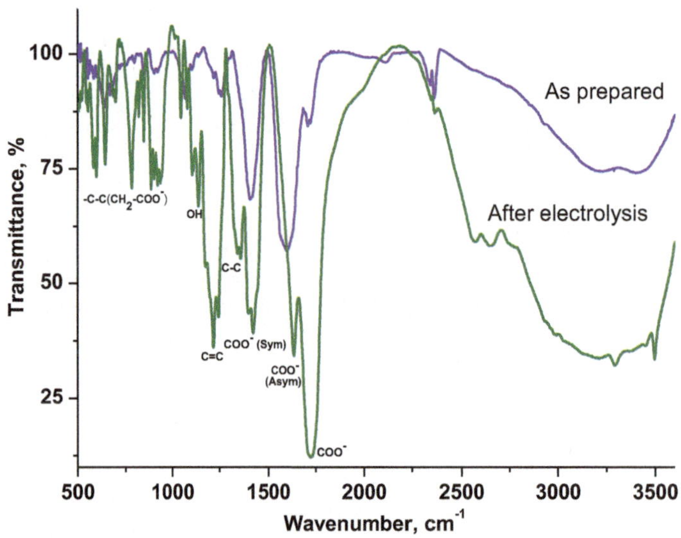

Figure 2. *FT-IR Spectra of electrolytic surface: (a) before and **(b)** after citric acid addition*

2. Effect of variation of CA

Figure 3 shows the polarization curves with the variation in the amount of CA at 30 °C. A current density of 0.40 Acm^{-2} was achieved with 0.1 wt % addition of CA compared to 0.32 A cm^{-2} achieved without the addition of CA at 1.0 V *vs.* Hg/HgO reference electrode. The maximum current density of 0.47 A cm^{-2} was obtained on the addition of 0.2 % CA at 1.0 V. Further increase in citric acid led to decrease in current density to 0.37 A cm^{-2} at 0.3 % CA addition. In alkaline solution without CA, $Ni(OH)_2$ precipitates on the surface and gets absorbed on the electrode, passivates the electrode decreasing the kinetics of the reaction. On the addition of CA, formation of $Ni(OH)_2$ on the surface decreases and a citrate complex of Ni is formed resulting in increase of current density. An increase in CA concentration leads to a strong citrate adsorption on the surface reducing the catalytic activity. Ni citrate particles present on the surface of the electrode may also act as nucleation sites for in situ deposition of metal thereby increasing the activity.

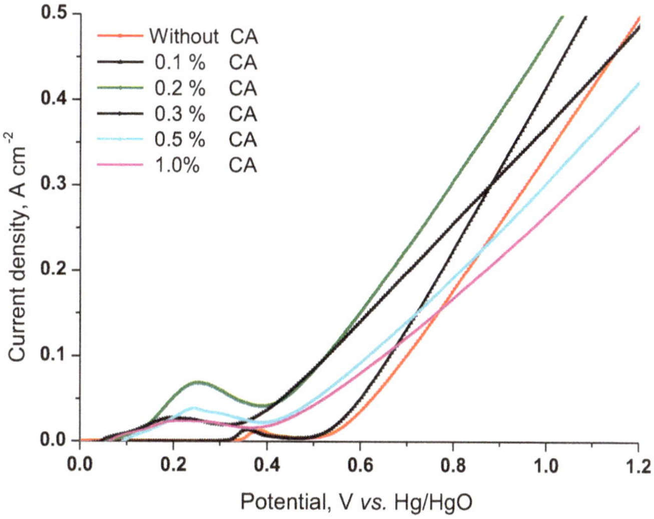

Figure 3. *Polarizations curve for alkaline water electrolysis with the different wt.% of CA at 30 °C.*

3. Cyclic voltammetry studies

Cyclic voltammograms obtained for the nickel alloy electrodes when scanned between 0 to 0.8 V *vs.* NHE at a scan rate of 5 mV s^{-1} in 30 % KOH solution are represented in Fig. 4. The oxide formation on surface of the nickel electrode is represented by the anodic current between the potentials of +0.37 and +0.58 V *vs.* Hg/HgO. The peak at potential of +0.16 V *vs.* Hg/HgO represents the oxide layer reduction upon current reversal. The Ni(II)- Ni(III) oxide transformations as represented by peaks acts as electrocatalyst for the OER[13,21]. Of all the additions, 0.2 wt. % of CA in 30 % potassium hydroxide solution showed the highest area.

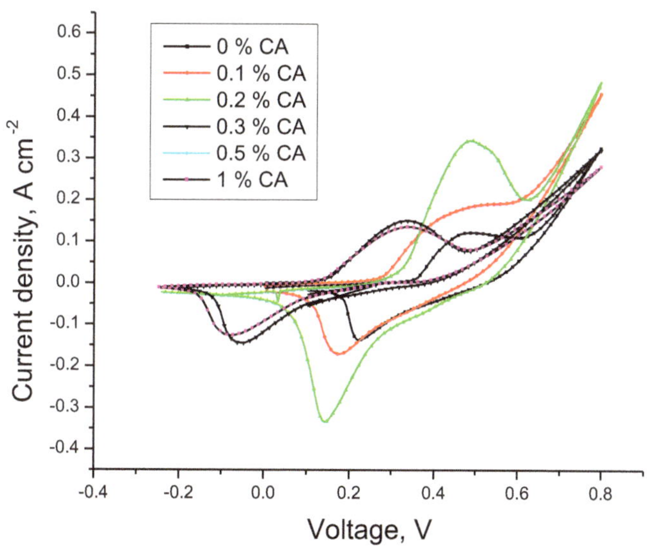

Figure 4. *Cyclic voltammograms of nickel alloy electrode with different wt.% of CA.*

The addition of 0.2 wt. % of CA enabled the current to increase more sharply when the potential reaches ~0.52 V. However, upon increasing the citric acid concentration above a 0.2 wt. % a further increase was not seen and the curve almost truncated. CA may reduce the surface tension of the electrolyte so the bubbles formed due to evolution of oxygen gas can leave the system easily. At higher concentration of CA, a strong absorption on the surface leads to decrease in catalytic sites and decrease in activity.

4. Effect of temperature

Figure 5a shows the steady state polarization curves of Ni-Zn-S alloy electrode at different temperatures on the addition of citric acid (0.2 wt. %) in alkaline water electrolysis. As can be seen from the polarization curves, performance increases with increase in temperature. A high current density of 0.78 A cm^{-2} was achieved at 1.0 V and 80 °C. With the introduction of citrate in the system, the overall efficiency increases and overvoltage also decreases for OER. The improved performance may be explained on the basis of an increase in the conductivity of the electrolyte at higher temperatures and also due to increase in catalytic activity with the increase in temperature. Figure 5b gives the long term stability of the cell, studied by holding the cell continuously at a potential of 1.0 V *vs.* Hg/HgO electrode and monitoring the current at the temperature of 30 °C. The current was constant indicating the stability of the cell with added citric acid.

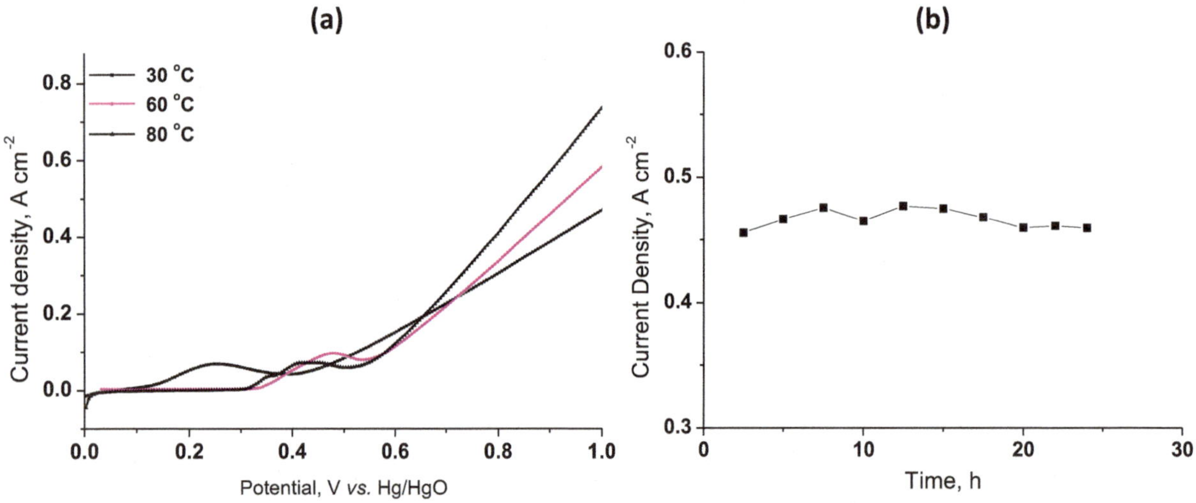

Figure 5.*Effect of temperature on the addition of CA (0.2 wt. %) in alkaline water electrolysis* ***(a)*** *and stability test with addition of CA (0.2 %) at 1.0 V* vs. *Hg/HgO at 30 °C* ***(b)***

5. SEM analysis

Scanning electron microscopy (SEM) of the nickel anode was performed before and after (24 h of continuous operation) the electrolytic process with the addition of the CA in the electrolyte. The typical SEM images are presented in Fig. 6. The morphology of deposits before electrolysis are shown in Figure 6 (a) and (b). In these micrographs homogenous globular particles are seen. The surface of the electrodes is also divided by coarser cracks and channels. The cracks are likely to be filled with the electrolyte when dipped in the solution. This increases the surface area of the catalyst in contact with the electrolyte and leads to increase in current density. Further, efficiency may also be increased due to convective replenishment of the electrolyte due to gas evolution. The EDAX data of the nickel alloy electrodes prepared has a composition of nickel (55 %), zinc (30 %) and sulphur (15 %).

The morphology of deposits after addition of the CA during the electrolytic process is shown in Fig. 6 (c) and (d). The surface of the electrodes is characterized by the presence of flakes covering the electrodeposited layers.

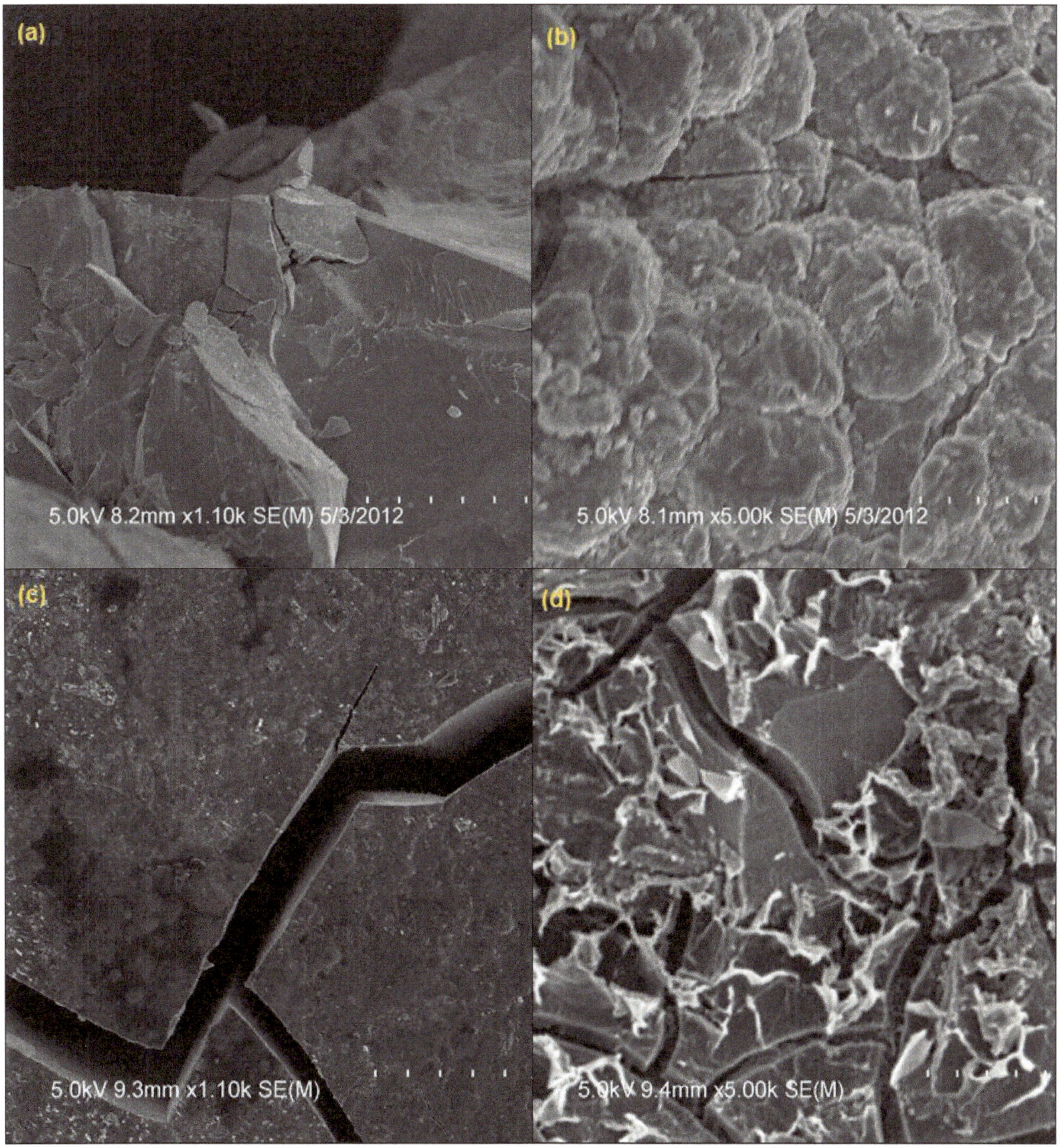

Figure 6. *SEM Images of nickel alloy electrodes:* ***(a)*** *and* ***(b)*** *after electrolysis with KOH,* ***(c)*** *and* ***(d)*** *After electrolysis with KOH and CA*

The deposits appear to be interspersed on the surface of the electrode and make the surface appear rough. Thus, it is evident that the addition of CA in the electrolyte has resulted in the deposition of Ni metal particle on the surface during the electrolytic process. Thus, an increase in surface area and an increase in possible active centers are revealed in the SEM micrographs of the sample after electrolysis. The citric acid complexing agent thus acts as a bridge for the transfer of electrolyte on the surface of the electrode and the in situ metal deposits with their high surface area and catalytic activity improve the performance of the electrode for OER. These results compare well with the powder deposits of Fe-Ni observed by use of citric acid by Lačnjavec *et al.* [22] in their studies on electrodeposition of Fe-Ni from citrate electrolytes.

6. XRD analysis

Figure 7 shows the XRD patterns of electrodeposited nickel electrodes before and after (24 h of continuous operation) electrolysis. The X-ray diffraction pattern of the before electrolysis sample has a broad peak around 44° corresponding to (111) position of Ni. Due to the high concentration of nickel, the intensity of these peaks is high and hence only these are seen. The Zn and S present are detected only by the EDAX analysis. In the after electrolysis curve the intensity of the nickel peak is very high and the peak is also very sharp. The average particle size was calculated using Debye-Scherrer relation. The particle size of samples after and before electrolysis was found to be between 11 and 16 nm respectively. This may be explained by the deposition of the in situ Ni particles from the Ni citrate complex formed. Such in situ deposited particles may be responsible for the increased electrochemical activity shown in the OER reaction. It shows that the size of Ni deposit particles can be modified by the addition of citric acid. Similar observations of metal deposition have been reported from the solution by Nikolić *et al.*[14].

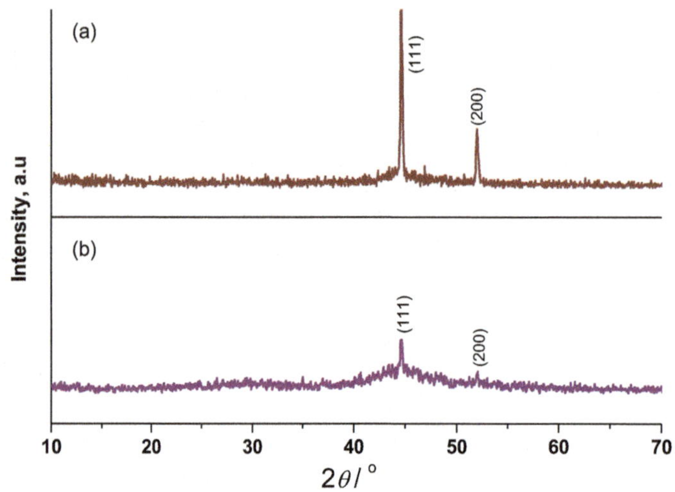

Figure 7. *XRD patterns of nickel surface: (a) after electrolysis; (b) before electrolysis*

The addition of 0.2 wt. % citric acid to the standard electrolyte as a complexing agent thus decreases the energy consumption during electrolysis of water. The addition resulted in the in situ deposition of Ni on the surface (as shown in SEM figures). The improved performance suggests that an extremely active surface center is created to improve the catalytic activity of the electrode. Further the formation of the nickel citrate complex may improve the wettability of the electrode and may act as a bridge for ion transfer between the electrolyte and the metal surface [23].

Conclusions

The results presented help in concluding that a deposit formed in-situ during the electrolytic process by using a complexing agent such as CA in the electrolytic solution is extremely active. An increase in the current density of about 25 % resulted at a temperature of 30 °C in the presence of 0.2 wt. % of CA at 1.0 V *vs.* Hg/HgO. The CA distributed throughout the nickel matrix as a complex created more organized transfer channels to produce high surface area electrodes for alkaline water electrolysis. SEM and XRD figures obtained after electrolysis confirm the deposition of the metal. This form of deposit is very active and improves the catalytic activity of the electrode. Improvement may also be explained due to increase in wettability and the bridging ability of the complexing agent to transfer ions between the electrolyte and metal surfaces. Thus, through simple processing, high current density and good efficiency can be obtained. We believe this type of complexing agent deserves further investigation and is a good candidate for use in alkaline water electrolysis.

Acknowledgement: *The authors would like to thank Director, ARCI for his constant encouragement and support and Department of Science and Technology (DST), Government of India for financial assistance. S. Seetharaman thanks ARCI for the Senior ARCI fellowship under which this work was carried out.*

References

[1] E. Rasten, G. Hagen, R. Tunold, *Electrochim. Acta* **48** (2003) 3945-3952
[2] P. Millet, D. Dragoe, S. Grigoriev, V. Fateev, C. Etievant, *Int. J. Hydrogen Energ.* **34** (2009) 4974-4982
[3] S. P. S. Badwal, S. Giddey, F. T. Ciacchi, *Ionics* **12**(2006)7-14
[4] J. Panek, A. Budniok, *Surf. Interface Anal.***40** (2008) 237-241
[5] A. A. Kamnev, B. B. Ezhov, V. Rusanov, V. Angelov, *Surf. Interface Anal.* **19**(1992) 577-580
[6] P. W. T. Lu, S. Srinivasan, *J. Electrochem. Soc.* **125** (1978) 1416-1422
[7] G. Zhang, Y. Qiu, X. Yang, X. X. Li, H. Pan, *Desalin. Water Treat.* **56** (2015) 905-911
[8] G. Sheela, M. Pushpavanam, S. Pushpavanam, *Int. J. Hydrogen Energ.* **27** (2002) 627-633
[9] M. S. Sadovnikova, V. M. Belikov, *Russ. Chem. Rev.* **47** (1978) 199-212
[10] B. H. Chen, L. Hong, Y. Ma, T. M. Ko, *Ind. Eng. Chem. Res.* **41** (2002) 2668-2678
[11] V. C. Kieling, *Surf. Coat. Tech.* **96** (1997) 135–139
[12] Q. Mao, L. Zhang, D. Huang , D. Wang, Y. Huang, H. Xu, H. Cao, Z. Mao, *Surf. Interface Anal.* **43** (2011) 903-912
[13] I. G. Casella, M. Gatta, *J. Electroanal. Chem.* **534** (2002) 31–38
[14] V. M. Nikolić, G. S. Tasić, A. D. Maksić, D. P.Šaponjić, S. M. Miulović, M. P. M. Kaninshi, *Int. J. Hydrogen Energ.* **35** (2010) 12369-12373
[15] M. P. M. Kaninski, A. D. Maksić, D. L. Stojić, S. C. Miljanić, *J. Power Sour.* **131** (2004)107-111
[16] S. M. Miulović, S. L. Maslovera, M. M.Šeović, B. B. Raduk, M. P. M. Kaninski, *Int. J. Hydrogen Energ.* **37** (2012) 16770-71675
[17] E. Gomez, S. Pane, E. Valles, *Electrochim. Acta* **51**(2005) 146-153
[18] A. Afshat, A. G. Dolati, M. Ghorbani, *Mater. Chem. Phys.* **77** (2003) 352-358
[19] S. Seetharaman, R. Balaji, K. Ramya, K. S. Dhathathreyan, M. Velan, *Ionics* **20** (2014) 713-720
[20] A. Seghiover, J. Chevalet, A. Barhoun, F. Lantelme, *J. Electroanal. Chem.* **442**(1998)113-123.
[21] M. C. Biesinger, B. P. Payne, L. W. M. Lau, A. Gerson, R. S. C. Smart, *Surf. Interface Anal.* **41** (2009) 324-332
[22] U. Lačnjevac, B. M. Jović, V. D. Jović, *Electrochim. Acta* **55** (2009) 535-543
[23] O. Gyliene, J. Aikaite, O. Nivinskiene, *J. Hazard. Mater.* **109** (2004) 105-111.

3

Monitoring of nickel(II) in biological samples using 4-(2-hydroxy-phenylethamino) benzene-1,3-diol by extractive differential pulse polarography

Dasari Rekha, Puthalapattu Reddy Prasad, Neelam Yughandhar Sreedhar✉

Department of Chemistry, Sri Venkateswara University, Tirupati-517502, A.P, India

✉Corresponding author: sreedhar_ny@rediffmail.com

Abstract

A novel analytical reagent, 4-(2-hydroxyphenylethamino)benzene-1,3-diol was synthesized for determination of nickel(II) in biological samples and plant materials by extractive differential pulse polarography. The analysis of nickel(II) was carried out by extracting the nickel(II)-(4-(2-hydroxyphenylethamino)benzene-1,3-diol) complex from chloroform prior to injecting into the instrument. The electrochemical behavior of nickel(II) complex was studied under optimum parameters like effect of pH, pulse amplitude, scan rate and choice of solvent. The calibration curve was constructed in the concentration range of 0.05-42 µg mL^{-1} at pH 4.0 (acetate buffer) with a correlation coefficient of 0.992, respectively. Interference of diverse ions was also investigated during the analysis of nickel(II) to optimize the conditions. The accuracy and validity of the proposed method were checked by the analysis of Certified Reference Materials which is distributed by the National Institute of Standard Technology

Keywords

Monitoring, nickel(II), Extractive differential pulse polarography

Introduction

Nickel(II) exposure into environment of more than 0.05 mg/kg has a toxic effect on human beings and animals [1,2]. Nickel(II) has a wide range of applications including but not limited to batteries, welding rods and wires, electronic equipment, pigment for paints, ceramics, consumer products, steel, magnets, and it can be the source of several diseases. Considering these toxicological considerations, monitoring of nickel(II) is of great importance in the field of the human health or environment and it is highly desirable to develop analytical methods at trace level.

Various analytical methods have been proposed for the monitoring of nickel(II) in environmental matrices such as spectrophotometry [3,4], X-ray fluorescence spectrophotometry

[5], atomic absorption spectrometry [6], inductively coupled plasma atomic emission spectrometry [7,8], inductively coupled plasma mass spectrometry [9], voltammetry [10,11] and chromatography [12]. Most of these techniques, however, are either very expensive, suffer from many interferences or require highly skilled technicians. On the other hand, differential pulse polarography (DPP) is an old but a conventional approach in trace analysis. In addition, the concentration of nickel is generally at very low level in environmental samples and it is not possible to determine in a simple complex sample concentration below the quantification limit of the instrument used. Hence, we have developed a simple, economic sensitive differential pulse polarography method, after extracting the metal complex from chloroform in biological samples and plant materials. The metal complex was formed by interaction of nickel(II) with newly synthesized analytical reagent 4-(2-hydroxyphenylethamino)benzene-1,3-diol at ordinary laboratory conditions. The various parameters have been evaluated. The developed method can be employed for efficient determination of nickel(II) at ppm levels. The results of analysis obtained were compared with those obtained by the well knowing method with certified values of nickel(II). The proposed method is found free from various interferences and successfully applied for determination of nickel (II) in biological and plant materials.

Experimental

Apparatus

An Elico CL-362 model polarographic analyzer was used for differential pulse polarography (DPP) measurements and the three–electrode system consisted of a hanging mercury drop working electrode, Ag/AgCl/KCl reference electrode and a platinum wire as the auxiliary electrode. The entire system was outfitted with the LX 300$^+$ X-Y recorder. Cyclic voltammetric measurements were carried out by the Metrohm 757 VA Computrace Model AFRDE 4 potentiostat and MSRX speed control unit supplied from Pine Instrument Company (USA) coupled with digital electronics 200X-y/t recorder. An Elico Li-129 Model glass-calomel combined electrode (Purchased from Elico Ltd, Chennai, India) was employed for measuring pH values. All experiments were performed at 25$\pm$1 °C.

Reagents

Analytical reagent grade chemicals and doubly distilled water were used throughout the experiments. 2,4-dihydroxy acetophenone, 2-amino phenol, organic solvents, HCl, HNO_3, H_2SO_4, $HClO_4$, CH_3CHOOH, $C_2H_3NaO_2$ were purchased from Merck Chemicals. A stock solution of nickel(II) (0.01 M) was prepared by dissolving 2.377 g of $NiCl_2$ $6H_2O$ in double distilled water in 100 mL volumetric flask. The working standard solution was freshly prepared by diluting the stock solution with double distilled water at the time of analysis. A stock solution of 0.001 M solution of 4-2-HPEDB-1,3,D was prepared by dissolving an appropriate amount of the reagent in 100 mL of methanol. The working solution was prepared by its appropriate dilution with the same solvent. Acetate buffer solutions of pH values ranging from 3.0–11.0 were prepared as appropriate ratios of 0.1 M acetic acid and 0.1 M sodium acetate (13.6 g/L).

Synthesis of 4-(2-hydroxyphenylethamino)benzene-1,3-diol (4-2-HPEDB-,3,D)

The equimolar ratio of 2,4-dihydroxy acetophenone (1.0 g) and 2-aminophenol (1.0 g) in methanol mixture was refluxed for three to four hours and the contents were cooled at room temperature creating an orange-red color precipitate. The resulting precipitate was filtered (Whatman filter paper No. 44) and washed with 50 mL ethanol. The crude solid was recrystallized from aqueous ethanol and dried on $CaCl_2$, with a Yield=97 % and M.P: 115-120 °C. Infrared absorption spectrum gave the following peaks (cm^{-1}, KBr pellet): 3375.3 (O-H), 2900 (C-H), 1601 (C=N) and 1465.9 and 1280 (C-O, O-H). The synthesis of analytical reagent and the metal complexation with the reagent are shown in the Scheme 1.

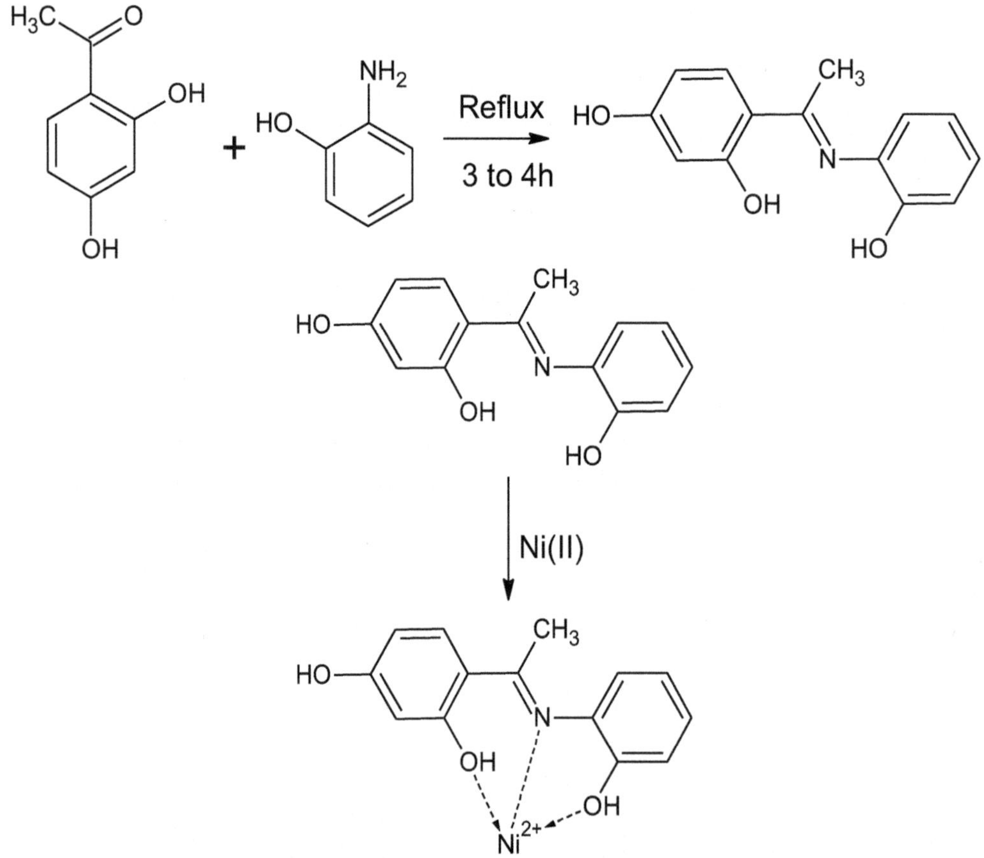

Scheme 1. *Synthesis and complexation of 4-2-HPEDB-1,3-D with nickel(II)*

General procedure

The aliquot of the working standard solution (0.001 M) containing 1–100 μL (0.2377 g) of metal ion was taken into 25 mL volumetric flask what was followed by addition of the solution containing 5 mL of acetate buffer solution (pH 4.0) and 2 mL of reagent (soluble in methanol) into the mixture. The mixture was shaken with 5 mL portions of chloroform for 30 s and allowed to stand for 5–10 min. The organic layer was collected and transferred into the polarographic cell, diluted with 9 mL of acetate buffer and then deoxygenated with nitrogen gas for 10 min. After recording the polarogram, a small increment (0.2 mL) of standard solution was added to the cell, treated for one min and the polarogram was again recorded under similar conditions. In the present study, the greatest precision was obtained at pH 4.0 having the correlation factor of 0.992 and RSD 0.7 % with a drop time of 2 sec, pulse amplitude of 50 mV, scan rate of 12 mV s^{-1} and the applied potential of – 615.0 mV.

Sample collection

Tirupati, city located in the south of India was chosen as the study area for the sample analysis. It is a rapidly growing city which is widely known at the global level. The arrival of approximately 50,000-60,000 pilgrims per day to visit the Lord Balaji Temple (Tirumala) is mainly responsible for threatening the natural beauty of this holy town. At the same time, several small and medium scale industries are situated in the suburban fringes of Tirupati town. For this reason, this holy pilgrim center has been chosen as the potential collection area for the present study. The biological and plant samples were collected from different locations of the study area in and around Tirupati. All necessary and possible precautions were taken at various stages starting from sample containers, sample collection and storage, processing and analyzing the samples.

Application

The hair samples were washed with acetone two to three times in a beaker with continuous stirring and dried in an electric oven at 70 °C for four hours. 2.0 g of the sample was weighed and taken in a beaker. Then, the (1:1) mixture of nitric acid and perchloric acid was added and the mixture was heated on a hot plate. The solution was evaporated to near dryness. The ash was taken up with 5 mL of HCl (35.5 %) and evaporated to dryness. The residue was taken up in 2 mL HCl (35.5 %) then filtered and made up to 25 mL with water. Suitable volumes of these solutions were taken for determination of nickel(II) as described above and the results are shown in Table 1.

***Table 1**. Determination of nickel(II) in biological sample*

Sample	Certified values	Reported method [18]	R.S.D, % (n=5)[a]	Present method Extractive-DPP	R.S.D, % (n=5)[a]
NIES No.5 Human hair: Pb(6.0), Cd(0.20), Sb(0.07), Zn(169), Al(240), Sc(0.05), Fe(225), Mg(208), Hg(4.4), Co(0.10), Rb(0.19), K(34), Mn(5.2), Cu(16.3), Ti(22), Ca(728), Cr(1.4), Ba(2.2), Se(1.4), Na(26), Sr(2.3).	1.80	1.90	0.12	1.96	0.65

[a]Mean values for five determinations

Analysis of plant material

The freshly collected *Pisum sativum* (Hulls), *Mangifera indica, Eucalyptus, Azadirachta indica* leaves (5.0 g each) samples were placed in a 250 mL beaker, and a solution of concentrated H_2SO_4/HNO_3 1:1 (v/v) (10 mL) was added. This mixture was heated until the solution was clear. The solution was filtered (Whatman filter paper No 44) off and concentrated in a porcelain bowl up to 5 mL, then cooled and diluted to 50 mL with deionized double-distilled water. The general procedure was then applied to 1 mL of this solution and the obtained results were presented in Table 2. In addition, plant tissues like Umbilicaria muhlenbergii has been examined in detail followed by the reference procedure [13]. The obtained results were compared with the reported method as shown in Table 3.

***Table 2**. Determination of nickel(II) in plant material*

Samples[a]	AAS method [19]	R.S.D, % (n=5)[b]	Present method	R.S.D, % (n=5)[b]
Pisum sativum (Hulls)	2.060	0.003	2.065	0.068
Mangifera indica leaves	2.150	0.004	2.153	0.072
Eucalyptus leaves	1.038	0.002	1.044	0.013
Azadirachta indica leaves	1.481	0.005	1.489	0.015

[a] Samples collected from Acharya N. G. Ranga Agricultural college Tirupati, A.P, India;
[b] Mean values for five determinations

The accuracy and precision of the present method were validated by analyzing the Certified Reference Materials (CRM's) which was distributed by the National Institute of Standard Technology (NIST) of Rice Flour (NIES-CRM-10A), Wheat Flour (ARC/CL-WF), and Rye Bread Flour (CSRM-12-2-05). Inter calibration was performed by using above materials and the analytical results obtained from the present method strongly agreed with the Certified Reference Materials (CRM's) and the data were mentioned in Table 5.

Table 3. *Determination of nickel(II) in plant tissues (Umbilicaria muhlenbergii) by differential pulse polarography.*

Samples	Nickel(II) content of zone, µg/g	
	Proposed method	Reported method(DPP) [13]
Set 1	12± 2	13± 3[a]
Set 2	14±6	12±5 [a]
Set 3	792±50	790±60
Set 4	635±40	640±50

[a]DPP+DMG analysis

Results and discussion

The analytical reagent 4-2-HPEDB-1,3,D is a ligand having chelating functional groups that form a complex with Ni(II) ions at pH 4.0 (acetate buffer). Figure 1 shows the typical FTIR spectrum of Ni(II)-(4-2-HPEDB-1,3,D) complex. In the FTIR spectra of the complex, a strong peak was observed in the region of 1610-1640 cm^{-1}, which is ascribed to involvement of nitrogen donor atoms of azomethine (-C=N-) in coordination to the Ni(II) ions [14]. The band at 524 cm^{-1} observed for the Ni(II)-(4-2-HPEDB-1,3,D) complex is assigned to v(M–O). The stretching vibration of the free ligands v(O-H), 3430-3464 cm^{-1} was not observed, suggesting deprotonation of hydroxyl groups and formation of Ni-O band [15]. The bands observed for the complexes between 720–620 and 564-495 cm^{-1} were metal sensitive and are assigned to v(M–O) and v(M–N) [16], respectively.

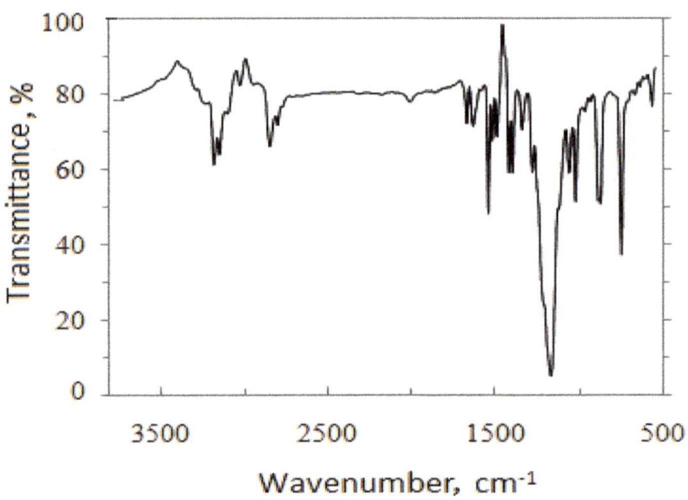

Figure 1. *FTIR spectra of Ni(II)- [4-2-HPEDB-1,3-D] complex*

Cyclic voltammetry is widely used for the characterization of electrochemically active systems. Figure 2 shows the cyclic voltammogram of nickel(II)-[4-2-HPEDB-1,3,D] system at a scan rate of 12 mV s^{-1} in the acetate buffer (pH 4.0). In the forward scan, the cathodic peak potential (E_{cp}) observed at ⍰0.62 V corresponds to reduction of nickel(II)-4-2-HPEDB-1,3,D complex. The anodic peak potential (E_{ap}) is observed at -0.12 V and corresponds to oxidation of nickel(II)-[4-2-HPEDB-1,3,D] complex generated in the anodic step. The anodic and cathodic peak potential difference (ΔE) is -0.50 V. The possible electrode mechanism can be described as follows:

Ni(II) (solution) + 4-2-HPEDB-1,3,D $_{(HMDE)} \rightarrow$ Ni-(4-2-HPEDB-1,3,D) $_{(HMDE)}$

Ni(II)-[4-2-HPEDB-1,3,D]$_{(HMDE)}$+ e^- $\rightarrow$ Ni(I)-[4-2-HPEDB-1,3,]$_{(HMDE)}$

Ni(I)-[4-2-HPEDB-1,3,D]$_{(HMDE)}$ $\rightarrow$ Ni(II) + e^-

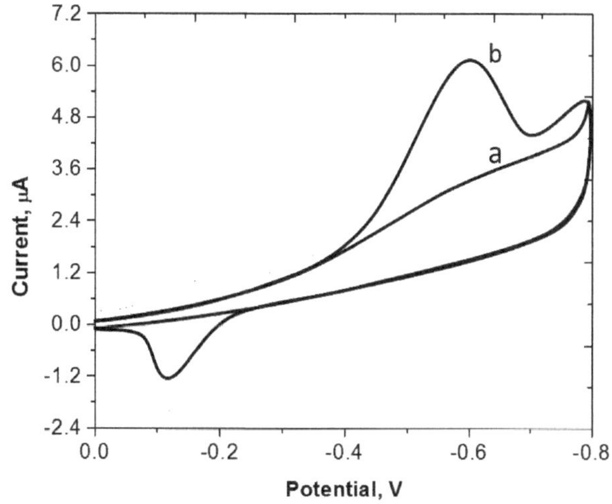

Figure 2. *Cyclic voltammogram of (a) Ni(II) and (b) Ni(II)-[4-2-HPEDB-1,3-D] complex at HMDE, acetate buffer (pH 4.0), scan rate 50 mV* s^{-1}

Differential pulse polarographic studies

Effect of pH, choice of solvent and scan rate

The effect of pH is one of the important parameter for the investigation of nickel(II) and [Ni-(4-2-HPEDB-1,3,D)] complex by differential pulse polarography [17]. The pH studied varied in the range of 2.0 to 10.0 for [Ni-(4-2-HPEDB-1,3,D)] complex. The maximum peak current was obtained with pH 4.0 at peak potential – 615 mV as shown in Figure 3. When pH was increased from 2.0 to 10.0 at constant potential, the concentration of metal complex decreased. Therefore, the acetate buffer of pH 4.0 was suitable for determination of nickel(II) for better sensitivity and selectivity in further investigations. The extraction of the nickel(II)-[4-2-HPEDB-1,3,D] complex was tested in various organic solvents, such as toluene, benzene, n-butanol, dimethyl formaldehyde, carbon tetrachloride, cyclohexane and chloroform. Among the various solvents studied, the chloroform was selected as the most suitable for extraction of nickel(II)-[4-2-HPEDB-1,3,D] complex. This was because of its maximum peak current and greater extraction ability, due to polarity of complex. In addition, the scan rate was also changed from 4 to 18 mV s^{-1}. The peak current values increased with increasing scan rate up to 12 mV s^{-1} and after that the peak current decreased. Therefore, the scan rate of 12 mV s^{-1} was selected for further studies.

Effect of foreign ions

The selectivity of the proposed method was enhanced by the study of diverse ions for determination of nickel(II) in biological samples, plant materials and the results are shown in Table 4. Foreign ions were added to the solution containing 10 μg mL^{-1} of nickel(II) and measurements were taken under optimized conditions. Many of diverse ions were tolerated up to the maximum level by using suitable masking agents with the error of ±2 % during the analysis of nickel(II). Though, some of the ions could be masked by addition of a sufficient amount of EDTA solution. Low concentrations of As^{3+} at 40 °C was also used to reduce the chromium ions. In addition, Ca^{2+}, Sr^{2+} interference was eliminated by prior extraction with KI/Isobutyl methyl ketone. Sodium, potassium, chlorides, nitrites, sulfates, bicarbonates, phosphates and other remaining ions do not interfere with determination of trace metals up to 50 μg mL^{-1}. Almost all the results were quantified in the presence of diverse ions to elevate the feasibility and selectivity of the proposed method at appropriate amounts of nickel(II) solution, followed by the general procedure.

***Table 4**. Tolerable limits of diverse ions in determination of nickel(II) with error of ±2 %*

Diverse ions	Tolerable limit, μg mL^{-1}
K^+, Mg^{2+}, NO_3^-	60,000
Al^{3+}, PO_4^{3-}, NO_2^-, ClO_4^-, SO_4^{3-},	40,000
Ca^{2+}, Sr^{2+}, B(III), ClO_3^-, BrO_3^-,	30,000
Mn^{2+}, Fe^{3+}, Ce(IV), Mo(VI),	20,000
V(V), Ti(IV), Cr(VI), Bi(III), U(IV), Cr(VI), Ba^{2+}	3,000
Cu^{2+}, Pd^{2+}, Zn^{2+}, Cd^{2+}, La^{3+}, Cr^{3+}, Cl^-, Zr(IV)	1,000
Pb^{2+}, Bi(III), Hg^{2+}, Ag^+, Th(IV), Sb^{3+}, Sn(IV),	500
Au^{3+}, Te(IV), Se(IV),	250
Co^{2+}	100

Calibration, detection limit and precision

The calibration curve was constructed based on the general procedure under optimized conditions at the concentration range of 0.05–42 μg mL^{-1} with the correlation coefficient of 0.992. The detection limit was found to be 0.15 μg mL^{-1} of the final solution by taking five individual replicates of nickel(II) solution which gives the relative standard deviation of 0.7 %. Typical calibration curve is shown in Figure 3b that is drawn based on the S.D. (standard deviation) values. The Method of Quantification (MOQ) value, calculated based on the calibration curve was found to be 7.940 μg mL^{-1} for nickel(II) in biological samples.

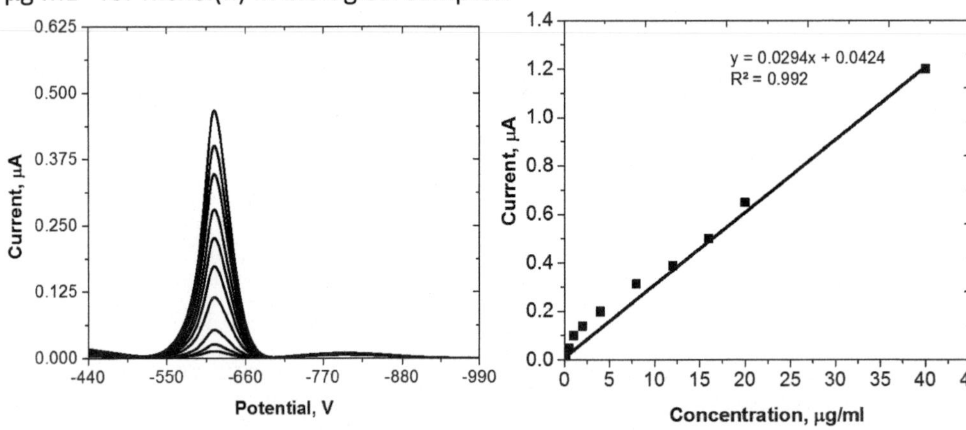

***Figure 3.** (left) Differential pulse polarogram of nickel(II), peak at – 615.0 mV, acetate buffer (pH 4.0), pulse amplitude 50 mV, scan rate 50 mV s^{-1}; (right) Calibration curve for nickel(II)*

The present method was successfully applied for determination of nickel(II) using the simple chloroform extraction procedure prior to DPP in biological samples, plant materials and the results are shown in Tables 1, 2 and 3. The validity of the present method was checked by analyzing the CRM's and the results are shown in Table 5. In addition, the obtained results are compared to several analytical reagents synthesized for the determination of nickel(II) in various environmental samples, presented in Table 6. The developed method shows good sensitivity in terms of solvent medium, techniques used for the determination of nickel(II) ion, concentration range and detection limits.

***Table 5.** Determination of nickel(II) in Certified Reference Material (CRM's)*

Sample	Standard values	Present method	RSD, % (n=5)[b]
Rice Flour (NIES-CRM-10A)	0.19	0.187	1.33
Wheat Flour (ARC/CL-WF)	0.153	0.155	0.62
Rye bread Flour (CSRM-12-2-05)	0.234	0.230	0.8

[a] CRM's distributed by NIST National Institute of Standard Technology [NIST (USA)]
[b] Mean values for five determinations

***Table 6.** Comparison of reagents for the determination of nickel(II) with other methods*

Reagents	Solvent/medium	Techniques	Concentration range	Limit of detection	Ref.
1-pyridylazo-2-naphthol	Amberlite XAD-resin	ICP-AES	0.10-275 μg L^{-1}	0.20 μg mL^{-1}	[8]
HNO_3	Surfactant Triton X-100	ICP-MS	up to 2.0 μg L^{-1}	0.11 μg L^{-1}	[20]
1-phenyl (3-methyl-3-benzoyl)5- -pyrozolone	Sufactant Triton X-100	GFAAS	up to 100 ng mL^{-1}	0.12 ng mL^{-1}	[21]
1-(2-Pyridilazo)-2-naphthol	-	FAAS	-	2.60 μg L^{-1}	[22]
2,(2-Quinolylazo) dimethyl dimethylaminoaniline(QADEAA)	sodium dodecyl sulfonate (SDS)	Spectro-photometry	0.01-0.4 μg mL^{-1}	0.20 μg mL^{-1}	[23]
5- Bromo-2- hydroxyl -3-metho-xybenzaldehyde-4-hydroxy benzoichydrazone	DMF	Spectro-photometry	0.117 to 2.64 μg mL^{-1}	0.016 μg mL^{-1}	[24]
2-pyridylazo 2-resorcinol (PAR)	Acetate buffer	HPLC	up to 50 μL	-	[25]
2-(2-Quinolylazo)-5-dimethylaminoaniline QADMAA	sodium dodecyl sulfonate (SDS)	Spectro-photometry	0.01-0.6 μg mL^{-1}	0.20 μg mL^{-1}	[26]
4-(2-hydroxy phenyl ethaminodiol), benzene-1.3-diol	Acetate buffer	Extractive-DPP	0.05-42 μg mL^{-1}	0.15 μg mL^{-1}	Present Work

Conclusions

A simple, sensitive, rapid, and eco-friendly solvent extractive differential pulse polarography method was designed for determination of nickel(II) in biological samples and plant materials. The preconcentration method following chloroform extraction procedure has been established. A new analytical reagent 4-2-HPEDB-1,3,D has been synthesized, which can be able to form stable complexes with nickel (II). The selective functional group of -C=N- in the complex, gives a good response and is easily reduced at the electrode surface. The reagent also suppresses the interference of diverse ions such as cations, anions, and other salts by a judicial use of masking agents.

References

[1] J. J. Scott-Fordsmand, P. H. Krogh, S. P. Hopkin, Ecotox. Environ. Safe. 43(1) (1999) 57-61.

[2] K. S. Kasprzak *Mutat. Res.* **533(1-2)** (2003) 67-97.

[3] E. Y. Hashem, M. S. Abu-Bakr, S. M. Hussain, *Spectrochem. Acta A* **59** (2003) 761-769.

[4] Y. Liu, X. Chang. S. Wang.Y. Guo, B. Din, S. Meng, *Talanta* **64 (1)** (2004)160-166.

[5] Z. Augustynski, B. Dziunikowski, J. Meitin, *J. Radioanal. Nucl. Chem.* **63 (2)** (1981) 325-334.

[6] J. L. Manzoori, A. Bavili-Tabrizi, *Microchim. Acta* **141(3-4)** (2003) 201-207.

[7] E. Vereda Alonso, A. Garcia de Torres, J. M. Cano Pavon, *Mikrochim. Acta* **110(1-3)** (1993) 41-45.

[8] S. L. C. Ferreira, C. F. de Brito, A. F. Dantas. N. M. Lopo de Araujo, A. C. S. Costa, *Talanta* **48(5)** (1999) 1173-1177.

[9] C. J. Park, S. A. Yim, *J. Anal. Atom. Spectrom.* **14** (1999) 1061-1065.

[10] M. Korolczuk, *Talanta*, **53(3)** (2000) 679-686.

[11] N. Y. Sreedhar, D. Rekhs, *Anal. Chem. Ind. J.* **8(2)** (2009) 209-212.

[12] X. Yang. Z. Qiao. W. Wei, S. Yao, *Talanta* **46(4)** (1998) 697-702.

[13] C. J . Flora, E. Nieboer. *Anal. Chem.* **52(7)** (1980) 1013-1020.

[14] L. Larai, Y. Harek, A. Reguig, M. M. Mostafa, *J. Serb. Chem. Soc.* **68(2)** (2003) 85-95.

[15] W. P. Griffith, S. I. Mostafa, *Polyhedron* **11(23)** (1992) 2997-3005.

[16] T. Priya Devi, R. K. Hemakumar Singh, *Rasayan Journal of Chemistry* **3(2)** (2010) 266-270.

[17] D. Rekha, N. Y. Sreedhar, P. Reddy Prasad, *Global Journal of Science Frontier Research* **10(2)** (2010) 21-29.

[18] A. A. Almeida. X. Jun, J. L. F. C. Lima, *Talanta* **50(2)** (1999) 253-259.

[19] A. Praveen Kumar, P. Raveendra Reddy, V. Krishna Reddy, *J. Autom. Method. Manag.* **1** (2007) 1-6.

[20] C. S. Kira, A. M. Sakuma, N. da Cruz Gouveia, *Journal of Applied Pharmaceutical Science,* **4(5)** (2014) 39-45.

[21] Z. Sun, P. Liang, Q. Ding, J. Cao, *Hazard. Mater. B,* **137(2)** (2006) 943-946.

[22] R. Galbeiro, S. Garcia, I. Gaubeur, *J. Trace. Elem. Med. Bio.* **28(2),** (2014) 160-165.

[23] Q. Hu, G. Yang, Z. Huang, J. Yin. *Anal. Sci.* **19(10)** (2003) 1449-1452.

[24] B. Saritha, T.S. Reddy, *IOSR J. Appl. Chem.* **7(3)** (2014) 22-26.

[25] H. Cyftci. A. Olcucu, T. Cyftci, *International Journal of Science and Technology* **2(2)** (2007) 105-108.

[26] Q. Hu, G. Yang, Z. Huang, J. Yin, *B. Kor. Chem. Soc.* **25(4)** (2004) 545-548.

4

Oxygen source-oriented control of atmospheric pressure chemical vapor deposition of VO_2 for capacitive applications

Dimitra Vernardou*,✉, Antonia Bei**, Dimitris Louloudakis*,***,
Nikolaos Katsarakis*,****,*****, Emmanouil Koudoumas*,*****

**Center of Materials Technology and Photonics, School of Engineering, Technological Educational Institute of Crete, 710 04 Heraklion, Crete, Greece*
***Department of Mechanical Engineering, School of Engineering, Technological Educational Institute of Crete, 710 04 Heraklion, Crete, Greece*
****Department of Physics, University of Crete 711 00 Heraklion, Crete, Greece*
*****Department of Electrical Engineering, School of Engineering, Technological Educational Institute of Crete, 710 04 Heraklion, Crete, Greece*
******5Institute of Electronic Structure & Laser, Foundation for Research & Technology- Hellas, P.O. Box 1527, Vassilika Vouton, 711 10 Heraklion, Crete, Greece.*

✉Corresponding Author: dvernardou@staff.teicrete.gr

Abstract
Vanadium dioxides of different crystalline orientation planes have successfully been fabricated by chemical vapor deposition at atmospheric pressure using propanol, ethanol and O_2 gas as oxygen sources. The thick a-axis textured monoclinic vanadium dioxide obtained through propanol presented the best electrochemical response in terms of the highest specific discharge capacity of 459 mAh g^{-1} with a capacitance retention of 97 % after 1000 scans under constant specific current of 2 A g^{-1}.

Keywords
Atmospheric pressure chemical vapor deposition; O_2 source; Vanadium dioxide; Electrochemical properties.

Introduction

Vanadium dioxide (VO_2) exists in more than 10 polymorphs, however, rutile, monoclinic (described as a slightly distorted rutile structure) and metastable phases have mainly attracted interest because of their interesting chemical and physical properties for catalytic and electrochemical applications [1-3]. The crystal structure of VO_2 consists of sheets of edge-sharing

VO_6 octahedra linked by corner sharing to adjacent sheets along the c-direction of the unit cell [4]. This corner-sharing structure strengthens the structural stability and the resistance to lattice shearing during cycling in lithium ion batteries [5]. In that context, VO_2 was found to show better electrochemical performance compared with the well-known V_2O_5 [6].

To date, various VO_2 forms have been synthesized such as nanocrystalline VO_2(B) [4], nanothorn VO_2(B) hollow microspheres [7], three dimensional hierarchical microflowers and microspheres [8,9] as well as VO_2/reduced graphene oxide [10,11], VO_2(B)/carbon nanotubes [12] and hydrogen treated VO_2 [13] composite powder materials with sufficient capacitive characteristics. But, most of the above mentioned materials showed poor cycling stability after 500 cycles. Hence, it is important to improve the long-term performance and high rate capability of VO_2.

Due to the small compositional differences between numerous phases of vanadium oxides, VO_2 preparation requires a stringent controlled process that provides desired oxygen stoichiometry and crystalline structure. In search of such a process, VO_2 has been grown using sol-gel [14], pulsed laser deposition [15], sputtering [16] and chemical vapor deposition (CVD) [17-20]. Among these techniques, CVD at atmospheric pressure (APCVD) is gaining attention because it does not require expensive vacuum systems, has fast deposition rates and can be easily integrated to float-glass production lines [21-24]. A range of precursors have been utilized including vanadium(IV) chloride (VCl_4) [22], vanadyl(IV) acetylacetonate ($VO(acac)_2$) [23-24], vanadium(V) oxytrichloride ($VOCl_3$) [25], vanadium(V) triisopropoxide oxide ($VO(OC_3H_7)_3$ [26-28]. Since highly oriented VO_2 films exhibited significant variation in conductivity as thermochromic layers [29-31], we were inspired to investigate the orientation effect on the APCVD coatings for capacitive applications.

In this paper, the APCVD of VO_2 on SnO_2-precoated glass substrate was studied using $VO(acac)_2$ and propanol, ethanol, O_2 gas as oxygen sources from the perspective of the potential application of such coatings as cathodes. We have demonstrated the differences in the resulting coatings obtained from a structure-orienting role played by the oxygen sources utilized and possible explanation of this role is examined.

Experimental

Preparation of VO_2 coatings

The APCVD reactor utilized in this work was an in-house design as also reported previously [3,23,24]. The vanadium source was the $VO(acac)_2$ (98 %, Sigma-Aldrich), which was placed in a bubbler at 200 °C, while the gas lines were kept at 220 °C to avoid any condensation or blocking. The carrier gas was N_2 (99.9 %), which was passed through the reactor during all depositions. The N_2 flow rate through the vanadium precursor bubbler was 1.4 L min^{-1}, for growth temperature and period of 500 °C and 7.5 min, respectively. Additionally, the flow rate of propanol (99.5 %, Sigma-Aldrich), ethanol (≥ 99.5 %, Sigma-Aldrich) and O_2 gas (99.9 %) was 0.8 L min^{-1}. Finally, the total N_2 flow rate was kept at 12 L min^{-1} in all CVD experiments.

The substrates were commercial SnO_2-precoated glass (Uniglass, Greece), all of dimensions 2×2×0.4 cm. Prior to deposition, they were all cleaned with H_2O and detergent, rinsed thoroughly with H_2O and deionised H_2O, and allowed to dry in air. Once the allotted time was complete, the reactor temperature was turned off and the substrate allowed cooling at 100 °C under an atmosphere of N_2. Then it was removed from the reactor, handled and stored in air.

Structural and morphological characterization of coatings

The structure of the coatings was examined in a Siemens D5000 Diffractometer for 2-theta = 15.00 – 60.00 °, step size 0.05 ° and step time 5 min/°. Additionally, their morphology was evaluated in a Jeol JSM-7000 microscope. In this case, coatings were over-coated with a thin film of gold prior to analysis to avoid charging.

Finally, the coating's thickness was estimated using a profilometer A-step TENCOR. A step was done by etching the vanadium oxide coatings off the SnO_2-precoated glass substrate in 1:3, H_2O_2 (30 %):HCl. Tin dioxide remained intact after this procedure and the thickness was obtained from the measured step height.

Electrochemical evaluation

Cyclic voltammetry experiments were performed using a three electrode electrochemical cell as reported previously [32-34] for a potential range of -1 V to +1 V, a scan rate of 10 mV s^{-1} and a number of scans up to 1000. In particular, Pt, Ag/AgCl and vanadium oxides on SnO_2-precoated glass substrates were used as counter, reference and working electrodes, respectively. All measurements were carried out in 1 M LiOH, which acted as electrolyte. The chronoamperometric measurements were done at -1 V and +1 V for a step of 200 s and a total period of 2000 s. Additionally, the chronopotentiometric curves were obtained at a constant specific current of 2 A g^{-1} and a potential range of +0.1 V to +0.6 V.

Results and discussion

Presented here are investigations into APCVD VO_2 coatings grown from $VO(acac)_2$ and three oxygen sources: propanol, ethanol and O_2 gas. It will be shown that a level of control can be exerted over orientation and morphology through the different sources. All coatings produced were stable in air for over six months and resistant to H_2O, acetone and toluene. Additionally, they passed the "scotch tape test"; a piece of sticky tape was placed on the coating and then removed without lifting off the coating.

Structure

Figure 1 presents the x-ray diffraction (XRD) patterns of the APCVD coatings using propanol, ethanol and O_2 gas. As shown in the case of propanol and ethanol, one diffraction peak is observed at 18.21 °, which matches to monoclinic a-axis textured VO_2 coatings [30,31,35]. This peak can be indexed to 100 plane showing the preferred orientation growth of these coatings. Furthermore, the XRD pattern of the as-grown coatings using O_2 gas show two peaks at 55.4 and 57.6 ° with Miller indices 022 and 220 due to monoclinic VO_2 phase indicating that is a 022-oriented single phase [36-38]. Finally, peaks at 26.5, 33.7, 37.1, 51.7 and 54.7 ° with respective Miller indices 110, 101, 211 and 220 (indicated with asterisk in Figure 1) are due to SnO_2–precoated glass substrate [39]. The preferential orientation of the VO_2 on SnO_2-precoated glass substrates at angles other than 27.8 ° is not clearly understood. Nevertheless, it seems to be more sensitive to O_2 source during the growth rather than to glass substrate. The reaction mechanism for the formation of VO_2 from $VO(acac)_2$ in the presence of O_2 gas has previously been studied in the literature [40]. The possible decomposition routes of $VO(acac)_2$ species involve a simple intramolecular rearrangement of the $VO(acac)_2$ precursor resulting in the release of two C_3H_4 molecules, followed by the decomposition of $VO(CH_3COO)_2$ to yield $(CH_3CO)_2O$ and VO_2. On the other hand, in the case of alcohols, the VO_2 deposits possibly act as heterogeneous catalytic sites for their oxidation to propanal (for propanol) and acetaldehyde (for ethanol). A similar behavior is also observed in the presence of methanol [41,42]. The active oxygen required for their oxidation comes from the $VO(acac)_2$ itself, since it can be regarded as a source of excess oxygen (there are 5 oxygen atoms to 1 V atom, while only 2 oxygen atoms are required for VO_2 formation).

Other researchers have also attempted to control the crystalline orientation of VO_2. Gary et. al. reported the a-axis textured VO_2 deposited on R-plane sapphire and suggested that the cause could be a stress developing on the interface between the substrate and the coating [30]. Muraoka *et. al.* studied the epitaxial growth of VO_2 001-oriented single phase on TiO_2 001 substrates and 110-oriented phase on TiO_2 110, respectively [29]. Ngom *et. al.* indicated that the crystalline orientation of the VO_2 thin films was drastically changed because of the formation of an interface layer between the VO_2 and the soda lime glass [35]. Chiu et. al. also attempted to grow VO_2 on glass using a 5 nm ZnO buffer layer. In the case of the direct VO_2 growth on amorphous glass, polycrystalline films formed, while only VO_2 011 peaks located at 27.90 ° were observed for the growth on ZnO [43].

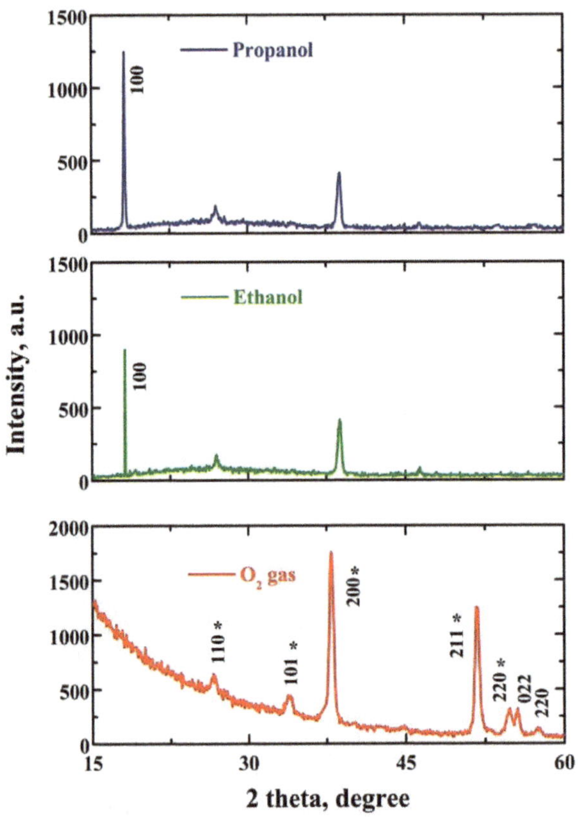

Figure 1. *XRD of APCVD vanadium oxides at 500 °C for 0.8 L min^{-1} flow rate of propanol, ethanol and O_2 gas.*

Morphology

Figure 2 presents the field-emission scanning electron microscope (FE-SEM) images of vanadium oxide coatings grown at 500 °C on SnO_2-precoated glass substrate for 0.8 L min^{-1} flow rate of propanol, ethanol and O_2 gas. For propanol and ethanol, compact grains of nearly round shape are mainly observed with their sizes being 160 nm and 40 nm, respectively. It was observed that thinner coatings grown using ethanol (95 nm) were denser, while as thickness increased for propanol (120 nm), the surface appeared less dense with growth at specific sites, suggestive of a Stranski-Krastanov type of growth mechanism [44]. Regarding the O_2 gas, agglomeration of grains forming rod-like structures is shown with thickness of 80 nm.

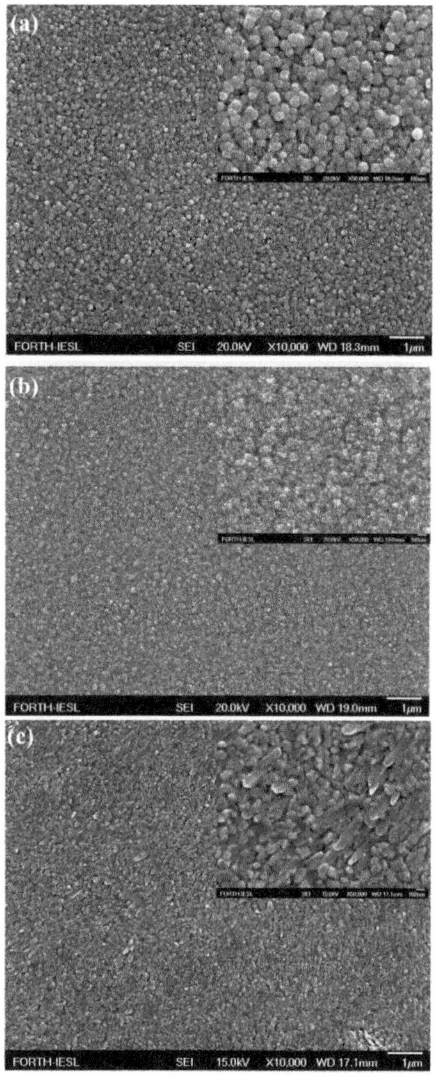

Figure 2. *FE-SEM of APCVD vanadium oxides at 500 °C for 0.8 L min^{-1} flow rate of propanol (a), ethanol (b) and O_2 gas (c).*

Electrochemical characteristics

In order to study the effect of oxygen source on the electrochemical performance of the coatings, cyclic voltammetry curves were obtained as indicated in Figure 3. The potential range was -1 V to +1 V at a scan rate of 10 mV s^{-1}. All curves are normalized to the mass of the working electrode. The mass was measured by a 5-digit analytical grade scale and found to be 0.00002 g, which was obtained by measuring the glass substrate before and after the growth. It can be observed that the as-grown vanadium oxide coatings using propanol present two anodic peaks at -0.05 V / +0.52 V and two cathodic peaks at -0.15 V / ≈+0.64 V (*vs.* Ag/AgCl), which are accompanied by color changes from green, blue to yellow and then yellow, blue to green. Since, the electrochemical cell is made up of glass, we have observed these color changes during the measurements. One may then assume that V^{+5} ions are reduced to V^{+4} and V^{+3}, since two anodic peaks are observed. A similar explanation can be given for the oxidation peaks, *i.e.* V^{+3} ions oxidize into V^{+4} and V^{+5}. These color changes are attributed to Li^+ intercalation and deintercalation [45]. On the other hand, the shape of the curve for the vanadium oxide coating using O_2 gas is different indicating one anodic peak at +0.11 V and one cathodic peak at -0.34 V accompanied by color changes from green to yellow and vice versa. This may be due to the existence of different VO_2

orientation planes compared with the one observed for alcohols. Furthermore, the specific current of the as-grown coatings using alcohols as oxygen source is the highest presenting an enhanced electrochemical activity. We then suggest that this is correlated to both the 100 plane and the increased thickness, which incorporate more active material for the insertion of Li^+.

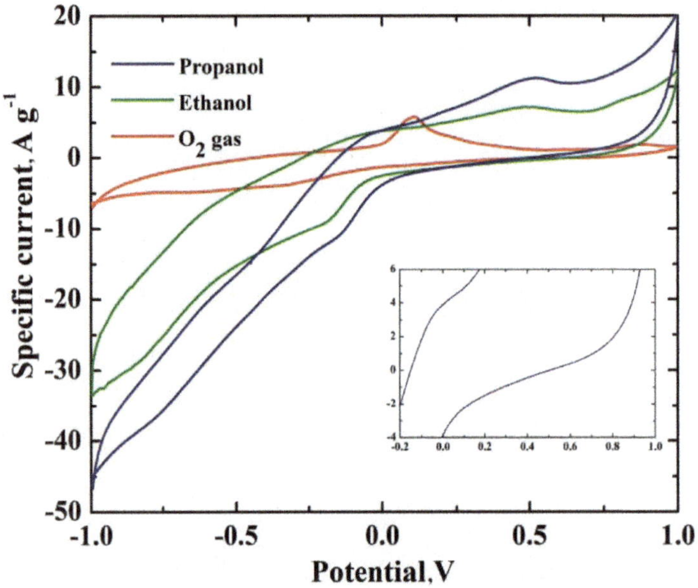

Figure 3. *Cyclic voltammograms of the first scan for the APCVD vanadium oxide coatings for 0.8 L min^{-1} flow rate of propanol, ethanol and O_2 gas and an electrode geometrical active area of 1 cm^2. Maximized cyclic voltammogram curve for the region of -0.2 V - +1 V of the grown vanadium oxide coating using propanol as inset.*

Chronoamperometry measurements were also performed to calculate the specific charge during Li^+ intercalation / deintercalation. It is estimated by integration of excess current measured upon switching the bias potential with time [32] as shown in Figure 4 for the as-grown vanadium oxide using 0.8 L min^{-1} of propanol. The amount of specific charge for propanol found to be 120 C g^{-1}, which is three times higher than that of O_2 gas.

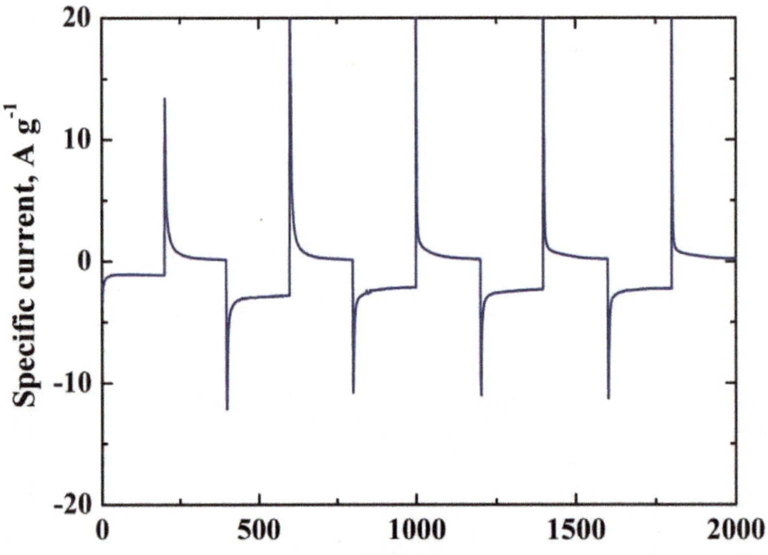

Figure 4. *The chronoamperometric response of the first scan recorded at -1 V and +1 V for an interval of 200 s of the as-grown coatings at 500 °C for 0.8 L min^{-1} flow rate of propanol.*

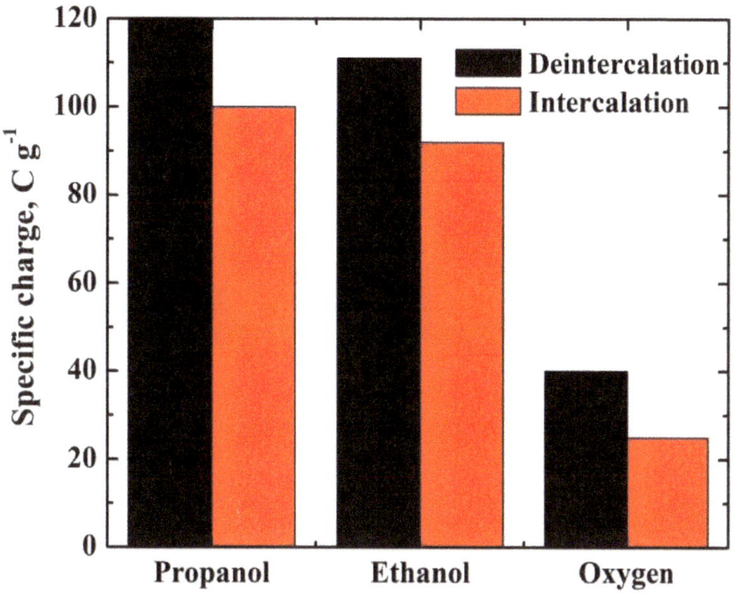

Figure 5. *Intercalated and deintercalated specific charge as a function with oxygen source utilized.*

Figure 6 presents the specific discharge capacities of the as-grown coatings at 500 °C for 7.5 in using 0.8 L min^{-1} flow rate of propanol, ethanol and O_2 gas under a constant specific current of 2 A g^{-1}. The propanol's curve indicates two plateaus at approximately 0.25 V and 0.5 V, which present the two-step Li^+ intercalation process as also observed in cyclic voltammetry analysis. The specific discharge capacity was 459 mAh g^{-1} with a capacitance retention of 97 % after 1000 scans (Figure 6 inset) keeping the staircase shape, which is promising for lithium ion batteries. The specific discharge capacity was higher than the APCVD metastable [3] and 022-oriented monoclinic VO_2 [24]. On the other hand, the ethanol and O_2 gas samples lack of staircase-like shape probably due to the less defined phase transition associated with Li^+. This result may arise due to the largest thickness of the propanol sample, which facilitates larger number of Li^+ within the vanadium oxide lattice.

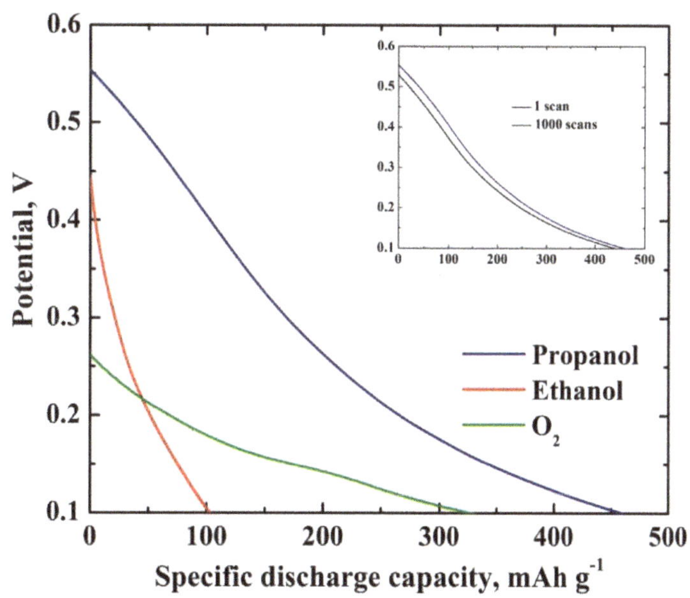

Figure 6. *The chronopotentiometric curves for the as-grown sample at 500 °C for 7.5 min using 0.8 L min^{-1} propanol, ethanol and O_2 gas under a constant specific current of 2 A g^{-1} and potential ranging from 0.1 V to 0.6 V. The 1000th scan of 0.8 L min^{-1} propanol is also included as inset.*

Conclusions

Vanadium dioxides of different crystalline orientation were grown by APCVD at 500 °C for 7.5 min using propanol, ethanol and O_2 gas. The a-axis textured monoclinic was enhanced with propanol and ethanol, while the 022-oriented single phase VO_2 was obtained with O_2 gas. Consequently, the samples with different orientations possessed different morphologies; a-axis textured coatings showed grains, while the 022-oriented phases presented agglomeration of grains forming rod-like structures. Electrochemical analysis has shown that the a-axis textured monoclinic VO_2 grown using propanol has several unique characteristics compared with the rest of the samples, which makes it promising electrode material for lithium ion batteries. These include the thickness, which facilitates Li^+ access to a large volume of active material and the specific discharge capacity of 459 mAh g^{-1} with capacitance retention of 97 % after 1000 scans under constant specific current of 2 A g^{-1} indicating both high rate performance and good stability.

References

[1] M. Rahman, J. Z. Wang, N. H. Idris, Z. Chen, H. Liu, *Electrochim. Acta* **56** (2010) 693-699.
[2] C. V. Subba Reddy, E. H. Walker Jr., S. Wicker Sr., Q. L. Williams, R. R. Kalluru, *Curr. Appl. Phys.* **9** (2009) 1195-1198.
[3] D. Vernardou, D. Louloudakis, E. Spanakis, N. Katsarakis, E. Koudoumas, *Int. J. Thin Film Sci. Tech.* **4** (2015) 187-191.
[4] E. Baudrin, G. Sudant, D. Larcher, B. Dunn, J. M. Tarascon, *Chem. Mater.* **18** (2006) 4369-4374.
[5] F. Wang, Y. Liu, C. Liu, *Electrochim. Acta* **55** (2010) 2662-2666.
[6] Y. Shi, S. -L. Chou, J. -Z. Wang, H. -J. Li, H. -K. Liu, Y. -P. Wu, *J. Power Sources* **244** (2013) 684-689.
[7] H. M. Liu, Y. G. Wang, K. X. Wang, E. Hosono, H. S. Zhou, *J. Mater. Chem.* **19** (2009) 2835-2840.
[8] H. Liu, Y. Wang, L. Li, K. Wang, E. Hosono, H. Zhou, *J. Mater. Chem.* **19** (2009) 7885-7891.
[9] S. Zhang, Y. Li, C. Wu, F. Zheng, Y. Xie, *J. Phys. Chem. C* **113** (2009) 15058-15067.
[10] L. Deng, G. Zhang, L. Kang, Z. Lei, C. Liu, Z. -H. Liu, *Electrochim. Acta* **112** (2013) 448-457.
[11] H. Wang, H. Yi, X. Chen, X. Wang, *J. Mater. Chem. A* **2** (2014) 1165-1173.
[12] L. Liang, H. Liu, W. Yang, *J. Alloys Compd.* **559** (2013) 167-173.
[13] X. Pan, Y. Zhao, G. Ren, Z. Fan, *Chem. Commun.* **49** (2013) 3943-3945.
[14] I. P. Parkin, R. Binions, C. Piccirillo, C. S. Blackman, T. D. Manning, *J. Nano Res.* **2** (2008) 1-20.
[15] M. Maaza, K. Bouziane, J. Maritz, D. S. McLachlan, R. Swanepool, J. M. Frigerio, M. Every, *Opt. Mater.* **15** (2000) 41-45.
[16] J. B. K. Kana, J. M. Ndjaka, P. O. Ateba, B. D. Ngom, N. Manyala, O. Nemraoui, A. C. Beye, M. Maaza, *Appl. Surf. Sci.* **254** (2008) 3959-3964.
[17] C. L. Choy, *Prog. Mater. Sci.* **48** (2003) 57-170.
[18] M. B. Sahana, M. S. Dharmaprakash, S. A. Shivashankar, *J. Mater. Chem.* **12** (2002) 333-338.
[19] M. B. Sahana, S. A. Shivashankar, *J. Mater. Res.* **19** (2004) 2859-2870.
[20] M. N. Field, I. P. Parkin, *J. Mater. Chem.* **10** (2000) 1863-1866.
[21] T. D. Manning, I. P. Parkin, M. E. Pemble, D. Sheel, D. Vernardou, *Chem. Mater.* **16** (2004) 744-749.
[22] D. Vernardou, M. E. Pemble, D. W. Sheel, *Thin Solid Films* **515** (2007) 8768-8770.
[23] D. Vernardou, M. E. Pemble, D. W. Sheel, *Chem. Vapor Depos.* **12** (2006) 263-274.
[24] D. Vernardou, M. Apostolopoulou, D. Louloudakis, N. Katsarakis, E. Koudoumas, *Chem. Vapor Depos.* **21** (2015) 1-3.

[25] T. D. Manning, I. P. Parkin, *Polyhedron* **23** (2004) 3087-3095.
[26] L. Kritikos, L. Zambelis, G. Papadimitropoulos, D. Davazoglou, *Surf. Coat. Tech.* **201** (2007) 9334-9339.
[27] S. Mathur, T. Ruegamer, I. Grobelsek, *Chem. Vap. Deposition* **13** (2007) 42-47.
[28] D. Vernardou, D. Louloudakis, E. Spanakis, N. Katsarakis, E. Koudoumas, *Sol. Energ. Mat. Sol. C.* **128** (2014) 36-41.
[29] Y. Muraoka, Z. Hiroi, *Appl. Phys. Lett.* **80** (2002) 583-585.
[30] G. Garry, O. Durand, A. Lordereau, *Thin Solid Films* **453** (2004) 427-430.
[31] Y. Zhang, R. Wang, Z. Qiu, X. Wu, Y. Li, *Mater. Lett.* **131** (2014) 42-44.
[32] D. Vernardou, M. Apostolopoulou, D. Louloudakis, N. Katsarakis, E. Koudoumas, *J. Colloid Interf. Sci.* **424** (2014) 1-6.
[33] D. Vernardou, A. Sapountzis, E. Spanakis, G. Kenanakis, E. Koudoumas, N. Katsarakis, *J. Electrochem. Soc.* **160** (2013) D6-D9.
[34] D. Vernardou, M. Apostolopoulou, D. Louloudakis, E. Spanakis, N. Katsarakis, E. Koudoumas, J. McGrath, M.E. Pemble, *J. Alloy. Compd.* **586** (2014) 621-626.
[35] B. D. Ngom, M. Chaker, A. Diallo, I. G. Madiba, S. Khamlich, N. Manyala, O. Nemraoui, R. Madjoe, A. C. Beye, M. Maaza, *Acta Mater.* **65** (2014) 32-41.
[36] X. Wu, Y. Tao, L. Dong, Z. Wang, Z. Hu, *Mater. Res. Bull.* **40** (2005) 315-321.
[37] J. Shi, S. Zhou, B. You, L. Wu, *Sol. Energ. Mat. Sol. C* **91** (2007) 1856-1862.
[38] J. B. Kana Kana, J. M. Ndjaka, B. D. Ngom, N. Manyala, O. Nemraoui, A. Y. Fasari, R. Nemutudi, A. Gibaud, D. Knoesen, M. Maaza, *Thin Solid Films* **518** (2010) 1641-1647.
[39] E. Elangovan, K. Ramamurthi, *Thin Solid Films* **476** (2005) 231-236.
[40] D. Vernardou, M. E. Pemble, D. W. Sheel, *Thin Solid Films* **516** (2008) 4502-4507.
[41] G. I. N. Waterhouse, G. A. Bowmaker, J. B. Metson, *Appl. Catal. A Gen.* **266** (2004) 257-273.
[42] P. Forzatti, E. Tronconi, A. S. Elmi, G. Busca, *Appl. Catal. A Gen.* **157** (1997) 387-408.
[43] T. –W. Chiu, K. Tonooka, N. Kikuchi, *Thin Solid Films* **518** (2010) 7441-7444.
[44] M. Ohring, *The Materials Science of Thin Films*, Academic Press, London, 1992.
[45] K. Lee, G. Cao, *J. Phys. Chem. B* **109** (2005) 11880-11885.
[46] D. Vernardou, I. Marathianou, N. Katsarakis, E. Koudoumas, I. I. Kazadojev, S. O' Brien, M. E. Pemble, I. M. Povey, *Electrochim. Acta* **196** (2016) 294-299.
[47] J. Qin, W. Lv, Z. Li, B. Li, F. Kang, Q. –H. Yang, *Chem. Commun.* **50** (2014) 13447-13450.
[48] Y. Wu, P. Zhu, X. Zhao, M. V. Reddy, S. Peng, B. V. R. Chowdari, S. Ramakrishna, J. *Mater. Chem. A* **1** (2013) 852-859.

5

New potentiometric electrode based on ion pair complex for determination of tropicamide in pure and pharmaceutical formulations

Mouhammed Khateeb✉, Basheer Elias*, Hazar Alksair*

Department of Basic Sciences, Faculty of Mechanical and Electrical Engineering, University of Al-Baath, Homs, Syria
**Department of Chemistry, Faculty of Sciences, University of Al- Baath Homs, Syria*

✉Corresponding author: drkhateeb2010@gmail.com, mkhateeb74@hotmail.com

---

Abstract

Construction and general performance of a novel modified carbon paste electrode (MCPE) for determination of tropicamide (TPC) in pure form and pharmaceutical formulations have been examined. Tropicamide-tetraphenylborate (TPC–TPB) ion pair has been prepared and used as electroactive material. The best MCPE electrode was composed of 7 % ion-pair, 46.5 % dioctylphthalat and 46.5 % graphite powder. The electrode shows stable potentiometric response for TPC in the concentration range 0.3–221.0 μM at 25 °C and pH range of 2.0–8.0. The electrode exhibits near Nernstian slope of 59.71±0.30 mV/decade and lower limit of detection of 0.09 μM with fast response time (less than 15 s). The selectivity of the electrode (TPC–TPB) was investigated with respect to some organic and inorganic cations. The MCPE was designed to have better mechanical resistance. The proposed method was successfully applied for determination of TPC in eye drop formulation.

Keywords

Tropicamide; Sodium tetraphenylborate; Ion pair complex; Carbon paste electrode; Pharmaceutical formulations

---

Introduction

Tropicamide (TPC), (R,S)-N-ethyl-3-hydroxy-2-phenyl-N-(pyrid-4-yl-methyl) propionamide (Fig. 1), is an antimuscarinic agent with short duration of mydriatic and cycloplegic effect. TPC is used for refractive examinations and preservative for optimal eye tolerance and activity [1]. The action of TPC is more rapid in onset and wears off more rapidly than most other mydriatics and its use is generally much the same as those described for other mydriatics [2,3].

Several methods have been reported for determination of TPC including spectrophotometry [4–8], high performance liquid chromatography (HPLC) [9–12] and gas chromatography–mass spectrometry (GC–MS) [13]. In the United States Pharmacopoeia (USP) [14] and British Pharmacopoeia (BP) [15] the non-aqueous titration method for determination of TPC in raw materials and the extractive spectrophotometric method for pharmaceutical preparations have been described, respectively.

In the literature survey, no potentiometric method has been reported yet for determination of TPC. Therefore, the objective of the present study was to develop an accurate and validated potentiometric method for determination of TPC in raw material and pharmaceutical dosage forms, that can also be used in quality control laboratories.

Since their innovation by Adams [16], carbon paste electrodes (CPEs) have widely been used in many fields such as voltammetry, amperometry and potentiometry [17]. In comparison with ion-selective electrodes based on polymeric membranes, chemically modified carbon paste electrodes (CMCPEs) possess some advantages. In addition to lower limits of detection with respect to electrodes with an internal reference solution, CMCPEs are ease for preparation and regeneration, have stable response and very low ohmic resistance [18].

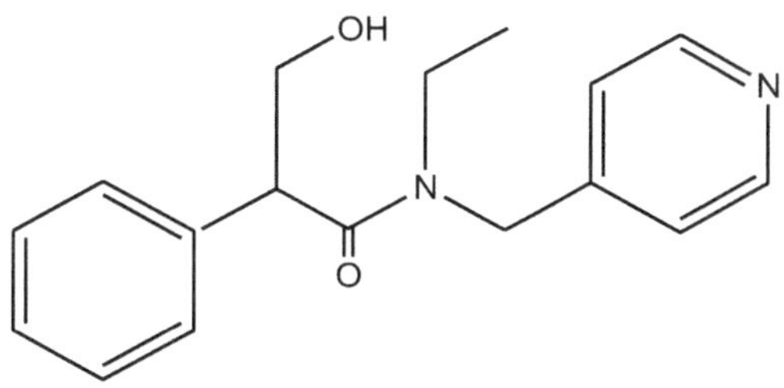

Figure 1. *Chemical structure of tropicamide.*

Experimental

Reagents and chemicals

All chemicals used were of analytical grade. Double distilled water was used throughout all experiments. Working reference standard of tropicamide (TPC $C_{17}H_{20}N_2O_2$, 284.35 g/mole) was obtained from Kunshan Chemical and Pharmaceutical Co., Ltd (India). Its purity was found to be 99.5 % according to the compendial testing method. Pharmaceutical preparation Mydriamed eye drop was provided by Medico Company for Pharmaceutical Industries (Homs, Syria). Sodium tetraphenylborate (Na–TPB) was from Merck, while dibutyl phthalate (DBP), dioctyl phthalate (DOP) and paraffin oil (p.Oil) were from BDH. Graphite powder was obtained from Aldrich.

Apparatus

All potentiometric measurements were made at 25±1 °C using a Sanwal potentiometer (pH meter, Ion meter model of DT9201A) with combined pH electrode (model 250A, Orion, USA) for pH measurements. Saturated calomel electrode (SCE) was used as the reference electrode. The electrochemical cell is represented as follows:

Hg, Hg_2Cl_2(s), KCl (sat.) ‖ sample solution | carbon paste electrode | Cu

Standard solution

Standard stock solution of 1.0×10^{-2} M TPC was prepared daily by dissolving the appropriate amount of drug in double distilled water. Standard working solutions 1.0×10^{-3} and 1.0×10^{-4} M were freshly prepared by suitable dilutions of the stock solution with double distilled water.

Calibration graph

Different amounts of 1.0×10^{-4} or 1.0×10^{-3} M TPC were added to 50mL of double distilled water to achieve the concentration range from 0.3 to 221.0 μM. The measured potential was recorded. The calibration graph was constructed by plotting the potential value versus pC_{TPC} (–logarithm concentration of TPC). The amount of drug was obtained from the regression equation.

Analysis of eye drop

0.1 mL of the commercial eye drop was transferred into a 100 mL volumetric flask and diluted with double distilled water. The general procedure was then followed in the concentration range mentioned above.

Preparation of tropicamide-tetraphenylborate (TPC–TPB) ion-pair

Sensing element used in the carbon paste electrode is the ion-pair compound made by interaction of TPC and Na–TPB. It was prepared by mixing 20 mL of 0.01 M acidic solution of TPC with 20 mL aqueous solution of 0.01 M Na–TPB. The resulting solution was then filtered, the precipitate washed with double distilled water and dried at room temperature [19].

Modified carbon paste electrodes (MCPEs)

MCPEs were prepared by thoroughly mixing various amounts of ion pairing agents with carbon powder and plasticizer in the mortar, until homogenization of this mixture was achieved. The resulting paste was then packed firmly into the hole of the electrode body. Electrical contact to the carbon paste was made with a copper wire. Fresh surface was obtained by applying manual pressure to the carbon paste and polished on a filter paper to a shiny surface (Fig.2).

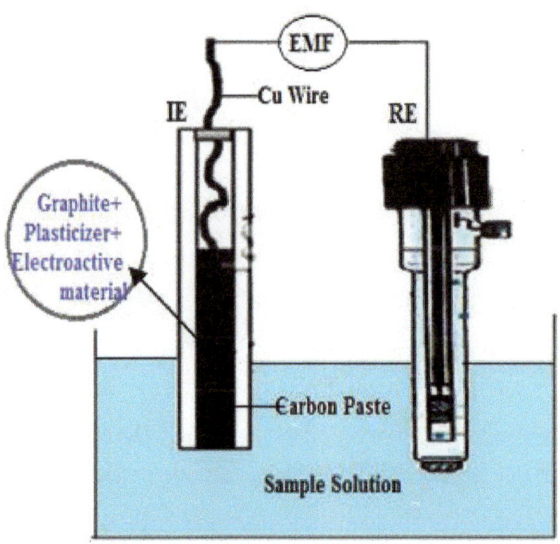

Figure 2. *Schematic illustration of electrochemical cell with MCPE as ~~an~~ indicator electrode (IE) and SCE as reference electrode (RE).*

Selectivity of the electrode

Potentiometric selectivity coefficient was evaluated using the matched potential method (MPM) [20]. According to the MPM, the activity of the analyte was increased from $\alpha_A = 9.90\times10^{-5}$ M (reference solution) to $\alpha'_A = 1.01\times10^{-4}$ M and the corresponding change in potential (ΔE) is measured. Then, 1.00×10^{-1} M solution of an interfering ion was added to ~~a~~ new 9.90×10^{-5} M analyte reference solution, until the same ΔE is recorded, determining thus the concentration of the added amount, α_B. The selectivity coefficient $K_{A,B}^{MPM}$ for each interfering ion was calculated using the following equation:

$$K_{A,B}^{MPM} = \frac{\alpha_A' - \alpha_A}{\alpha_B}$$

Results and discussion

Calibration graph and effect of ion pair (IP) percentage on electrode potential

It is known that sensitivity and linearity of an electrode depend significantly on the amount of the ion–pair (IP) in the carbon paste composition. Thus, influence of the TPC–TPB IP percentage in the carbon paste composition was investigated first. Preliminary experiment showed that carbon paste electrode without IP modifier has no response towards the analyte. For this purpose, ten electrodes were prepared containing the IP modifier percentage from 1 to 10 % and the results are summarized in Fig. 3 and Table 1.

Increase-of IP percentage in the paste is found to increase the electrode response and stability of potentiometric readings, as well as slopes (S) of linear parts of calibration graphs defined by equation $E = f(\mathrm{p}C_{\mathrm{TPC}})$, reaching S = 59.714 mV dec^{-1} at 7 % TPC–TPB (Fig. 4). At percentages above 7 %, a decrease of slope and reduction of linearity range where $E = f(\mathrm{p}C_{\mathrm{TPC}})$ of the MCPE electrode are observed due to the kinetics of IP within the paste.

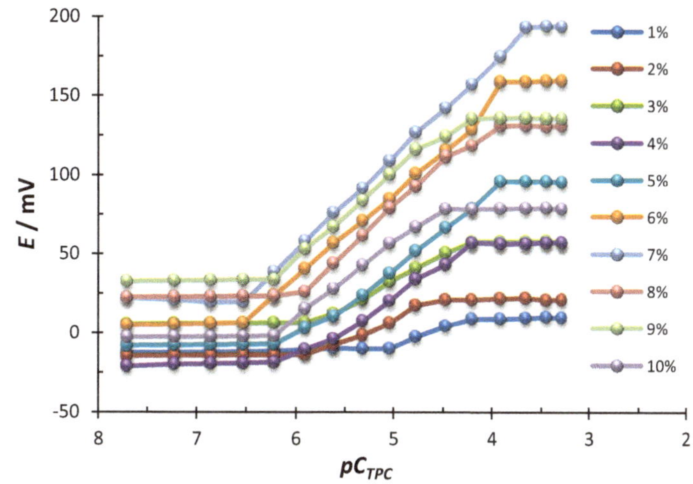

Figure 3. *Effect of IP percentage in the MCPE on the calibration graph of TPC*

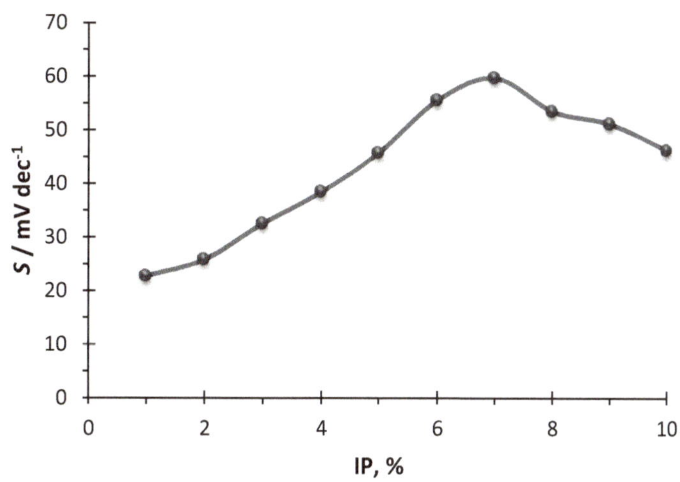

Figure 4. *Effect of IP percentage in MCPE on the slope (S) of the linear, E =f(pC_{TPC}), range of the calibration graph of TPC.*

Table 1. *Potentiometric data for MCPE with different IP percentage obtained from linear equations.*

Composition, %			S / mV dec^{-1}	Intercept	Linear range, µM	r
Graphite	Plasticizer	IP				
49.5	49.5	1	22.77	105.17	9.27-64.70	0.9873
49.0	49.0	2	25.83	136.41	1.25-33.40	0.9810
48.5	48.5	3	32.53	190.24	1.25-64.70	0.9942
48.0	48.0	4	38.43	215.31	0.62-64.70	0.9868
47.5	47.5	5	45.73	268.98	0.62-122.00	0.9906
47.0	47.0	6	55.53	366.42	0.62-64.70	0.9912
46.5	46.5	7	59.71	409.40	0.30-221.00	0.9995
46.0	46.0	8	53.49	344.94	1.25-64.70	0.9904
45.5	45.5	9	51.17	354.48	0.62-64.70	0.9922
45.0	45.0	10	46.22	286.89	0.62-33.40	0.9971

Response characteristics and statistical data

The characteristics performance of the best electrode (shadowed in blue in Table 1) was determined and the results are summarized in Table 2. The best proposed MCPE electrode shows nearly Nernstian response over the concentration range 0.3-221.0 µM.

Table 2. *Response characteristics of the best MCPE-TPC electrode.*

Parameter	TPC-TPB electrode
IP, %	7
Regression equation	$E = -59.7\text{pC}_{\text{TPC}} + 409.40$
Correlation coefficient (r)	0.9995
Linear range, µM	0.3–221.0
LOD, µM	0.09
LOQ, µM	0.300
Response time, sec	≤15
Life time, day	70
Working pH range	2.0–8.0

The slope of calibration graph for the best MCPE is 59.7 mV/decade for TPC concentrations in the range of 0.3–221.0 µM, with standard deviation of ±0.3 mV after five replicate measurements. Limit of detection (LOD) was calculated from the intersection of two extrapolated segments of the calibration graph [21]. LOD and limit of quantification (LOQ) were 0.09 and 0.3 µM, respectively.

Effect of plasticizer on the potential response

In this study, three plasticizers, di-octylphthalate (DOP), di-butylphthalate (DBP) and paraffin oil (p. Oil) were used to examine possible optimization of the paste. Contents of examined plasticizers were 46.5, 46.0 and 45.0 w %, contents of graphite powder were 46.5, 46.0 and 45.0 w % and contents of electroactive compound (TPC–TPB IP) were 7, 8 and 10 w %, respectively. The sum of percentages of all three components was always adjusted to 100 %. The results obtained showed that response performances of prepared pastes are rather different, depending on the kind of plasticizer, proportion of the plasticizer towards graphite and amount of electroactive compound (Table 3). Typical potential responses of electrodes constructed with three plasticizers are given in Fig.5. As shown in Fig. 5, the DOP-graphite electrodes were superior to DBP-graphite and p.Oil-graphite electrodes in both the response slope and linear concentration range. So, DOP was selected as the plasticizer of the carbon paste. The best paste composition of the DOP-graphite electrode was 46.5 % graphite, 46.5 % DOP and 7 % IP.

Table 3. *General characteristic of different plasticizers of MCPE-TPC electrode.*

Composition of the plasticizer, %	IP, %	S / mV dec^{-1}	Linear range, µM	LOD, µM	Response time, s
DOP (46.5)	7	59.71	0.30 – 221.00	0.09	15
DBP (46.0)	8	52.18	0.62 – 64.70	0.31	20
p. Oil (45.0)	10	44.88	2.48 – 122.00	0.89	30

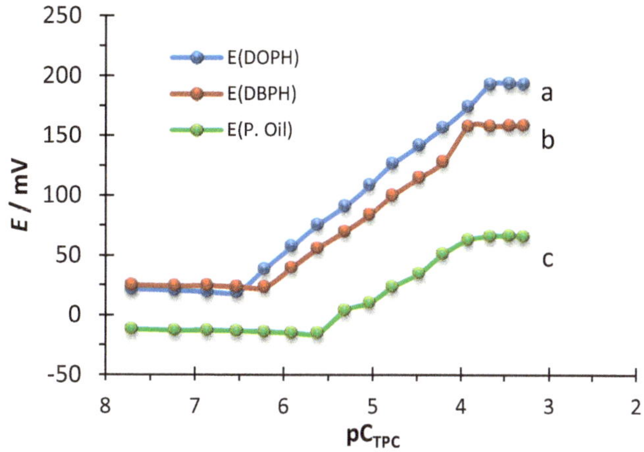

Figure 5. *Optimization of plasticizers with CPE compositions:**(a)** DOP 46.5 %, graphite 46.5 %, IP 7.0 %, **(b)** DBP 46.0%, Graphite 46.0 %, IP 8.0 %), **(c)** p. Oil 45.0 %, graphite 45.0, IP 10.0 %.*

Dynamic response time

Dynamic response time is the required time for the electrode to achieve values within ±1 mV of the final equilibrium potential after successive immersions in the sample solutions [22]. Its calculation involved the variation and the recording of the TPC concentration in a series of solutions from 0.5 to 50.0 µM. The electrode was able to quickly reach its equilibrium response in the whole concentration range. This time for the MCPE was about 15 seconds in the concentrated solutions ≤221.00 µM (Fig 6).

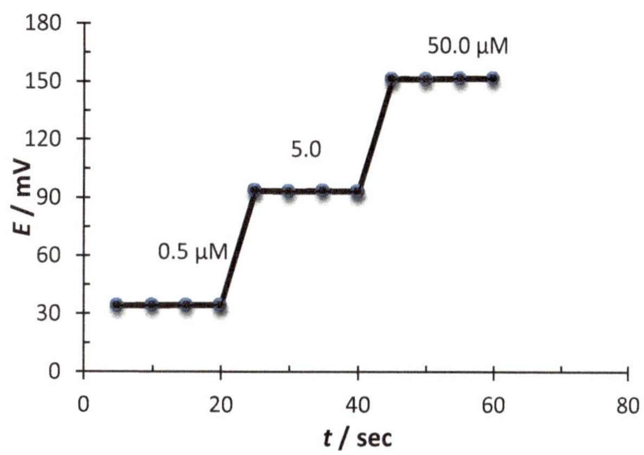

Figure 6. *Response time of MCPE for different concentrations of TPC.*

Selectivity of the electrode

Influence of various basic substances on the response of MCPE-TPC electrode was investigated by measuring the potentiometric interference from different kinds of sugars, inorganic cations and certain alkaloids. Selectivity coefficients were evaluated by the matched potential method (MPM). Table 4 showed that the proposed MCPE is highly selective towards TPC. The electrode showed no response to a number of potentially interfering ions usually used in the manufacturing of pharmaceutical preparations, such as starch and lactose. The inorganic cations did not interfere due to differences in their mobilities and permiabilities as compared with TPC cation.

Table 4. *Selectivity coefficients of some interfering ions*

Interfering ion	Log K_{MPM}	Interfering ion	Log K_{MPM}
Na^+	-3.51	Ba^{2+}	-2.51
K^+	-3.38	Fructose	-
Ca^{2+}	-2.97	Glucose	-
Mg^{2+}	-3.60	Sucrose	-
NH_4^+	-2.43	Maltose	-
Mn^{2+}	-2.71	Lactose	-
Zn^{2+}	-2.43	Starch	-

Effect of pH

To examine the effect of pH on the MCPE-TPC electrode response, the potential was measured at specific concentration of the TPC solution (50.0, 5.0 and 0.5 μM) having pH value of 1.0 up to 14.0 (concentrated NaOH or HCl solutions were employed for pH adjustments). The results showed that the potential remained constant despite pH change in the range of 2.0-8.0, indicating applicability of this electrode in the specified pH range (Fig. 7).

Relatively prominent fluctuations in the potential vs. pH behavior took place below and above the formerly stated pH limits. In detail, the fluctuations above the pH value of 8.0 might be justified by removing the positive charge on the drug molecule. Fluctuations below the pH value of 2.0 may be due to interference of hydronium ion.

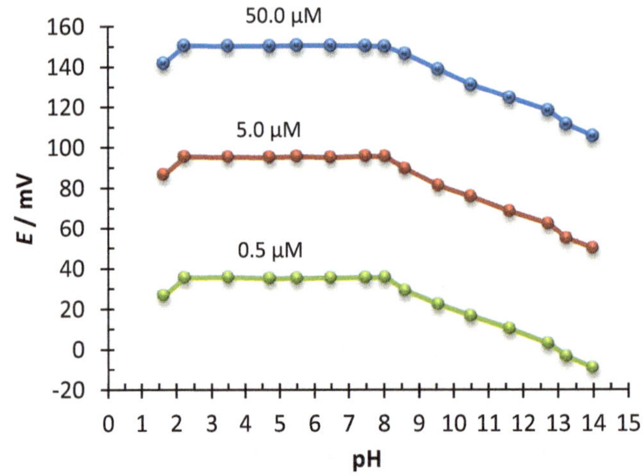

Figure 7. *The pH effect on potential response of the MCPE-TPC electrode.*

Effect of temperature

To investigate thermal stability of the electrode, the effect of temperature of TPC solution on the MCPE was studied at different temperature values (20-70 °C). The MCPE exhibited good Nernstian behavior in the temperature range of (20-60 °C). The standard cell potentials (E^o_{cell}) were determined at different temperatures and used to determine the isothermal temperature coefficient ($dE°/dt$) of the cell with the aid of the following equation [23]:

$$E°_{cell} = E°_{cell,25\,°C} + (dE°/dt)_{cell}\, t$$

Plotting t *vs.* $E°_{cell,25ºC}$ produced the straight line (Fig. 8). The slope of this line was taken as the isothermal coefficient of the cell which was found to be -1.96×10^{-3} V/°C. The values of isothermal coefficient of the electrode revealed relatively high thermal stability within the investigated temperature range.

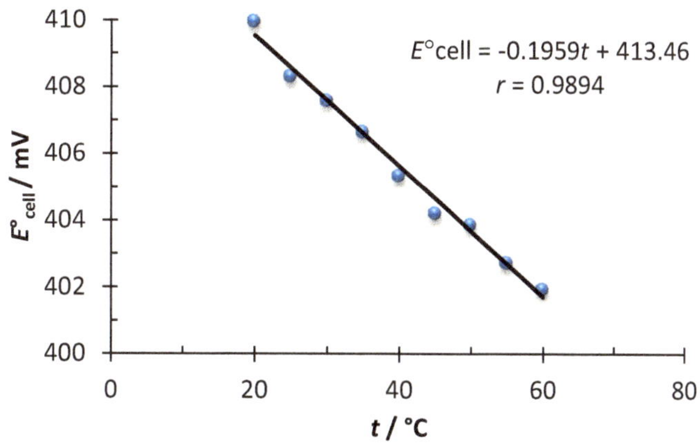

Figure 8. *Variation of the* $E°_{cell}$ *with temperature for the MCPE-TPC electrode.*

Potentiometric titration

Representative titration curve for determination of the investigated drug is shown in Fig.9. This MCPE electrode can be used successfully as indicator electrode in potentiometric titrations of the investigated drug. It also indicates that 1:1 [TPC]:[TPB] is formed as seen from the curve.

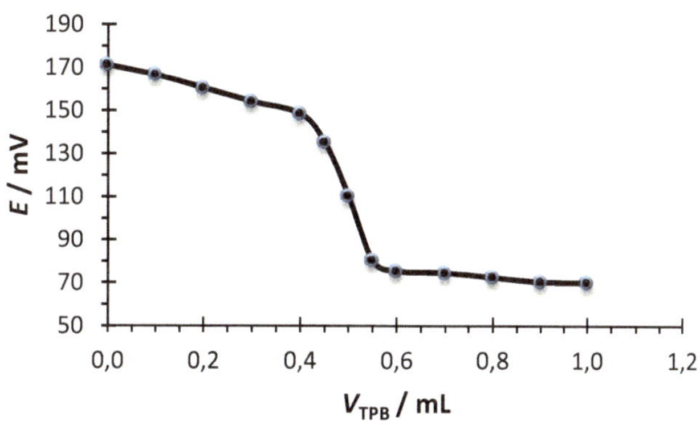

Figure 9. *Typical potentiometric titration of 50 mL of* 1×10^{-4}*M TPC with* 1×10^{-2} *M TPB using MCPE*

Life-time study

The MCPE electrode life time was estimated with the calibration graph, periodical tests with standard solutions (0.30-221.0 µM) and calculation of its response slope. For this estimation, the electrode was employed extensively (1 hour per day) for 80 days. As it can be seen from Table 5, in the case of MCPE this time is 70 days which shows the long-term stability of this kind of electrode. In MCPEs the surface of the electrode is renewable and can be used for longer time.

Table 5. *Life time of MCPE electrode*

Time, day	S / mV dec^{-1}	Linear range, µM	Time, day	S / mV dec^{-1}	Linear range, µM
1 / 24	59.71	0.30-221.00	45	59.65	0.30-122.00
1	59.35	0.30-221.00	50	60.02	0.62-122.00
2	59.50	0.30-221.00	55	59.22	0.62-122.00
5	59.96	0.30-221.00	60	59.11	0.62-221.00
10	59.60	0.30-221.00	65	59.09	0.30-122.00
15	59.41	0.30-221.00	70	58.65	1.25-122.00
20	59.75	0.30-221.00	75	58.31	1.25-122.00
25	59.18	0.30-221.00	80	57.76	2.48-64.70
30	59.15	0.30-221.00			
35	59.31	0.30-221.00			
40	59.55	0.30-221.00			

Accuracy and precision

The precision and accuracy of the method were also evaluated. The standard deviation, relative standard deviation and recovery of different TPC amounts were determined and recorded in Table 6. The accuracy of the method is indicated by excellent recovery (99.80-101.40 %) and precision is supported by low standard deviation.

Table 6. *Accuracy and precision for determination of TPC in pure form by the proposed method.*

C_{TPC}, μM		RSD, %	Recovery, %
Taken	Found ± SD*		
0.5	0.499±0.01	2.00	99.80
5.0	5.07±0.08	1.58	101.40
50.0	50.10±0.58	1.16	100.20
100.0	99.80±0.82	0.82	99.80

*Average of five replicates

Analytical application

The application of the proposed method using calibration graph gives good results as shown in Table 7. The results were compared with the official method [14] and shown that the MCPE has good efficiency as regard of sensitivity, index of retrieving and repetition.

Table 7. *Determination of TPC in eye drop by here proposed and official methods.*

Formulation	Label claim	Recovery, % ± SD[a]	
		Proposed method	Official method [14]
Mydriamed	0.5 % 5 mg/mL	99.60±0.10 t^b= 0.45 F^c=1.23	99.80±0.09 t^b=0.25
	1.0 % 10 mg/mL	100.90±0.11 t^b=1.83 F^c=1.21	100.60±0.10 t^b=1.34

[a] Average of five replicates; [b] Tabulated *t*-value at 95 % confidence level is 2.776. [c] Tabulated *F*-value at 95 % confidence level is 6.26

Conclusion

The proposed chemically modified carbon paste electrode (MCPE) demonstrated advanced performance with fast response time and long stability, and shows high sensitivity, reasonable selectivity, and applicability over wide concentration range without sample pretreatment. The proposed procedure can be used for the routine analysis of TPC in bulk and eye drop. The sample recoveries from all samples were in good agreement with their respective label claims.

References

[1] J. G. Hardman, L. E. Limbird, P. B. Molinoff, R. W. Ruddon, *The Pharmacological Basis of Therapeutics Goodman and Gilman's, 9th edition*, Mc Graw-Hill, New York, 1996, 156-7, 1635.

[2] J. G. Hardman, A. Goodman Gilman, L. E. Limbird, *The pharmacological pasis of therapeutics, 10th ed.* New York: Mc Graw-Hill, 2001, 1673-85.

[3] J. N. Delgabo, W. A. Remers, (editors.), Wilson and Gisvold's textbook of organic medicinal and pharmaceutical chemistry. 10th ed. New York: Lippincott-Raven Publ., 2004, 585.

[4] Z. A. Shoaibi, A. A. Gouda, *Journal of Young Pharmacists* **4** (2012) 42-48.

[5] A. Massoud, S. Effat, I. Shiva, F. Hassan, *Iranian Journal of Pharmaceutical Sciences* **3(1)** (2007) 43-50.

[6] S. Effat, A. Massoud, S. H. Shiva, B. Mitra, *Iranian Journal of Pharmaceutical Sciences* **6(3)** (2010) 171-178.
[7] S. Swati, R. C. Mashru, *Journal of the Chilean Chemical Society* **55(4)** (2010) 43-50.
[8] N. Y. Hasan, M. A. Elkway, B. E. Elzeany, N. E. Wagieh, *Journal of Pharmaceutical and Biomedical Analysis* **28** (2002) 87-97.
[9] M. J. Galmier, A. M. Frasey, S. Meski, E. Beyssac, J. Petit, J. M. Aiache, C. Lartigue, *Biomedical Chromatography* **14(3)** (2000) 202–204.
[10] M. Amanlou, G. Asmardi, P. C. Andalibi, N. Javadi, F. Khodadady, Z. B. Omarny, *Journal of Chromatogaphy* **1(2)** (2005) 136-139.
[11] B. P. Harshit, C. H. Payal, D. P. Bhavesh, K. S. Samir, *International Journal of Institutional Pharmacy and Life Sciences* **4(5)** (2014) 20-27.
[12] M. Jovovic, N. Kostic, B. Janc, J. Stojanovic and A. Malenovic, *Journal of Liquid Chromatography and Related Technologies* **35** (2012) 231–239.
[13] Z. Türkmen, M. Selda, S. Bağdadi, S. Cengiz, *Nobel Medicus* **33** (2015) 50-54.
[14] United States Pharmacoepia 29. Rockville: *United States Pharmacoepial Convention*, 2005, 295, 2223.
[15] British Pharmacopoeia. London: *Her Majesty's, Stationery Office*, 2005, 2028.
[16] R. N. Adams, *Analytical. Chemistry* **30** (1958) 1576.
[17] I. Svancara, K. Vytras, K. Kalcher, A. Walcarius, J. Wang, *Electroanalysis* **21** (2009)7.
[18] K. Kalcher, J.M. Kauffmann, J. Wang, I. Svancara, K. Vytras, C. Neuhold and Z. Yang, *Electroanalysis* **7** (1995) 5.
[19] M. R. Ganjali, S. Aghili, M. P. Hamedani, P. Norouzi, *International Journal of Electrochemical Science* **7** (2012) 10394–10403.
[20] Y. Umezawa, P. Bühlmann, K. Umezawa, K. Tohda, S. Amemiya, *Pure and Applied Chemistry* **72 (10)** (2000) 1851-2082.
[21] L. Meites, P. Zuman, H.W. Nurnberg, Recommended terms, *Pure and Applied Chemistry* **57 (10)** (1985) 1491-1505.
[22] P. L. Baily, M. Riley, *Analyst (London)* **100 (1188)** (1975) 145-156.
[23] A. J. Debethune, T. S. Licht, N. Swendeman, *Journal of Electrochemical Society* **106** (1959) 616-625.
[24] N. L. Teradal, S. N. Prashanth, J. Seetharamappa, *J. Electrochem. Sci.* Eng. **2** (2012) 67-75.

6

Cyclic voltammetric study of tin hexacyanoferrate for aqueous battery applications

Denys Gromadskyi✉, Volodymyr Chervoniuk, Sviatoslav Kirillov

Joint Department of Electrochemical Energy Systems, 38A Vernadsky Ave, 03680 Kyiv, Ukraine

✉Corresponding author: d.gromadskyi@gmail.com

Abstract

A hybrid composite containing 65 mass % of tin hexacyanoferrate mixed with 35 mass % of carbon nanotubes has been synthesized and its electrochemical behavior as a negative electrode in alkali metal-ion batteries has been studied in 1 mol L^{-1} aqueous solution of sodium sulfate. The specific capacity of pure tin hexacyanoferrate is 58 mAh g^{-1}, whereas the specific capacity normalized per total electrode mass of the composite studied reaches 34 mAh g^{-1}. The estimated maximal specific power of an aqueous alkali-metal ion battery with a tin hexacyanoferrate electrode is ca. 3.6 kW kg^{-1} being comparable to characteristics of industrial electric double-layer capacitors. The maximal specific energy accumulated by this battery may reach 25.6 Wh kg^{-1} at least three times exceeding the specific energy for supercapacitors.

Keywords

Aqueous alkali-metal-ion battery; Tin hexacyanoferrate-based composite

Introduction

Nowadays, the list of electrochemical energy storage devices is quite wide, from electric double-layer capacitors (the so-called supercapacitors) to various rechargeable batteries (lead-acid, lithium-ion and others). Their specific energy and power characteristics significantly vary depending on electrode materials, electrolyte compositions, *etc.* Supercapacitors demonstrate very high specific power values (up to 6 kW kg^{-1} at 95 % efficiency) but their specific energy is immeasurably low compared to any commercially available battery (1–8 Wh kg^{-1} vs. 26–170 Wh kg^{-1}) [1,2].

In recent years, alkali metal-ion batteries, supercapacitors and their hybrids with neutral aqueous electrolytes gain popularity due to their relatively low cost, fire safety and environmental friendliness as opposed to devices with nonaqueous organic solutions which decompose under applied voltage at the presence of a small amount of water impurity and, inter alia, may provoke the corrosion of metallic cell components [3–7]. As electrode materials for the aqueous rechargeable batteries, multivalent metal complexes (*e.g.* barium, cobalt, copper, iron, manganese, nickel, titanium and zinc hexacyanoferrates) [8–15] are commonly utilized. All these compounds having a perovskite-like structure similar to the Prussian blue ($A_xM_y[Fe(CN)_6]$, where A is an alkali metal and M is a multivalent metal) are good hosts for alkali and alkaline earth ions [16]. Their electrochemical properties arise from the process of alkali metal ion intercalation/deintercalation [17]:

$$M_y[Fe(CN)_6] + xA^+ + xe^- \leftrightarrow A_xM_y[Fe(CN)_6] \quad (1)$$

Hexacyanoferrate anion is relatively low-toxic (LD_{50} oral-rat is 1600 mg kg^{-1}) [18] because it does not tend to release free cyanide, and the total toxicity of a salt should be defined by the harmfulness of the multivalent metal. If the data on lethal doses (LD_{50}, oral-rat) and prices for raw materials used for the synthesis of hexacyanoferrates (mainly nitrates and chlorides of reagent grade or p.a. quality) are compared, Table 1, one can see that tin chloride has a low value of LD_{50} and is quite inexpensive [18,19]. This makes tin hexacyanoferrate (SnHCF) an interesting candidate for electrochemical studies. Its theoretical specific capacity $Q_{sp.(theor.)}$ in the case of insertion/deinsertion of sodium ion,

$$Sn_{1.5}[Fe^{III}(CN)_6] + Na^+ + e^- \leftrightarrow NaSn_{1.5}[Fe^{II}(CN)_6] \quad (2)$$

calculated as

$$Q_{sp.(theor.)} = \frac{nF}{M} \quad (3)$$

equals to 64.9 mAh g^{-1}. In Eq. (3), n is the number of electrons taking part in the electrode half-reaction, F = 26801 mAh mol^{-1} is the Faraday constant, and M is the molecular mass of $NaSn_{1.5}[Fe(CN)_6]$.

***Table 1** Toxicity and cost of some multivalent metal chlorides and nitrates.*

	Chlorides						
	$BaCl_2$	$CoCl_2$	$CuCl_2$	$MnCl_2$	$NiCl_2$	$SnCl_2$	$ZnCl_2$
LD_{50}, mg kg^{-1}	118	80	584	1484	105	700	350
Price, € kg^{-1}	45	1272	174	199	904	148	122
	Nitrates						
	$Ba(NO_3)_2$	$Co(NO_3)_2$	$Cu(NO_3)_2$	$Mn(NO_3)_2$	$Ni(NO_3)_2$	$Sn(NO_3)_2$	$Zn(NO_3)_2$
LD_{50}, mg kg^{-1}	355	691	794	300	1620	N/A	1190
Price, € kg^{-1}	122	458	300	100	1590	N/A	38

Despite of the large number of publications referred to hexacyanoferrate-based batteries no research works exist directed towards the use of SnHCF in energy storage applications, although it serves as a material for electrochemical sensors [20]. This paper aims at filling this gap, i.e. at exploring potentialities of a SnHCF based composite as an electrode material for aqueous alkali-ion batteries.

Experimental

1. Composite synthesis

A composite containing SnHCF and multi-walled carbon nanotubes (CNTs) as a conductive additive was synthesized via room-temperature precipitation technique according to the following equation,

$$3SnCl_2 + 2K_3[Fe(CN)_6] = 2Sn_{1.5}[Fe(CN)_6] + 6KCl \quad (4)$$

FloTube™ 9000 CNTs from C-Nano were treated in nitric-sulfuric acid mixture as proposed in Ref. [21], dispersed in water, and then homogenized by a mechanical stirrer for 15 min. Tin chloride dihydrate (reagent grade, ReaKhim) was added into the reaction vessel and mixed together for next 15 min. After that, the 0.3 mol L^{-1} solution of potassium hexacyanoferrate (reagent grade, ReaKhim) was added dropwise to avoid fast precipitation and coagulation of SnHCF particles. A SnHCF-CNT suspension was stirred for 1 h, filtered, washed with deionized water to neutral pH and dried at 80 °C for 12 h. It should be noted that the amount of CNTs in the SnHCF/CNT composite was varied from 20 to 80 mass %.

2. Electrode preparation and characterization

In order to prepare the electrodes a SnHCF/CNT composite was homogenized in ethanol where polytetrafluoroethylene (60 mass % dispersion in water, Sigma Aldrich) as a binding agent was added in advance. The mass ratio of the SnHCF/CNT composite and the polytetrafluoroethylene binder was 9 to 1. Pre-electrode slurries were formed by manual mixing in a mortar and alcohol evaporation. Ready-to-use electrodes were obtained after densification by rolling and drying at 80 °C for 12 h. Their porosity analysis was performed at 77 K by means of Quantachrome® Autosorb Station 3.

3. Cell assembling and electrochemical testing

Electrochemical measurements were carried out in a flat three-electrode cell by means of cyclic voltammetric technique on a home-made potentiostat at room temperature. The working electrode was the SnHCF/CNT composite with binding agent (mass 2 mg, geometric surface area 0.09 cm^2). The counter electrode with the geometric surface area of 0.5 cm^2 was made of a commercial activated carbon (Kuraray YP50F). These electrodes were separated by two layers of a porous paper membrane (TF4030, Nippon Kodoshi), and a silver sulfate reference electrode fabricated as described in Ref. [22] was located between them. Its potential was +0.137 V vs. Ag/AgCl at 25 °C. After assembling, the cell was impregnated by an electrolyte, 1 mol L^{-1} Na_2SO_4 aqueous solution.

Cyclic voltammetric testing is suitable for determining key characteristics of any energy storage material, namely its operation potential range, reversibility of charge/discharge processes as well as specific capacity [23]. The latter was calculated as

$$Q_{sp.} = \frac{I\Delta U}{3.6 \times vm} \quad (5)$$

where $Q_{sp.}$ is the specific capacity (mAh g^{-1}), I is the current (A), ΔU is the operating potential range (V), v is the scan rate (V s^{-1}) and m is the electrode mass.

Results and discussion

The electrochemical behavior of SnHCF-based electrodes depending on the CNT content is shown in Fig. 1. Cyclic voltammograms (CVs) demonstrate cathodic and anodic peaks characteristic of electrodes where reversible or quasi-reversible electrochemical reactions take place. Adding CNTs leads to changing the profile of CV making it typical for electrodes with electric double-layer capacitance.

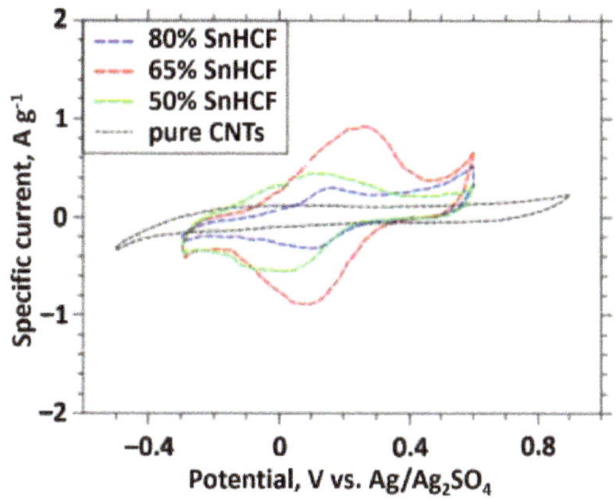

Fig. 1. *CVs of SnHCF-based electrodes containing different amount of CNTs recorded at 5 mV s*$^{-1}$.

The heights of the peaks responsible for the intercalation/deintercalation of Na^+ ion into/from the SnHCF structure are different. This means that respective specific capacities are also different and, moreover, non-linearly vary with the SnHCF/CNT ratio. Upon growing the amount of CNTs the specific capacity increases rapidly reaching a maximal value (~45 mAh g^{-1}) at 35 mass % of CNTs and then decreases. Such behavior presumably reflects the contribution of CNTs to the structure formation of the composites. CNTs not only play the role of a conductive additive; they create a structured framework facilitating the access of ions to electroactive hexacyanoferrate species deposited in mesopores with pore radii greater than 12 nm. This is seen from Fig. 2(a) where pore size distributions calculated according to the density functional theory (DFT) is shown. Besides, SnHCF-based composites have a slightly higher volume V of micropores of about 1 nm radii r than CNTs possibly due to oxidizing the CNT surface by potassium hexacyanoferrate. Similar behavior is observable in Fig. 2(b) for pore size distributions obtained by the Barrett-Joyner-Halenda (BJH) method, which is less sensitive to microporosity being better suitable for mesoporous materials.

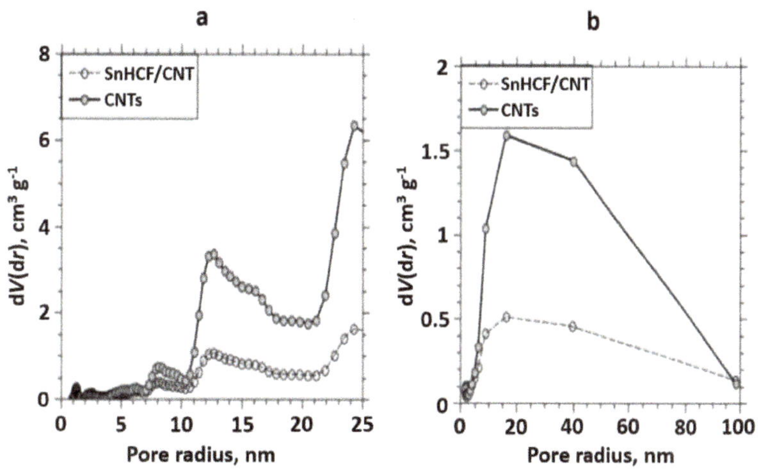

Fig. 2. (a) *Pore size distributions for SnHCF electrode with 35 mass % CNT and pure CNT electrode calculated by means of* ***(a)*** *DFT and* ***(b)*** *BJH techniques.*

As the SnHCF-based material containing 35 mass % of CNTs demonstrates the best electrochemical performance, all following experiments were conducted with it. It should be also mentioned that in performance estimates, some authors account only for the mass of the electroactive material and ignore the masses of a conductive additive and a binder [24–26]. In this work, we prefer to operate with the mass of the electrode as a whole since just this value is of practical importance. The specific capacity of the electrode material without the contribution of CNTs and polytetrafluoroethylene can be calculated via division by 0.585 in accord with the mass ratio of electrode components (see Subsections 1 and 2 in Experimental). This gives $Q_{sp.}$ of about 77 mAh g^{-1}, while the specific capacity[1] of other insoluble hexacyanoferrates in aqueous electrolytes varies from 36 to 59 mAh g^{-1} for individual compounds [10,15,26]. The $Q_{sp.}$ value obtained is 12 % higher than the theoretical one (64.9 mAh g^{-1}), probably due to the impact of unwanted faradaic side processes (e.g. the decomposition of electrode and/or electrolyte components). Due to this fact in what follows we confined ourselves with a narrower potential range (-0.25...+0.50 V).

The CV curves demonstrate a significant dependence on the scan rate, Fig. 3(a), and respective peak heights linearly vary with the square root of the scan rate, Fig. 3(b). This indicates the diffusion control of the Na^+ intercalation/deintercalation upon the charge/discharge of the SnHCF-based electrode. An analysis of the CV for the SnHCF-based electrode at the slowest scan rate (5 mV s^{-1}) shown in Fig. 4 gives the peak height ratio I_a/I_c of almost unity characterizing the electrochemical process as a reversible one in spite of the deviation of the peak shift U_a/U_c from the theoretical value of 59 mV that may be caused by interactions between intercalated species as proposed by Laviron [28].

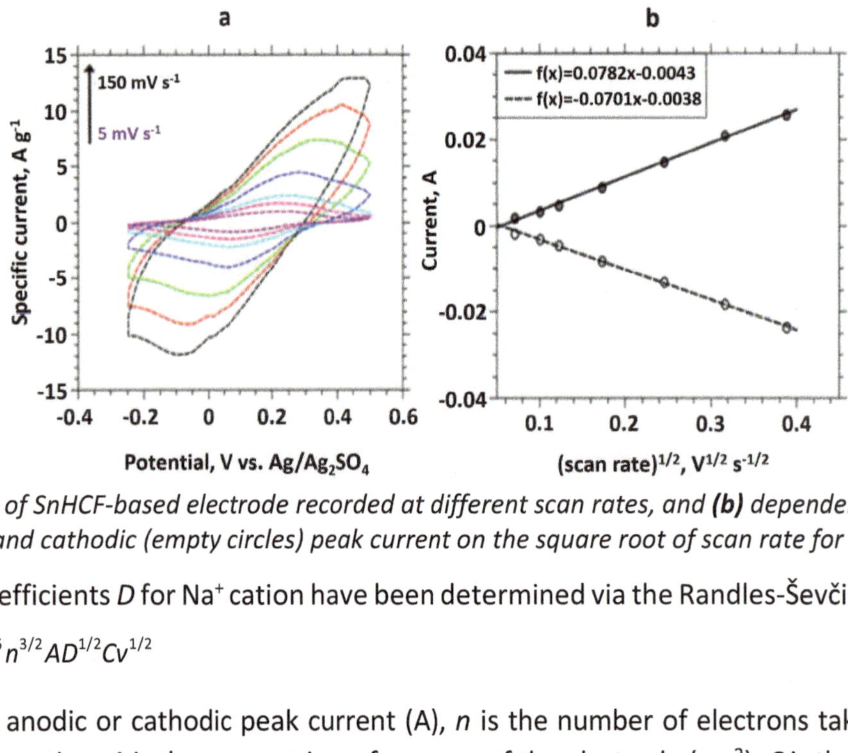

Fig. 3 (a) *CVs of SnHCF-based electrode recorded at different scan rates, and* ***(b)*** *dependence of anodic (filled circles) and cathodic (empty circles) peak current on the square root of scan rate for this electrode.*

Diffusion coefficients D for Na^+ cation have been determined via the Randles-Ševčik equation [29]

$$I_p = 2.69 \times 10^5 n^{3/2} A D^{1/2} C v^{1/2} \quad (6)$$

where I_p is the anodic or cathodic peak current (A), n is the number of electrons taking part in the electrode half-reaction; A is the geometric surface area of the electrode (cm^2), C is the concentration of Na^+ ions in the electrolyte solution (mol cm^{-3}); v is the scan rate (V s^{-1}). They are $2.1 \cdot 10^{-6}$ cm^2 s^{-1} and $2.6 \cdot 10^{-6}$ cm^2 s^{-1} for the intercalation and deintercalation processes, respectively.

[1] In some articles authors consider insoluble hexacyanoferrates, which are typical battery-type materials, as supercapacitor-type ones. Consequently, they characterize them by specific capacitances instead of specific capacities that is fundamentally flawed as described in Ref. [27]. Therefore, in order to compare our data with results obtained by other researchers we transfer the specific capacitance into the specific capacity through its multiplication by operating potential range and division by 3.6.

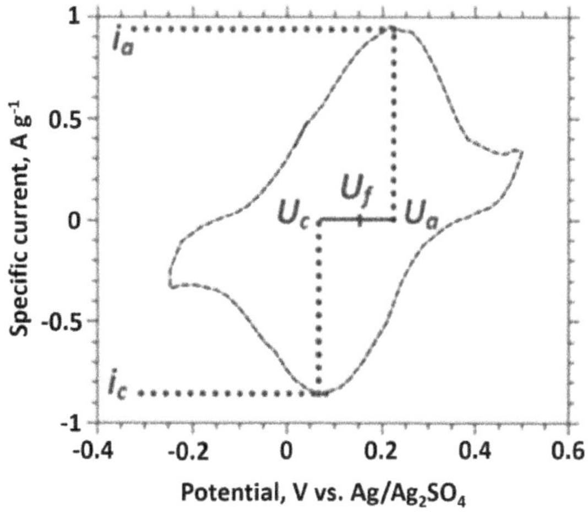

Fig. 4. *CV of SnHCF-based electrode recorded at 5 mV s^{-1}*

The *D* values obtained are significantly higher than diffusion coefficients in transition metal oxides commonly employed as electrodes in 'traditional' lithium-ion batteries, being rather comparable with effective diffusion coefficients obtained in supercapacitor electrodes produced from commercial activated carbons [30–33]. Such behavior can be explained by an effect of CNTs on the structure of the synthesized composite, where SnHCF particles are located in 12 nm pores formed by a CNT network. These nanosized particles are better accessible to electrolyte, and all advantages of finely subdivided electrode materials [34] are applicable to them.

The maximal specific capacity of the composite electrode (34 mAh g^{-1}) is observed at 5 mV s^{-1}, Fig. 5. Using a 0.585 factor as described above, the capacity related to the mass of neat SnHCF is determined as 58 mAh·g^{-1} thus representing 89 % of the theoretical value.

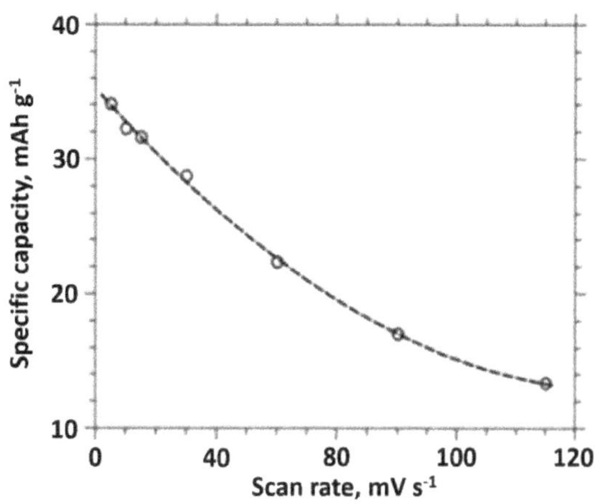

Fig. 5. *Dependence of specific capacity of SnHCF-based electrode on the scan rate*

Another important requirement for energy storage devices in addition to high specific capacity is their charge/discharge cycling stability. According to data shown in Fig. 6 the SnHCF-based electrode demonstrates an excellent cycling behavior. No capacity decrease is observed after 500 charge/discharge cycles. Furthermore, $Q_{sp.}$ even increases a little (by ~5 %) at the first cycles, possibly due to deeper ion intercalation into the hexacyanoferrate structure (pretreatment), and becomes stable reaching the value of 23 mAh g^{-1}. This is confirmed by changes in the anodic and cathodic peak heights collected in Table 2, which are leveling on growing the cycle number. The insignificant drift of the formal potential on cycling (10^{-5} V cycle^{-1}) should also be considered as an additional verification of the good reversibility of the electrochemical system studied. Furthermore, it is asserted [35] that the constancy of the formal potential signifies right mass balancing of the negative and positive electrodes in the cell.

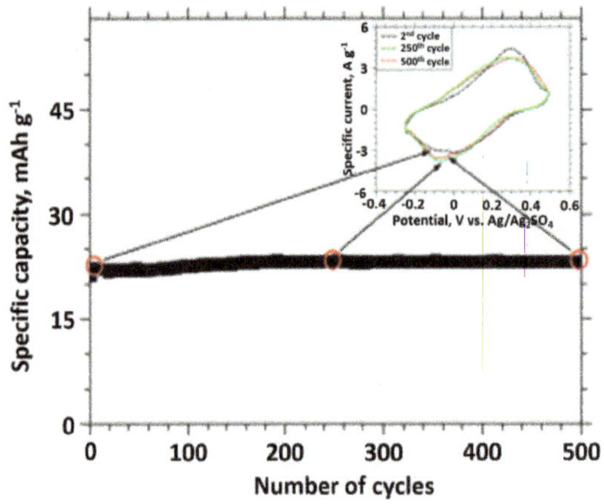

Fig. 6. *Changes in the specific capacity of SnHCF-based electrode during voltammetric cycling at 50 mV s*$^{-1}$

Table 2. *CV parameters of SnHCF-based electrode obtained from cyclic voltammograms given in Fig. 6.*

Cycle number	i_a, A g^{-1}	I_c, A g^{-1}	I_a/I_c	U_a, V	U_c, V	U_f, V	$Q_{sp.}$, mAh g^{-1}
2	4.43	3.19	1.39	0.28	0.0024	0.141	22.0
250	3.68	3.80	0.97	0.29	-0.0050	0.143	23.3
500	3.70	3.80	0.97	0.30	-0.0075	0.146	23.3

Using the data from Fig. 5 recalculated into the specific energy $E_{sp.}$ and power $P_{sp.}$ by means of Eqs. 7 and 8, one can build the so-called Ragone plot (Fig. 7), which helps to better understand the application field of the synthesized composite [36],

$$E_{sp.} = Q_{sp.}\Delta U \tag{7}$$

$$P_{sp.} = \frac{E_{sp.}}{\tau} \tag{8}$$

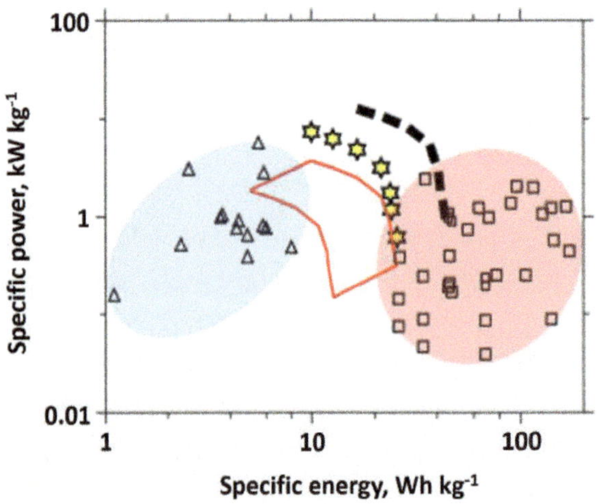

Fig. 7. *Ragone plot for pure SnHCF (dashed line), SnHCF-based electrode (stars), and a 'virtual' aqueous alkali metal-ion battery with SnHCF-based electrode (an area within the red polygon) in comparison with industrial supercapacitors (triangles limited by a blue colored area) and various rechargeable batteries (squares limited by a pink colored area) [1,2].*

In Eq. (8), τ is the time of charging or discharging the electrode in h.

In Fig. 7 we also plot the data on specific power and energy recalculated per mass of the pure electroactive material as well as per the total mass of a battery with the SnHCF-based composite as a negative electrode and additional cell elements (our approach to this calculation is given in footnote[2]). This clearly demonstrates how the characteristics are changing depending on the way of their presentation (material, electrode or device). As has been already mentioned, 'the truest' data are those normalized per electrode mass. Despite of fundamental differences in the charge storage mechanism (intercalation/deintercalation instead of adsorption/desorption), the SnHCF-based electrodes and hence energy accumulating devices with such electrodes demonstrate the specific power comparable with electric double-layer capacitors, and their specific energy values are close to lead-acid batteries. For instance, such composite electrodes can be charged up to half-capacity in a fast mode (by 7.5 seconds) with specific power of 6.1 kW kg^{-1}, while in order to charge them fully (up to 25.6 Wh kg^{-1}), less than 3 min is needed. We expect that a battery pack with SnHCF/CNT electrodes will be able to achieve the specific power and energy of (1.8–3.6) kW kg^{-1} and (12.8–25.6) Wh kg^{-1}, respectively.

The formal potentials of some multivalent metal hexacyanoferrates taken from literature sources [40–50] are collected in Fig. 8 where they are contingently divided into two groups representing negative and positive electrodes. The relatively low formal potential U_f of the SnHCF-based electrode determined as $U_f = (U_a+U_b)/2$ (+0.143 V *vs.* Ag/Ag_2SO_4 or ca. +0.28 V *vs.* Ag/AgCl) signifies an opportunity of employing the synthesized composite as a negative electrode (anode) for aqueous alkali metal-ion batteries. The best counter electrode for SnHCF that allows for getting a maximal operating voltage for a battery is VHCF and InHCF, but the extra-high cost of vanadium and indium raw materials makes CuHCF preferable for large-scale manufacturing [19]. Our further efforts will be directed towards both cathode and anode HCF materials. SnHCF-based composites mixed with other multivalent metal hexacyanoferrates (e.g. FeHCF and/or TiHCF due to their low U_f values) would be of great interest as such mixtures have been shown to have much higher specific capacity compared to the individual components [25,51].

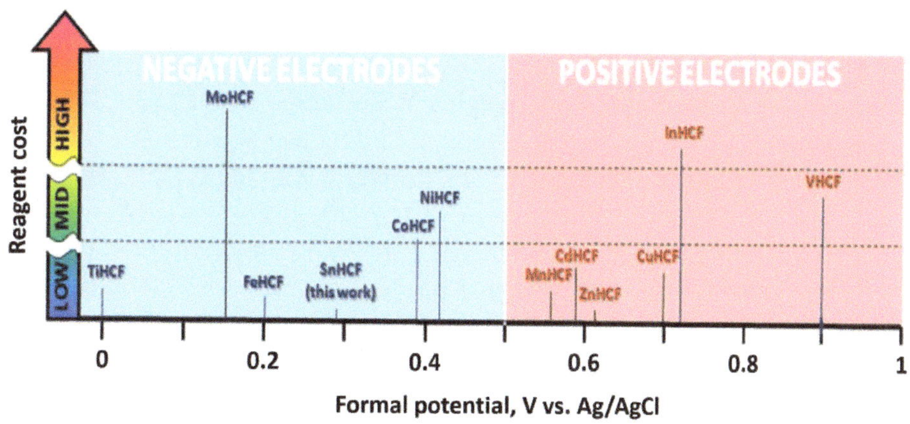

Fig. 8 *Formal potential of SnHCF-based electrode compared with formal potentials of other hexacyanoferrates in the Ag/AgCl scale.*

[2] Following to this approach the specific power of a single SnHCF-based electrode should be doubled if compared to the specific power of a two-electrode cell, whereas the specific energy remains almost unchanged in the case if the cell is assembled in an anode-limited design and both (negatively and positively charged) electrodes have the same working potential ranges *vs.* a nominal zero point [37]. Further, in our calculations, we account for the mass of the whole cell, since according to Refs. [38,39], the mass contribution of two electrodes to the total value of specific energy is of about 25–50 %.

Conclusions

A novel SnHCF-CNT composite has been fabricated by means of a room-temperature precipitation technique. The specific capacity of this electrode determined via cyclic voltammetry reaches 34 mAh g^{-1} at the slowest scan rate. Excellent electrochemical stability during charge/discharge cycling as well as fast ionic diffusion ($D \approx 2 \cdot 10^{-6}$ cm^2 s^{-1}) makes the synthesized SnHCF/CNT composite an attractive electrode material for low-cost and environmental friendly aqueous alkali metal-ion batteries, which can compete with supercapacitors in high-power applications.

References

[1] A. Burke, *Proc. IEEE* **95(4)** (2007) 806–820.

[2] A. Burke, *Int. J. Energy Res.* **34** (2010) 133–151.

[3] H. Kim, J. Hong, K. Y. Park, H. Kim, S. W. Kim, K. Kang, *Chem. Rev.* **114(23)** (2014) 11788–11827.

[4] A. J. Stevenson, D. G. Gromadskyi, D. Hu, J. Chae, L. Guan, L. Yu L, G. Z. Chen, *Supercapattery with hybrids of redox active polymers and nanostructured carbons*, in: X.L. Feng (Ed.), *Nanocarbons for Advanced Energy Storage*, Wiley-VCH, Weinheim, Germany, 2015, pp. 179–210.

[5] D. G. Gromadskyi, J. H. Chae, S. A. Norman, G. Z. Chen, *Appl. Energy* **159** (2015) 39–50.

[6] D. H. Hromadśkyi, Y. F. Fateev, N. H. Stryzhakova, Y. A. Maletin, *Mater. Sci.* **46**(3) (2010) 412–417.

[7] D. G. Gromadskyi, Y. F. Fateev, Y. A. Maletin, *Corros. Sci.* **69** (2013) 191–196.

[8] P. Padigi, G. Goncher, D. Evans, R. Solanki, Potassium barium hexacyanoferrate – a potential cathode material for rechargeable calcium ion batteries, J. Power Sources **273** (2015) 460–464.

[9] F. Zhao, Y. Wang, X. Xu, Y. Liu, R. Song, G. Lu, Y. Li, *ACS Appl. Mater. Interfaces* **6(14)** (2014) 11007–11012.

[10] C. D. Wessells, R. A. Huggins, Y. Cui, *Nat. Commun.* **2** (2011) Article No. 550.

[11] Y. You, X. Yu, Y. Yin, K.W. Nam, Y.G. Guo, *Nano Res.* **8(1)** (2015) 117–128.

[12] L. Wang, Y. Lu, J. Liu, M. Xu, J. Cheng, D. Zhang, J. B. Goodenough, *Angew. Chem. Int. Ed.* **52** (2013) 1964–1967.

[13] C. D. Wessells, S. V. Peddada, R. A. Huggins, Y. Cui, *Nano Lett.* **11** (2011) 5121–5425.

[14] M. Xie, Y. Huang, M. Xu, R. Chen, X. Zhang, L. Li, *J. Power Sources* **302** (2016) 7–12.

[15] H. Lee, Y.I. Kim, J.K. Park, J.W. Choi, *Chem. Commun.* **48** (2012) 8416–8418.

[16] C. Ling, J. Chen, F. Mizuno, *J. Phys. Chem.* C. **117** (2013) 21158–21165.

[17] J. Yun, J. Pfisterer, A.S. Bandarenka, *Energy Environ. Sci.* **9** (2016) 955–961.

[18] MSDS database, URL: *http://ccinfoweb.ccohs.ca/msds/search.html* (viewed on 10.02.2016).

[19] Sigma Aldrich catalog, URL: *http://www.sigmaaldrich.com/catalog/search* (viewed on 10.02.2016).

[20] H. Razmi, A. Taghvimi, *Int. J. Electrochem. Sci.* **5** (2010) 751–762.

[21] M. S. P. Shaffer, X. Fan, A. H. Windle, *Carbon* **36(11)** (1998) 1603–1612.

[22] D. G. Gromadskyi, *J. Chem. Sci.* **128** (2016) 1011– 1017.

[23] S. Zhang, N. Pan, *Adv. Energy Mater.* **5(6)** (2015) DOI: 10.1002/aenm.201401401.

[24] J. Chen, K. Huang, S. Liu, *Electrochem. Commun.* **10** (2008) 1851–1855.

[25] Y. Yang, Y. Hao, J. Yuan, L. Niu, F. Xia, *Carbon.* **84** (2015) 174–184.

[26] Y. Yang, H. Hao, X. Wang, Q. Yan, J. Yuan, Y. Shao, L. Niu, *Electrochim. Acta* **167** (2015) 364–371.

[27] A. Laheäär, P. Przygocki, Q. Abbas, F. Béguin, *Electrochem. Commun.* **60** (2015) 21–25.

[28] E. Laviron, *J. Electroanal. Chem. Interfacial Electrochem.* **52** (1974) 395–402.

[29] P. Zanello, *Inorganic electrochemistry: theory, practice and application*, Royal Society of Chemistry, Cambridge, UK, 2003, p. 615.
[30] C. K. Park, S. B. Park, S. H. Oh, H. Jang, W. Cho, *Bull. Korean Chem. Soc.* **32** (2011) 836–840.
[31] H. Hia, L. Lu, G. Ceder, *J. Power Sources* **159** (2006) 1422–1427.
[32] S. B. Tang, M. O. Lai, L. Lu, *Mater. Chem. Phys.* **111(1)** (2008) 149–153.
[33] Y. Maletin, V. Strelko, N. Stryzhakova, S. Zelinsky, A. B. Rozhenko, D. Gromadsky, V. Volkov, S. Tychina, G. Gozhenko, D. Drobny, *Energy Environ. Res.* **3** (2013) 156–165.
[34] A. S. Aricó, P. Bruce, B. Scrosati, J.M. Tarascon, W. Schalkwijk, *Nat. Mater.* **4** (2005) 366–377.
[35] Z. G. Dai, C. Peng, J. H. Chae, G. Z. Chen, *Sci. Rep.* **5** (2015) Article No. 9854.
[36] W. G. Pell, B. E. Conway, *J. Power Sources* **63** (1996) 255–266.
[37] X. Wu, Y. Cao, X. Ai, J. Qian, H. Yang, *Electrochem. Commun.* **31** (2013) 145–148.
[38] C. Liu, Z. Yu, D. Neff, A. Zhamu, B.Z. Zhang, *Nano Lett.* **10** (2010) 4863–4868.
[39] A. Du Pasquier, J. A. Shelburne, I. Plitz, F. Badway, A. S. Gozdz, G. G. Amatucci, *Proc. 11th Int. Semin. Double Layer Capacitors and Similar Energy Storage Devices*, Deerfield Beach, USA. 2001, pp. 1–10.
[40] Z. Gao, G. Wang, P. Li. Z. Zhao, *Electrochim. Acta* **36**(1) (1991) 147–152.
[41] C. H. Luangdilok, D. J. Arent, A. B. Bocarsly, R. Wood, *Langmuir.* **8(2)** (1992) 650–657.
[42] L. M. Siperko, T. Kuwana, *Electrochim. Acta* **32(5)** (1987) 765–711.
[43] K. Itaya, K. Ataka, S. Toshima, *J. Am. Chem. Soc.* **104**(18) (1982) 4767–4772.
[44] S. Dong, Z. Jin, *Electrochim. Acta* **34**(7) (1989) 963–968.
[45] P. Wang, Z. Jing, W. Zhang, G. Zhu, *J. Solid State Electrochem.* **5**(6) (2001) 369–374.
[46] S. Dong, Z. Jin, *J. Electroanal. Chem.* **256(1)** (1988) 193–198.
[47] M. H. Pournaghi-Azar, H. Razmi-Nerbin, *J. Electroanal. Chem.* **456(1–2)** (1998) 83–90.
[48] M. Jiang, X. Zhou, Z. Zhao, *J. Electroanal. Chem.* **292(1–2)** (1990) 289–296.
[49] S. Dong, F. Li, J. *Electroanal. Chem.* **210(1)** (1986) 31–44.
[50] J. Joseph, H. Gomathi, G. P. Rao, *J. Electroanal. Chem.* **431(2)** (1997) 231–235.
[51] A. Safavi, S. H. Kazemi, H. Kazami, *Electrochim. Acta* **56** (2011) 9191–9196.

7

Preparation of NiO-graphene oxide nanosensor for adsorptive stripping voltammetric determination of dinoterbon in food samples

Kasaram Roja, Puthalapattu Reddy Prasad, Punyasamudram Sandhya*, Neelam Yugandhar Sreedhar✉

Electroanalytical Lab, Department of Chemistry, Sri Venkateswara University, Tirupati-517 502, A. P., India
**Department of Chemistry, Sri Padmavati Mahila Visvavidyalayam, Tirupati-517 502, A. P., India*

✉Corresponding Author: sreedhar_ny@rediffmail.com

Abstract

Graphene oxide (GO) modified NiO electrochemical nanosensor was developed for the determination of the dinoterbon in food samples using adsorptive stripping voltammetry. The modified nanosensor characterized by TEM, XRD, cyclic and adsorptive stripping voltammetry. Dinoterbon pesticide exhibited a single well-defined cathodic peak at pH 4.0 at Britton–Robinson buffer (-810.0 mV). The voltammetric characterization of the pesticide residues is evaluated and the parameter such as the effect of pH, scan rate, pulse amplitude, deposition potential and deposition time were optimized. The current–concentration plot obtained using this peak was straight-lined over the range from 0.05 to 50.00 μg mL^{-1} with limit of detection (LOD) 0.028 μg mL^{-1}. The proposed method was efficiently applied to the determination of dinoterbon in food samples. The mean recoveries of the pesticide 97.40 to 99.88 % with a relative standard deviation of 0.114 % in food samples respectively.

Keywords
Nanosensor, pesticide, voltammetry

Introduction

Wide-range usage of pesticides in agriculture, which leads to accumulation of pesticide residues in soil, water and food, has imposed a serious risk to human health and the environment worldwide [1]. Carbamates are one of the major classes of synthetic pesticides and due to their broad biological activity, these compounds are used on a large scale around the world [2]. The most commonly used pesticide with acaricides and fungicides activity is dinoterbon (2-*tert*-butyl-4,6-dinitrophenyl ethyl carbonate), see Fig. 1. It has been used extensively by farmers on major crops and other field crops [3]. However, it is used to prevent the larvae of the pests from growing up. Dinoterbon, like all fungicide pesticides exhibits toxicity to humans, including carcinogenicity, reproductive, developmental toxicity, neurotoxicity and acute toxicity [4].

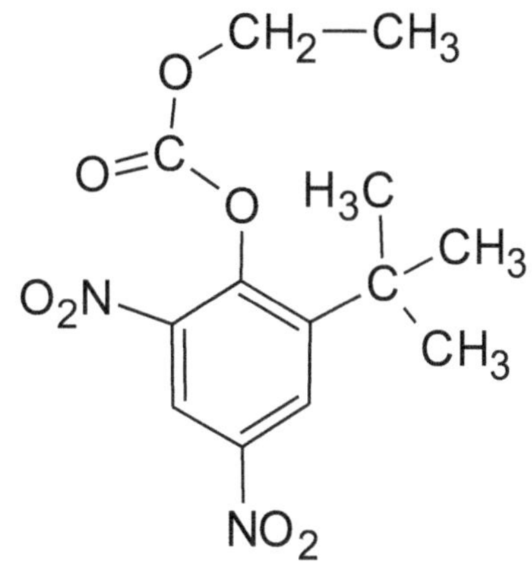

Figure 1. *Chemical structure of dinoterbon*

Graphene oxide a two-dimensional sheet of sp^2-bonded carbon atoms arranged in a honeycomb lattice, has attracted increasing attention since it was first isolated from three-dimensional graphite by mechanical exfoliation [5]. Due to its extraordinary thermal, mechanical, and electrical properties, graphene is usually considered as a competitive candidate for next generation electronic application devices such as super capacitors [6,7], batteries [8], fuel cells [9], solar cells [10], sensors [11], biosensors [12], energy storage [13] and catalysts [14]. However, many researchers have reported that the pure graphene actually exhibit unsatisfactory electrical conductivity because of the inevitable aggregation [15]. On the contrary, functionalized graphene sheets are easier to disperse in organic solvents, which can improve the dispersion and the homogeneity of the graphene with in aqueous solutions and yield novel types of electrically conductive nanocomposites [16-18]. Also, some of the useful and unique properties of graphene can only be realized after it is functionalized with organic groups such as hydroxyl, carboxyl and amino groups [19]. However, nanostructural metal oxide semiconductors possess a high surface area, nontoxicity, good biocompatibility, catalytic activity and chemical stability. Among the metal oxide semiconductors, nickel oxide (NiO), a p-type semiconductor with a wide band gap of 4.0 eV at 300 K, has been investigated for various applications such as solar cells, electrochemistry nanosensors [20,21]. Electrochemical nanosensors based voltammetric techniques have become extremely useful for the monitoring of pesticides [22,23].

In the present paper authors developed a novel NiO/GO electrochemical sensor prior to determination of dinoterbon pesticide residues at low concentration levels in food samples. The developed NiO-GO/GCE nanocomposite characterized by XRD, TEM, cyclic voltammetry and adsorptive stripping voltammetry (AdSV). Under the optimized operational conditions, the developed electrochemical nanosensor showed a specific and excellent performance with a good sensitivity, selectivity and wide dynamic range toward the quantification of dinoterbon. The results implicate the applicability of NiO/GO nanocomposite for rapid, sensitive and selective analysis of dinoterbon.

Experimental

Instrumentation and reagents

The electrochemical measurements were carried out in Autolab, three electrode systems consisting of modified glassy carbon electrode as a working electrode, Ag/AgCl (salt KCl) as a reference electrode and a platinum wire as an auxiliary electrode. Transmission electron microscopy (TEM) micrographs were performed a JEOL JEM 200CX operating at 200 kV. All reagents used were of analytical reagent grade. Ultrapure water was used throughout the experiment. The technical grade samples of dinoterbon fungicide in the form of 50 % wet-table powders were obtained from Bayer India Ltd. India.

Synthesis of nickel oxide nanocomposite

The NiO prepared from $NiNO_3{\cdot}6H_2O$ precursor by drop wise addition of 0.1 mol L^{-1}, KOH to a 0.1 M $NiNO_3{\cdot}6H_2O$ solution was kept vigorously stirred until the pH becomes 10.0. The precursor was filtered and rinsed with ultrapure water for twice and with ethanol once. Wet cake obtained was dried in oven at 100 °C overnight and was heated at 400 °C for 4 h to form black NiO nanoparticles.

Fabrication of the NiO-GO modified GCE

Before modification, the bare GCE (diameter 3 mm) was prudently polished to a mirror-like surface with 0.3, 0.05 M alumina slurries in sequence, then sonicated in ethanol and distilled water for 3 min respectively and dried with nitrogen. The 15 µL of NiO/GO suspension, prepared by simply mixing of NiO and GO suspensions with an appropriate volume ratio (v/v), was then dropped onto the clean electrode. Then the electrode was dried under room temperature, making the GO/GCE, NiO/GCE, GO and NiO-GO modified GCE.

Recommended analytical procedure for the determination of dinoterbon

An aliquot of working standard solution containing 5.0 µg mL^{-1} of dinoterbon pesticide is taken into 25 mL volumetric flask. To this 5 mL of Britton-Robinson buffer of (pH 4.0) added and transferred into electrolytic cell and diluted with 9.0 mL of supporting electrolyte and then deoxygenated with nitrogen gas for 5 min. The pesticide residue was treated with electrochemical NiO-GO nanocomposite on the surface of glassy carbon electrode. The dinoterbon pesticides were determined by cyclic and AdSV mode. Electrolysis was done at +0.8 to -1.20 V *vs.* SCE, accumulation time 80 s, pulse amplitude of 25 mV, scan rate 20 mV s^{-1} and pH 4.0. The maximum voltammetric peaks appearing for sample is at -0.810 V for dinoterbon.

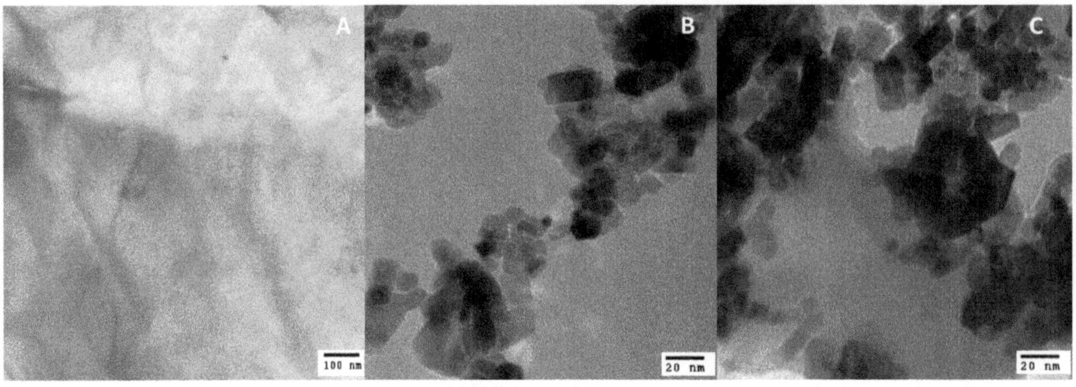

Figure 2. *TEM images of GO (A), NiO (B) and NiO/GO (C) modified glassy carbon electrode.*

Results and discussion

The structure of the GO fabricated NiO nanocomposite was examined by TEM and XRD. The morphological structure of the resulting NiO/GO was investigated by TEM (Fig. 2). The TEM image of single layer graphene oxide as shown in Fig. 2A. Fig. 2B, shows the TEM image of many NiO nanoparticles with sphere-like morphology are homogeneously dispersed. The many NiO nanoparticles with sphere-like morphology are homogeneously anchored on the surface of the GO sheets as shown in Fig 2C. The revealing sphere-like NiO nanoparticles with an average size of 28.6 nm. The XRD was used to investigate the phase structure of the resulting hybrids (Fig. 3). Fig. 3a shows the X-ray diffraction spectrum of NiO samples. The XRD patterns exhibit five prominent peaks at 2θ = 37.4°, 43.3°, 63.5°, 75.4° and 79.6° can be readily indexed as (111), (200), (220), (311) and (222) crystal planes of the bulk NiO, respectively. Fig. 3b shows the X-ray diffraction spectrum of GO fabricated NiO nanocomposite. The XRD pattern of the NiO/GO hybrid exhibits crystalline NiO diffraction peaks, which are in good agreement with the standard NiO (JCPDS No. 04-0835). The as-prepared GO displays a characteristic (002) peak at 25.4 which is in good agreement with previous reports [24]. By applying the Scherer formula to the XRD peaks, the average crystal size was calculated to be 22.4 nm for the nickel crystalline particles. This result indicates that the intermediate products was completely converted to NiO, which can be indexed as disordered stacked graphitic sheets. This finding indicates that GO is similar to previous reports [25].

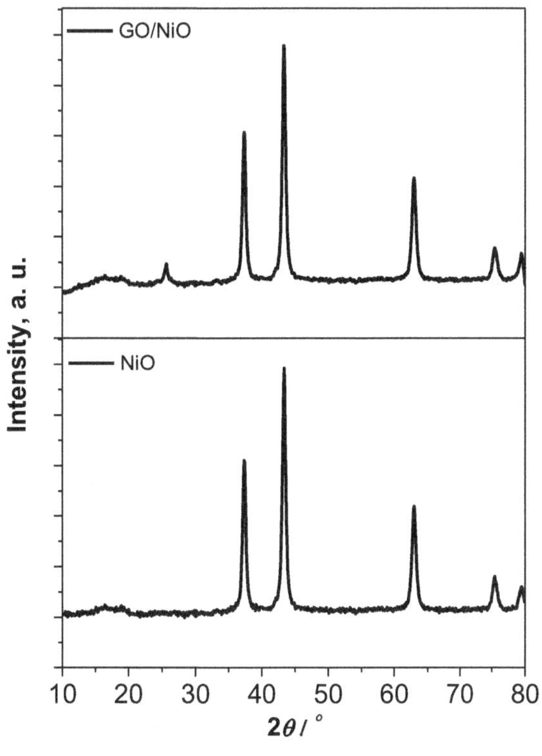

Figure 3. *XRD patterns of NiO and GO-NiO nanocomposite*

Voltammetric response of dinoterbon at various electrodes

The electrochemical performance of various electrodes was first investigated by cyclic voltammetry. Fig. 4, showed the cyclic voltammetry of the bare GCE, GO/GCE, NiO/GCE and NiO-GO/GCE in 0.1 M pH 4.0 BR buffer solution in the presence of 5.0 µg mL^{-1} dinoterbon. There was no obvious peak observed at bare GCE. However, a cathodic peak response at −0.810 V was observed at the NiO/GCE and GO/GCE, respectively. According to the currently accepted mechanism [24,25], the reduction peak should be attributed to a four-electron transfer reduction of the each nitro group (-NO_2) to give the hydroxylamine derivative. It can be reoxidized to the nitroso compound at a more positive potential. Under subsequent cycling, the nitroso group was reversibly reduced to respective hydroxylamine. Consequently, the redox behaviors of dinoterbon at the NiO-GO electrode is shown in scheme. 1.

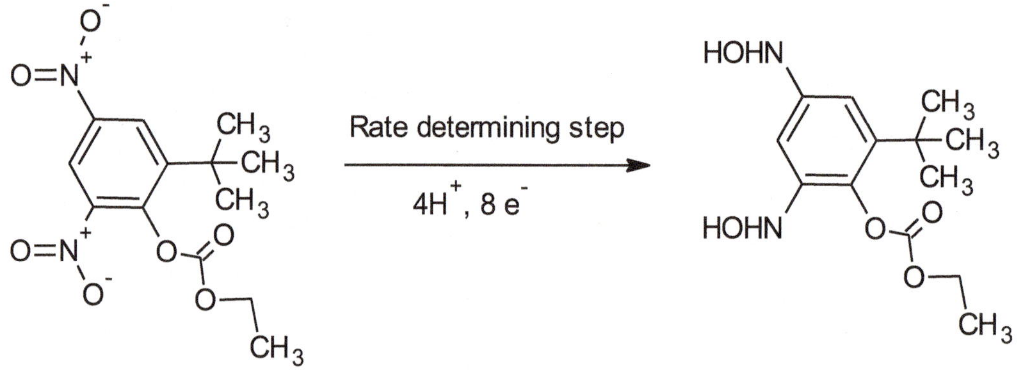

Scheme 1. *Reduction mechanism of dinoterbon*

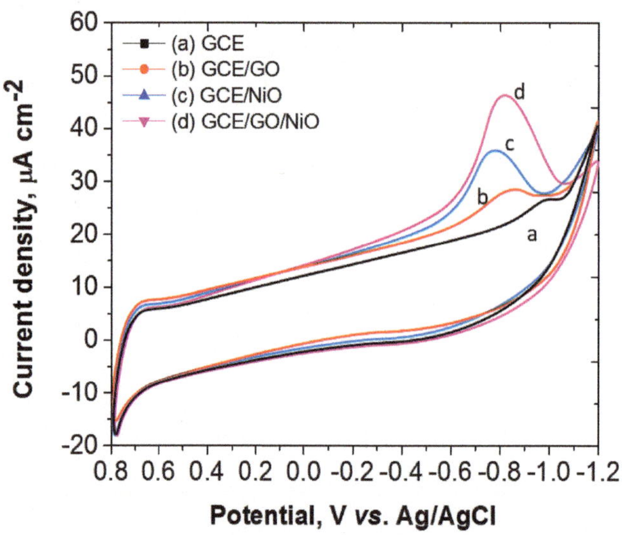

Figure 4. *Typical cyclic voltammogram of dinoterbon for an accumulation time of 80 sec at GCE (a); GO/GCE (b); NiO/GCE (c); NiO-GO/GCE(d); rest time: 10 s; scan rate: 20 mVs*$^{-1}$*; concentration: 5.0 µg mL*$^{-1}$*; pH: 4.0 (BR buffer); pulse amplitude: 25 mV.*

The NiO-GO/GCE strong cathodic peak current response at -0.810 V, which was more prominent than those obtained at the bare GCE, NiO/GCE and GO/GCE. Furthermore, one can see that the incorporation of NiO into GO nanocomposite possesses more prominent peaks, indicating that the use of GO can significantly enhance the electron transfer between NiO-GO and the GC electrode. The normalized signal response of the different electrodes to 5.0 µg mL^{-1} dinoterbon were calculated. The dinoterbon response at the NiO-GO/GCE is more than the normalized response at the NiO/GCE and GO/GCE. These showed the NiO/GO/GCE gave the highest normalized signal response. The enhanced performance of the NiO-GO/GCE nanocomposite can be attributed to the excellent affinity of dinoterbon with NiO and the enhanced electron transfer, which could amplify the interaction between dinoterbon and NiO due to the formation of entangled NiO-GO structure.

The difference in the electrochemical behaviour of dinoterbon at bare GCE, GO/GCE and NiOGO/GCE were also evaluated by AdSV in 0.1 mol L-1 BR buffer pH 4.0, containing 20 µg mL-1 dinoterbon, at scan rate of 20 mV s–1 (Fig. 5).

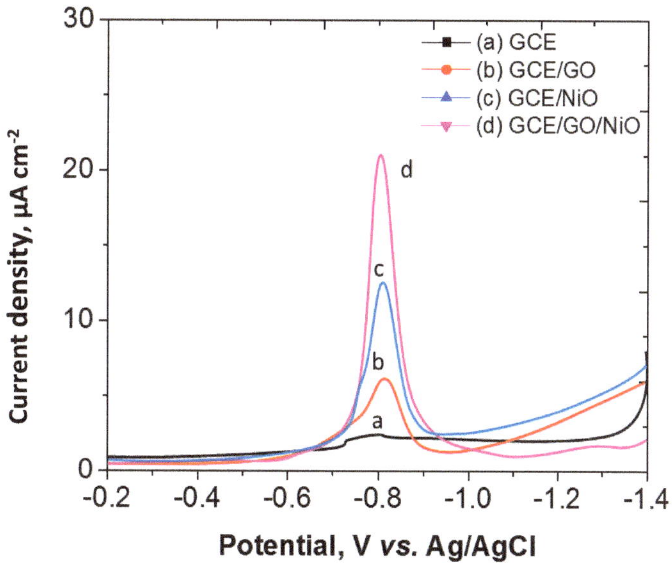

Figure 5. *Adsorptive stripping voltammogram of dinoterbon at bare GCE (a), GCE/GO (b), GCE/NiO (c), NiO-GO/GC electrode(d); pH 4.0 (BR buffer) accumulation time: 80 s.; stirring rate: 1500 rpm; scan rate:20 mVs^{-1}; pulse amplitude: 25 mV.*

No characteristic peak related to the dinoterbon reduction was observed for the bare glassy carbon electrode which indicates GCE does not exhibit electrocatalytic activity for dinoterbon. However, when the determination was performed at the GO/GCE, NiO/GCE and NiO-GO/GCE electrodes, there was a peak for dinoterbon at -0.713 and -0.688 V, respectively. These peaks indicate that both GO/GCE and NiO-GO/GCE electrodes exhibit electrocatalytic activity and can identify the fungicide dinoterbon. However, the electrode modified with the hybrid material, NiO-GO/GCE, showed a well electrocatalytic response with higher catalytic current and the potential was less positive compared to the other modified electrodes. It can be explained due to the GO physically incorporated into the NiO nanocomposite were oriented in such a way that their extremities were more susceptible to reacting with the fungicide dinoterbon.

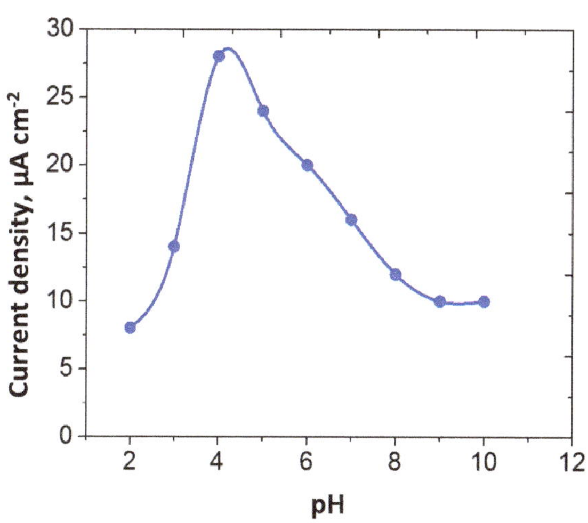

Figure 6. *Effect of pH on dinoterbon at NiO-GCE/GCE; accumulation time: 80 s.; rest time: 10 s., stirring rate: 1500 rpm; scan rate: 20 mVs^{-1}; concentration: 5.0 µg mL^{-1}; pulse amplitude: 25 mV.*

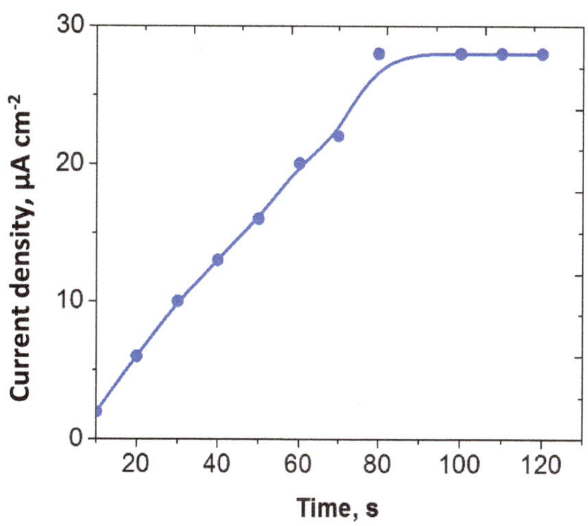

Figure 7. *Effect of accumulation time on the AdSV response of dinoterbon at NiO-GO/GCE; rest time:10 s; stirring rate: 1500 rpm; scan rate: 20 mVs^{-1}; concentration: 5.0 µg mL^{-1}; pH: 4.0 (BR buffer); pulse amplitude: 25mV.*

Optimization parameters for dinoterbon detection at NiO-GO/GCE

The peaks of AdSV for dinoterbon (20 μg mL^{-1}) with NiO-GO/GCE were compared in different supporting electrolytes, namely, 0.1 mol L^{-1} Britton–Robinson buffer, 0.2 mol L^{-1} sodium acetate–acetic acid buffer, 0.1 mol L^{-1} phosphate buffer, carbonate buffer, and borate buffer solution. The highest peak current was obtained with 0.1 mol L^{-1} Britton–Robinson buffer as the electrolyte. Thus, 0.1 mol L-1 Britton–Robinson buffer was chosen as the analytical medium, in which the peak shape was well defined. As shown in Fig. 6, the effects of pH on the AdSV peak current of dinoterbon (20 μg mL^{-1}) with NiO-GO/GCE was also studied in 0.1 M Britton–Robinson buffer (pH 2.0–10.0). The maximum current appeared at pH 4.0 for the determination of dinoterbon. In the following experiment, pH 4.0 was selected.

Further investigation of electroactivities using the various ratios of NiO to GO was preliminarily studied to optimise the catalytic performance of NiO-GO/GCE toward dinoterbon. The mass ratio of NiO:GO was varied from 1:0.25 to 1:0.5, 1:1, 1:2, and 1:4; the highest peak current was obtained at a mass ratio of 1:2. The catalytic activity of the as-prepared NiO-GO hybrid for dinoterbon reduction was under the synergistic effect of the primary component (NiO) and the support component (GO). The combination of GO with NiO is an ideal strategy of improving the catalytic performance of NiO by enhancing its charge transfer efficiency. However, excessive GO in the composite can also decrease the loading level of NiO. Thus, the mass ratio of NiO-GO has an important influence on the catalytic oxidation reactions of dinoterbon. The optimum catalytic performance was achieved from NiO-GO with a mass ratio of 1:2. The excellent adsorption capacity of GO, (2) the larger effective surface area of NiO-GO/GCE, and (3) the synergistic catalytic effect of GO, and NiO toward dinoterbon, as detailed above.

The AdSV peak current of dinoterbon increases with accumulation time increasing from 0 to 80 s, as shown in Fig. 7. But when it exceeds 80 s the peak current remains almost constant for a 5.0 μg mL^{-1} dinoterbon solution, meaning that an accumulation/or extraction equilibrium is achieved at the electrode/solution interface. The influence of accumulation potential is examined from +0.3 to −0.5 V. The results showed that the peak current of dinoterbon is almost independent of accumulation potential. This is due to the neutral nature of dinoterbon under this condition. Thus, an accumulation is performed under open-circuit.

The effect of pulse height variation on the peak current of dinoterbon voltammograms was studied in the range of 15–100 mV. The obtained results showed that increasing the pulse heights up to 25 mV will cause an increase in peak current. Pulse heights more than 25 mV cause broadening of the dinoterbon voltammogram and so decreasing the peak current intensity of the analyte. So the optimum pulse height value was selected to be 25 mV.

Table 1. *Tolerance limits of matrix substances for determination of dinoterbon by the AdSV method*

substances	Tolerance limit, μg mL^{-1}	substances	Tolerance limit, μg mL^{-1}
Na^+, K^+, NH_4^+	1200	Cu^{2+}, Cr^{6+}	60
Ca^{2+}, Mg^{2+}	1000	Phenols and nitrophenols	50
Cl^-, NO_3^-, HCO_3^-	800		
PO_4^-, SO_4^{2-}	750		
Al^{3+}, Zn^{2+}, Fe(II)	400		

Interference study

Prior to the application of the developed method on food samples it was vital to investigate the effect of some of the interfering ions on the recovery percentage of dinoterbon. The AdSV determination of dinoterbon was tested in the presence of spiked known amounts of interfering ions and molecules. The tolerance limit was defined as the amount of the foreign substance causing a change of ±5 % in the peak current intensity reading. The tolerable limits of interfering substance are given in Table 1. The results showed that most of the investigated substances do not interfere in the AdSV determination of dinoterbon in food samples

Determination of dinoterbon pesticide in food samples

The food samples namely, Onion (*Alium sepa*), Cauliflower (*Brassica olera. var. Botrytis*), Lady's Finger *(Abelmoschus esculentus)*, Cucumber (*Cucumis sativus*), Sweet Potato (*Ipomoea*) and Tomato (*Lycopersicum esculentum*) were collected from the Tirupati local market, A.P, India. Afterwards, they were taken in small mesh and dried in an oven at 90 °C to constant weight. In order to digest the samples, 1.0 g of food sample was digested with 10 mL concentrated HNO_3 (65 %) and 3.0 mL H_2O_2 (30 %) in microwave system, then again evaporated to near dryness. After evaporation, 10 mL of deionized water was added and the sample was mixed. The resulting mixture was filtered through filter paper. The filtrate was diluted to 25 mL with deionized water. All the samples were stored in polyethylene bottles. For sample analysis spiked with dinoterbon standards at 10.0, 50.0, 100 μg mL^{-1}. Then 10 μg mL^{-1} of the digested sample solution was dissolved in 25 μL BR buffer solution at pH 4.0 for determination. Samples were also analyzed by GC methods. The recovery rates in food samples exhibited a range of 97.40 to 99.88 % (average of five determinations) with less than 1.72 % of RSD and precision data are reported in Table 2.

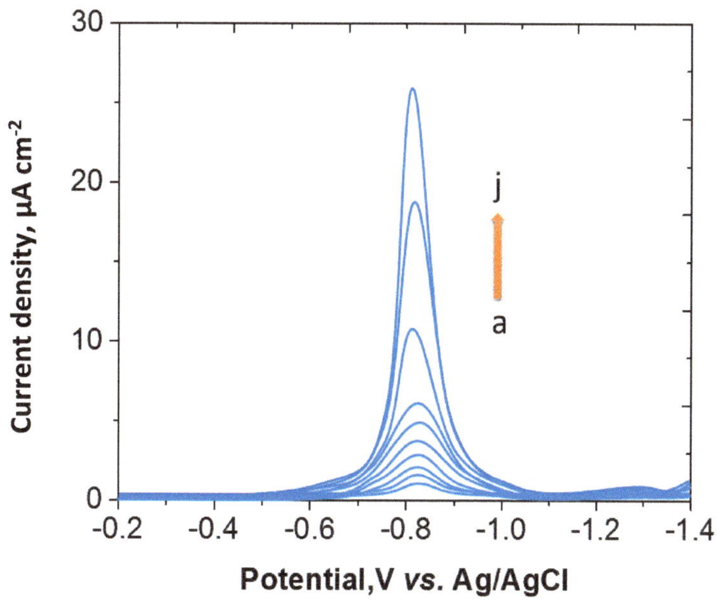

Figure 8. *AdSV of the NiO-GO/GCE by (a) 0.2, (b) 0.4, (c) 0.8, (d) 1.0 (e) 2.0 (f) 4.0, (g) 8.0, (h) 16, (i) 32 (j) 50 μg mL^{-1} dinoterbon; accumulation time of 80 s, stirring rate:1500 rpm; scan rate: 20 mVs^{-1}; pH: 4.0 (BR buffer); pulse amplitude:25 mV.*

Adsorptive stripping voltammetric quantification of dinoterbon

Due to the good sensitivity of adsorptive stripping voltammetry was applied for further electrochemical detection of dinoterbon under the optimized operating conditions. Fig. 8, displayed the AdSV of the NiO-GO/GCE in 0.1 mol L^{-1} pH 4.0 BR buffer solution at a potential range from −0.4 V to 0.2 V in the presence of various dinoterbon concentrations. Well-defined peaks, proportional to the concentration of the corresponding dinoterbon, were observed in plots. The corresponding calibration plot was presented in Fig. 9, indicating that the response was gradually saturated at a higher dinoterbon concentration. Under the optimal experimental conditions, NiO–GO/GCE was used to detect dinoterbon by AdSV. It can be seen from the figure that well-defined AdSV responses from adsorbed dinoterbon were observed and increased gradually with the increase of the dinoterbon concentration. The good linear relationship between oxidation current and log *C* dinoterbon was obtained from 0.05 μg mL^{-1} to 50.0 μg mL^{-1} with the regression equation of i / μA = 0.4742x+0.1283 (R = 0.9992). A LOD of 0.0283 μg mL^{-1} was calculated according to the formula LOD = 3 σ/S, where σ is the mean of standard deviation of five measurements taken from the signal obtained from the blank, S is the slope of the calibration curve, and the number 3 comes from the required ~98 % level of confidence in the difference between the observed signal and the blank response.

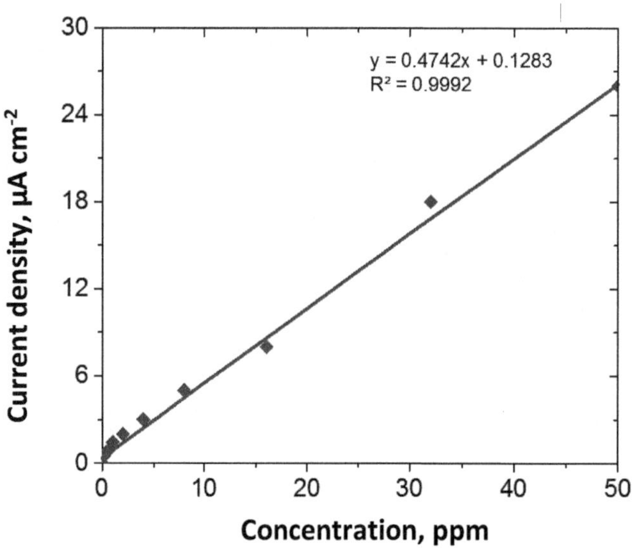

Figure 9. *Calibration curve on the AdSV response of dinoterbon at NiO-GO/GCE; 5.0 μg mL^{-1}; accumulation time of 80 s, stirring rate: 1500 rpm; scan rate: 20 mV s^{-1}; pH: 4.0 (BR buffer); pulse amplitude: 25 mV.*

Table 2. *AdSV determination of pesticide in food samples (no. of determinations =5)*

Name of the food samples	Amount added, μg mL^{-1}	Amount found, μg mL^{-1}	Recovery, %	R.S.D*
Onion	5.0	4.98	99.60	0.11
	10.0	9.92	99.20	0.08
	25.0	24.96	99.84	0.16
Cauliflower	5.0	4.87	97.40	0.02
	10.0	9.94	99.40	0.12
	25.0	24.82	99.28	0.14
Cucumber	5.0	4.88	97.60	0.22
	10.0	9.94	99.40	0.04
	25.0	24.96	99.84	0.16
Sweet Potato	5.0	4.91	98.20	0.12
	10.0	9.89	98.90	0.05
	25.0	24.97	99.88	0.20
Tomato	5.0	4.99	99.80	0.16
	10.0	9.93	99.30	0.08
	25.0	24.89	99.56	0.06

*R.S.D: Relative standard deviation

The inter-assay precision was estimated at six different NiO-GO a nanocomposite modified electrodes for the determinations in 0.1 mol L^{-1} BR buffer (pH = 4.0) containing 5.0 μg L^{-1} dinoterbon.

Similarly, the intra-assay precision was evaluated by assaying one working electrode for five replicate determinations under unvarying conditions. The relative standard deviation values of inter-assay and intra-assay were found to be 6.8% and 4.6%, respectively, indicating acceptable precision and reproducibility. In addition, the developed electrochemical metal oxide nanosensor (NiO-GO) was very stable at room temperature. No obvious decrease in the electrochemical response was observed in the 5 days and over ~90% of the initial response remained after four weeks, indicating that is acceptable stability.

Conclusions

In this paper, combining the advantageous characteristics of dinoterbon at NiO/GCE, GO/GCE and the NiO-GO/GCE nanocomposite have been prepared. The NiO-GO/GCE nanocomposite with excellent electrocatlytic and between NiO and GO modified electrode. It's improved the absorptivity and charge transfer properties on the surface of nanocomposite and improve the stability. The constructed NiO-GO/GCE sensor exhibited many advantages such as low applied potential, good fabrication reproducibility, acceptable stability, fast response and low detection limit. The NiO-GO/GCE sensor has potential application in monitoring of dinoterbon in food samples.

References

[1] R. C. Gupta, *Toxicol. Mechan. Methods*, **14** (2004) 103-143

[2] W. T. Song, Y. Q. Zhang, G. J. Li, H. Y. Chen, H. Wang, Q. Zhao, D. He, C. Zhao, L. Ding, *Food Chem.* **143** (2014) 192-198

[3] P. Lautala, M. Kivimaa, H. Salomies, E. Elovaara, J. Taskinen, *Eur. J. Pharm. Sci.* **4** (1996) S167-168

[4] X. Guo, Z. Wang, S. Zhou, *Talanta* **6** (2004) 135-139

[5] K. S. Novoselov, A. K. Geim, S. V. Morozov, D. Jiang, Y. Zhang, S. V. Dubonos, I. V. Grigorieva, A. A. Firsov, *Science* **306** (2004) 666-669

[6] M. D. Stoller, S. J. Park, Y. W. Zhu, J. H. An, R. S. Ruoff, *Nano Letters* **8** (2008) 3498-3502

[7] A. K. Mishra, S. Ramaprabhu, *Journal of Physical Chemistry C*, **115** (2011) 14006-14013

[8] E. Yoo, J. Kim, E. Hosono, H. S. Zhou, T. Kudo, I. Honma, *Nano Letters* **8** (2008) 2277-2282

[9] J. Xiao, D. Mei, X. Li, W. Xu, D. Wang, G. L. Graff, W. D. Bennett, Z. Nie, *Nano Letters* **11** (2011) 5071-5078

[10] X. Wang, L.J. Zhi, N. Tsao, Z. Tomovic, J.L. Li, K. Mullen, *Angew and te Chemie International Edition* **47** (2008) 2990-2992

[11] S. J. Guo, S. J. Dong, *Journal of Materials Chemistry* **21** (2011) 18503-18516

[12] J. D. Qiu, J. Huang, R. P. Liang, *Sensors and Actuators B* **160** (2011) 287-294

[13] D. Chen, L. H. Tang, J. H. Li, *Chemical Society Reviews* **39** (2010) 3157-3180

[14] S. S. Han, H. Kim, N. Park, *Journal of Physical Chemistry* **C115** (2011) 24696-24701.

[15] D. Li, R. B. Kaner, *Science* **320** (2008) 1170-1171

[16] S. Stankovich, D. A. Dikin, G. H. B. Dommett, K.M. Kohlhaas, E.J. Zimney, E.A. Stach, R.D. Piner, S.T. Nguyen, R.S. Ruoff, *Nature* **442** (2006) 282-286

[17] D. Li, R. B. Kaner, *Science* **320** (2008) 1170-1171.

[18] K. Wang, Q. Liu, L.N. Dai, J.J. Yan, C. Ju, B.J. Qiu, X.Y. Wu, *Analytica Chimica Acta,* **695** (2011) 81-84

[19] D. Zheng, S. K. Vashist, K. Al-Rubeaan, J.H.T. Luong, F.S. Sheu, *Talanta,* **99** (2012) 22-28

[20] T. Seto, H. Akinaga, F. Talano, K. Koga, T. Orii, M. Hirasawa, *J.Phys. Chem.B* **109** (2005)13403-13405

[21] P. Lunkenheimer, A. Loidl, C. Ottermann, H. Bange, *Phys.Rev.B:Condens. Matter* **44** (1991) 5927-5930

[22] C. N. Reddy, P. R. Prasad, and N.Y. Sreedhar, *Journal of Analytical Methods in Chemistry* **2013** (2013) 1-7

[23] D. Rekha, N. Y. Sreedhar, P.R. Prasad, *Global journal of frontier Science research*, **10** (2010) 21-29

[24] Y. Li, B. Zhang, J.X. Zhao, Z. Ge, X. Zhao, L. Zou, *Applied Surface Science* **279** (2013) 367-373

[25] H. Zhang, H. Huang, H. Ming, H. Li, L. Zhang, Y. Liu and Z. Kang, *J. Mater. Chem.*, **22** (2012) 10501-10506

8

Electrochromism in tungsten oxide thin films prepared by chemical bath deposition

Julijana Velevska[1,✉], Nace Stojanov[1], Margareta Pecovska-Gjorgjevich[1], Metodija Najdoski[2]

[1]*Institute of Physics, Faculty of Natural Sciences and Mathematics, University Sts Cyril and Methodius, Arhimedova 3, 1000 Skopje, Republic of Macedonia*
[2]*Institute of Chemistry, Faculty of Natural Sciences and Mathematics, University Sts Cyril and Methodius, Arhimedova 5, 1000 Skopje, Republic of Macedonia*

✉Corresponding author: julev@pmf.ukim.mk

Abstract

*Tungsten oxide (WO_3) thin films were prepared by a simple, economical, chemical bath deposition method onto fluorine doped tin oxide (FTO) coated glass substrates. The electrochemical properties of the films were characterized by cyclic voltammetry. The obtained films exhibited electrochromism, changing color from initially colorless to deep blue, and back to colorless. Visible transmittance spectra of (WO_3) films were recorded in-situ in their both, bleached and colored states. From those spectra, absorption coefficient (α) and the optical energy gaps were evaluated. The dependence of the optical density on the charge density was examined and the coloration efficiency (η) was calculated to be 22.11*cm^2C^{-1}*. The response times of the coloring and bleaching to an abrupt potential change from -2.5 V to +2.5 V and reverse, were found to be 9.3 and 1.2 s respectively. The maximum light intensity modulation ability of the films, when the AM1.5 spectrum is taken as an input, was calculated to be about 50 %.*

Keywords

Optical materials; cyclic voltammetry; response time; solar light modulation

---

Introduction

Electrochromism is a unique property of the material to change reversibly its optical properties when it is electrochemically reduced or oxidized [1]. Electrochromic materials exhibit color change between the clear transparent state and a darkened colored state, or between two colored states. At the same time, there are materials that exhibit multiple colored states and are described as polyelectrochromic. The classification of the electrochromic materials is related to the potential at which the coloration process occurs. Materials with cathodic coloration exhibit coloration at negative potential, *i.e.* they darken upon reduction (charge insertion). Anodic materials, on the other hand, exhibit coloration at positive potential, *i.e.* they darken upon oxidation (charge extraction). When the electrochromic material is integrated in device, it could modulate the reflectance/transmittance of the incident illumination [2]. Electrochromic materials are currently attracting much interest in academia and industry for both their spectroelectrochemical and commercial applications [3]. Electrochromism is known to exist in many types of materials, both organic and inorganic. Common inorganic electrochromic materials are transition metal oxides and metal hexacyanometallates, while viologens, phthalocyanines, conducting polymers and metallopolymers are common organic and polymer electrochromic materials.

Transition metal oxides have attractive technological importance for electrochromic applications because they show considerable variation in stoichiometry, and can be quite easily deposited in a form of thin film which is appropriate for device manufacturing. Among them, tungsten oxide (WO_3) is of intense interest and has been extensively investigated due to its appreciable electrochromic properties in the visible and infrared region. It exhibits large optical modulation, good durability, low power consumption, less stress for the viewer's eyes, and relatively low price [4].

Eectrochromic WO_3 thin films have been prepared by a large number of techniques, such as thermal evaporation, electrodeposition, spray pyrolysis, chemical vapor deposition, electron beam evaporation, magnetron sputtering, sol-gel methods [5 - 16] *etc.* Among these techniques, chemical bath deposition methods have many advantages: they do not require sophisticated expensive equipment, various substrates including metals, semiconductors or insulators can be used, the starting chemicals are commonly available and cheap, and the preparation parameters are easily controlled [17 - 21]. These methods have benefit of being easily realizable from the point of view of industrialization, especially on large area devices, with the required electrochromic properties [22, 23]. The electrochromic properties of WO_3 thin films, like the transmittance modulation (ΔT), coloration efficiency (η), switching time (τ), and cyclic durability, strongly depends on their structural, morphological and compositional characteristics [1, 2], which, in turn, depends directly on the deposition method and deposition conditions.

The target of this research is to investigate electrochromic properties of WO_3 thin films prepared by a simple chemical bath deposition method [24] and their possible application for solar light modulation.

Experimental

Tungsten oxide films were deposited onto fluorine doped tin oxide (FTO) coated glass substrates commercially available with transparency of about 80 % for visible light, and sheet resistance of about 10 – 20 $\Omega\square^{-1}$. Before the deposition, the substrates were immersed in acetone and ethanol to be degreased in an ultrasonic bath, and then rinsed in deionized water and dried in air. WO_3 films were deposited from a chemical bath with optimized composition and process conditions.

The bath solution was prepared by dissolving 1.65 g $Na_2WO_4 \cdot 2H_2O$ in 90 ml deionized water. The substrates were immersed in the beaker filled with deposition solution, and vertically supported against its walls. Then, the whole system was heated slowly, up to 95°C with continuous stirring. The preparation of the bath solution and the deposition of the thin films have been explained in details in Ref. [24]. The thickness of the films depends on the deposition time. In this work the deposition time was 20 min, and the thickness of the films was 150 nm.

The electrochemical properties of WO_3 films were characterized by cyclic voltammetry measurements performed using Micro AUTOLAB II equipment (Eco-Chemie, Utrecht, Netherlands) in one compartment three electrodes electrochemical cell with WO_3 film as working, platinum wire as counter, and saturated calomel electrode (SCE) as a reference electrode. The cycling was carried out in 1 moldm^{-3}KCl aqueous solution as an electrolyte. The voltage scan rate was 10 mV s^{-1}, and the film working area was 1 cm^2.

Electrochromic investigations were performed *in situ* in an electrochromic device (ECD) consisted of: home-built glass cell (4 ×2.5×4 cm), WO_3 film deposited on FTO substrate as working electrode (WE), blank FTO substrate as counter electrode (CE), and 1 moldm^{-3}KCl aqueous solution as an electrolyte. The distance between the electrodes was about 1.5 cm, the volume of the electrolyte was about 20 ml, and the active surface area of the electrodes was approximately 6 cm^2.

The optical transmittance spectra were recorded by using Varian CARY 50 Scan UV-Visible spectrophotometer in the wavelength range from 300 to 900 nm, in both, the completely colored and bleached states of the film. An electrochromic cell with two clean FTO substrates filled with electrolyte was measured as 100 % background. Coloration and bleaching of the film were performed with -2.5 V and +2.5 V respectively. Spectra were recorded 3 min after the voltage was applied. In order to obtain intermediate states of coloration, the film was also colored with coloration potentials of -1.5 V and – 2 V.

The visible transmission spectra were used for evaluation of the optical band gaps E_g of the WO_3. For that purpose, the absorption coefficient (α) was evaluated from the transmittance data (T) and the film thickness (t), using the equation [25]:

$$\alpha = \frac{1}{t}\ln\frac{1}{T} \tag{1}$$

The optical band gaps of the film were evaluated from the absorption coefficient by fitting the data to the relation [26 - 28]:

$$\alpha h\nu = A\left(h\nu - E_g\right)^n \tag{2}$$

where, A is a constant, $h\nu$ is the energy of the incident photon, E_g is the optical energy gap, and n is a number which determines the type of electron transition causing the absorption. The value of n is 1/2 for direct allowed, 3/2 for direct forbidden, 2 for indirect allowed, and 3 for indirect forbidden transitions.

The coloration efficiency (η) of the WO_3 was calculated from the optical density change (ΔOD) at a wavelength of 700 nm, and the charge density ($\Delta Q/S$) during coloration after the films were fully bleached:

$$\eta = \frac{S\log\left(T_b/T_c\right)}{\Delta Q} \tag{3}$$

where T_b and T_c are the transmittance of the fully bleached and colored states respectively, ΔQ is the injected charge, determined by the applied current and the time of its application, and S is the active area of the electrochromic film.

The time needed for the electrochromic film to reach some fraction (usually above 70 %) of its maximum colored or bleached state (response time, τ) was examined as a change in the transmittance at 700 nm due to abrupt voltage change between +2.5 V and -2.5 V.

For an electrochromic material to be practically successful, it must have the ability to switch between its bleached and colored states frequently, whilst maintaining other important features consistently. Cycle life is defined as the number of cycles completed before the material fails, and measures material stability. The cycling behavior of chemically deposited WO_3 films during electrochromic switching was directly observed spectroscopically by the *in situ* measurements of the transmittance at 700 nm of the fully bleached and colored states of the film after some number of cycles. The cycling was performed by alternatively applying potential of ±2.5 V.

The ability to switch between two states (bleached and colored) in a relatively short response time makes the tungsten oxide films a possible candidate for transmittance modulation device. Taking the solar irradiance spectrum AM1.5 for a normal incident illumination on tungsten oxide based electrochromic device (ECD = glass/FTO/WO_3/electrolyte/FTO/glass) and the absorption coefficient spectra of the WO_3 film in its bleached and colored states, the output integral of the spectral intensity and the integral of the spectral modulation could be calculated [29].

Results and discussion

The WO_3 films investigated in this work exhibited good electrochromic behavior. They could be repeatedly colored and bleached by alternative application of a negative and positive potential respectively, versus a counter electrode. WO_3 is cathodically coloring material which means that it possesses a reduced colored state. It is transparent in oxidized state (positive potential), and has a deep blue color in reduced state (negative potential). The X-ray diffraction (XRD) analysis showed that the films were crystalline [24].

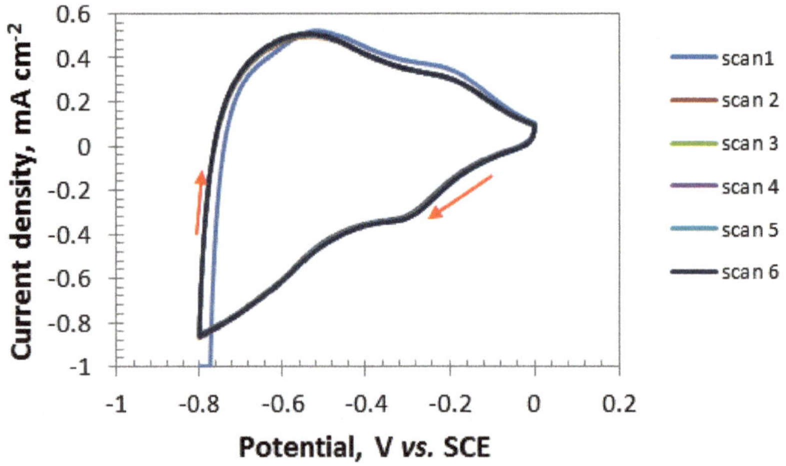

Figure 1. *Cyclic voltammetric curves (six cycles) of chemically deposited WO_3thin film.* Arrows indicate direction of the potential scan.

The electrochromic behavior of the films was examined by cycling voltammetry. The cyclic voltammetric (CV) curves were obtained by sweeping the potential in the range of - 0.8 V to 0 V *vs.* SCE at scanning rate of a 10 mV s^{-1}. In Fig. 1 are presented six CV curves of the WO_3 thin film. As can be seen all CV curves (except the first one) have a same shape, which means that the films exhibited good stability. The CV curves also showed an increase of the cathodic current density to -0.86 mA cm^{-2} at -0.8 V, due to the reduction process occurring in the film and its switching to the blue color, whereas the anodic peak is observed at around - 0.541 V due to the oxidation process and bleaching of the film. The coloring process is followed by reduction of W^{VI} ions and double injection of potassium ions and electrons, and the bleaching process is followed by oxidation of W^{V} ions and double extraction of potassium ions and electrons. The following equation can express the coloring/bleaching process:

$$WO_3(\text{transparent}) + xK^+ + xe^- \leftrightarrow K_xWO_3(\text{blue}) \quad (4)$$

It was observed [30] that during the coloration process, the XRD peaks change in the position and in the intensities indicating a structural transition associated with intercalation of ions. When the ions were deintercalated by applying reverse potential, the crystalline structure reverts to the initial lattice. However, this phenomenon remains to be studied in our future research.

The optical transmittance of the films was recorded to understand the type of electron transition and it was used to evaluate optical energy gap. By analyzing the optical transmission spectra, it is also possible to determine whether the optically induced transition is direct or indirect, and allowed or forbidden.

The optical transmission spectra of the WO_3 film in the wavelength range from 350 to 900 nm in both, bleached and colored states, taken *in situ*, are presented in Fig. 2. One can see significant transmittance difference (more than 60%) that occurs in the red region of the visible spectrum with tendency to continue in the near infrared (NIR) region.

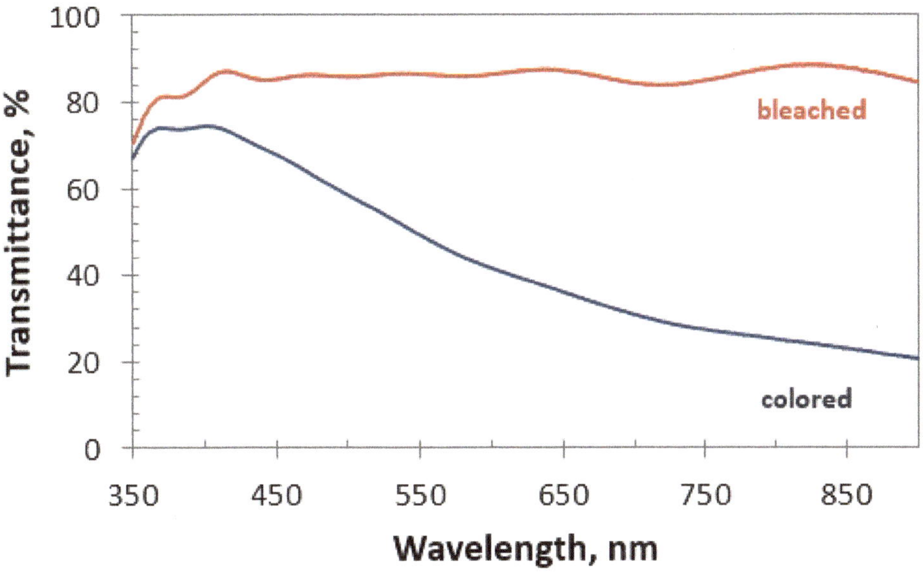

Figure 2. *In-situ visible transmittance spectra of chemically deposited WO_3 thin film in the bleached and colored states.*

Fig. 3 shows the optical transmittance of the ECD constructed by using chemically deposited WO_3 film in its bleached and colored states for an applied bleaching potential of +2.5 V and coloring potentials of -1.5 V, -2 V, and -2.5 V. The recorded transmission spectra for the device in the bleached state for positive potentials lower than +2.5 V are indistinguishable from the spectrum of the device bleached at +2.5 V, and hence are not shown. Also, no difference was observed when negative potentials higher than -1.5 V were applied. The film started to change its color at -1.5 V, and one can see that the spectra recorded at -1.5 V, -2 V and -2.5 V are distinguishable from each other, and from the spectrum of the bleached state of the film. No significant difference was observed between the spectra recorded at coloration potentials of -2.5 and -3 V, which means that the film is wholly reduced at -2.5 V. Actual photographs of the device in the bleached (+2.5 V) and colored (-1.5 V, -2 V, and -2.5 V) states are presented in Fig. 4. From these photographs one could clearly notice the different shades of blue obtained by different coloring potentials.

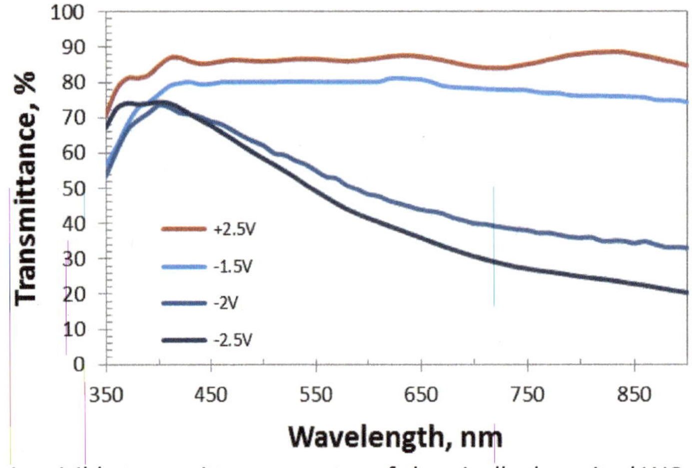

Figure 3. *In-situ visible transmittance spectra of chemically deposited WO_3 thin film bleached at +2.5 V and colored -1.5 V, -2 V, and -2.5V.*

A B C D

Figure 4. *Photographs of ECD* ***A:*** *bleached at +2.5 V,* ***B:*** *colored at -1.5 V,* ***C:*** *-2 V and* ***D:***-*2.5 V.*

The optical energy gaps of the films were evaluated utilizing the transmittance data and the equations (1) and (2). The plots of $(\alpha h\nu)^2$ versus $h\nu$ for the chemically deposited WO_3 thin film in both, bleached and colored states, are shown in Fig. 5. The film showed a better fit for $n = 0.5$ which shows the direct electron transition mechanism in both states (bleached and colored) of chemically deposited WO_3 films. The energy gaps were calculated from the linear parts in Fig. 3 as intercepts with the photon energy axis. The evaluated band gaps for the film in its bleached and colored states were 3.38 eV and 3.32 eV respectively. These values are in good agreement with the reported values on tungsten oxide thin films [31].

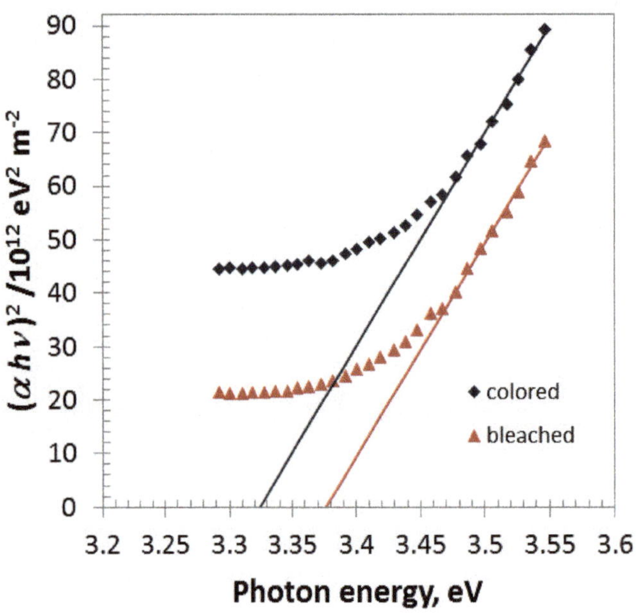

Figure 5. *The plots of* $(\alpha h\nu)^2$ vs. $h\nu$ *for WO_3 thin films prepared by chemical bath deposition.*

To investigate in more detail the optoelectrochemical properties of the chemically deposited WO_3 thin films, the optical density change (ΔOD) was plotted against the charge density change ($\Delta Q/S$), and displayed in Fig. 6. The coloration efficiency η at 700 nm was extracted as the slope of the line fitted to the linear region of the curve. The calculated η value was found to be 22.11 $cm^2\ C^{-1}$. The η value obtained in this work is higher when compared to the values obtained for electrodeposited and sol-gel coated WO_3 thin films [32-35], but lower compared to the values obtained for the sputtered WO_3 [36].

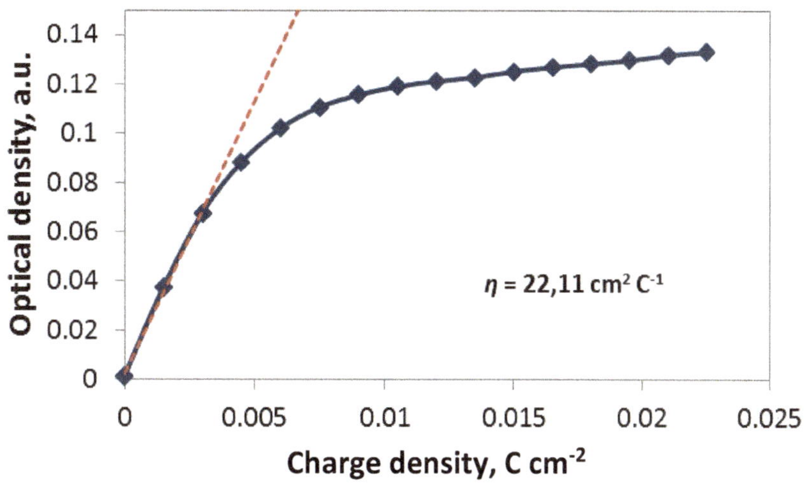

Figure 6. *Optical density variationwith respect to the charge density measured at 700 nm*

In order to investigate the transition response time between coloration and bleaching, the transmittance was measured *in-situ* through the ECD. The applied potential was switched between 2.5 V (transparent state) and - 2.5 V (blue state). Fig. 7 shows the dynamic coloration/bleaching characteristics of the ECD, recorded at the wavelength of 700 nm. The coloration and bleaching times (τ_c and τ_b), defined as time required for achieving 70 % of the total transmission change [37, 38] was found to be 9.3 s and 1.2 s respectively, which means that the coloring kinetics is slower than the bleaching one. The faster bleaching time is due to the good conductivity of the tungsten bronze (K_xWO_3) and the conductor (K_xWO_3) to semiconductor (WO_3) transition. On the other hand, the slower coloration time is due to the higher resistance during WO_3 to K_xWO_3 transition [39].

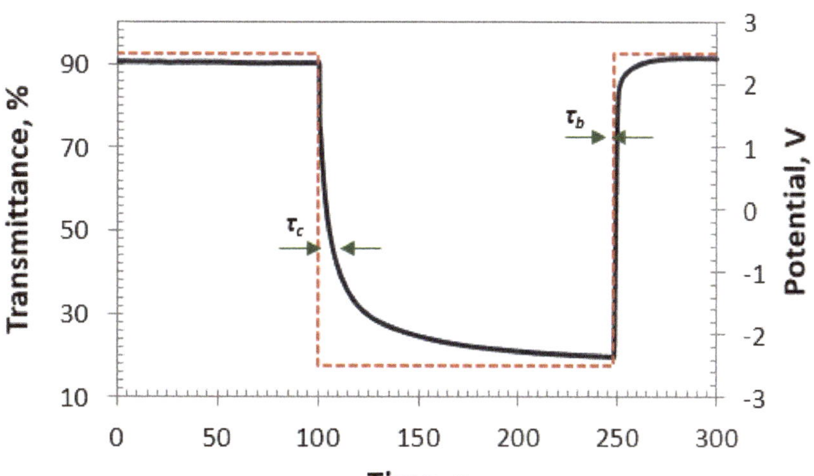

Figure 7. *Switching time characteristics (at 700 nm) between the colored and bleached states* for ECD, measured at ±2.5 V.

Fig. 8 shows the transmittance at 700 nm of the ECD in the bleached and colored states up to 10000 color-bleach cycles. As can be seen from Fig. 8, the transmittance shows insignificant variations which means that the electrochemically deposited WO_3 films are stable and could be electrochemically switched for 10^4 cycles without serious deterioration. Unfortunately, after the 10^4 cycles the optical transmittance change rapidly decreased, so we could say that the device durability is up to 10^4 cycles of bleaching and coloring.

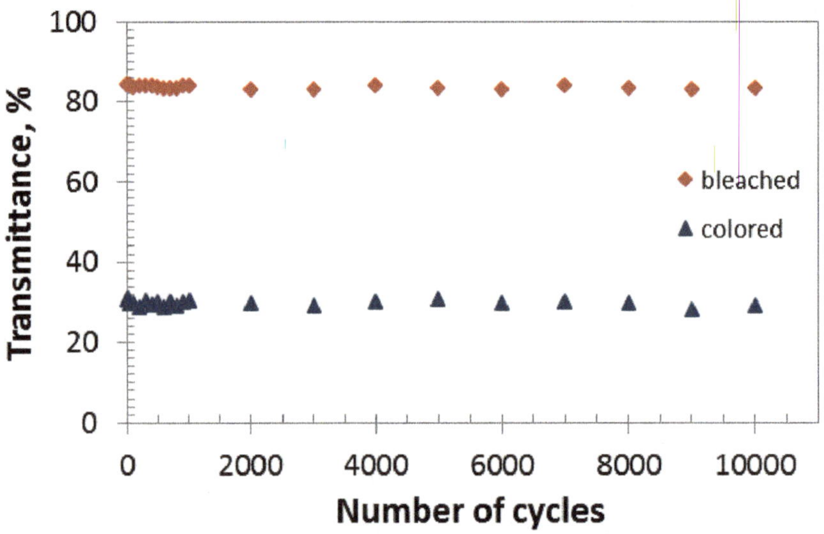

Figure 8. *In-situ transmittance (at 700 nm) of chemically deposited WO_3 thin film in the bleached and colored states vs. number of cycles.*

Finally, the irradiance of the solar spectrum AM 1.5 [40] and the absorption coefficient spectra (calculated from the transmittance spectra) of the chemically deposited WO_3 films in their bleached and colored states (Fig. 9), were taken as input parameters. The output spectral intensities transmitted across the WO_3 films were calculated and presented in Fig. 10. The results of the numerical integration for the spectral intensity within the visible region (350 – 900 nm) are presented in Table 1.

Table 1. *Integral transmitted intensity from 350 to 900 nm(I_t) through the WO_3 films in their bleached and colored states.*

State	I_t/ W m^{-2}
Bleached	506
Colored	253

The relative change of the integrated intensity (the visible transmitted intensity and the light modulation) could be calculated by the equation:

$$\text{Modulation} \approx \frac{I_t(\text{bleached}) - I_t(\text{colored})}{I_t(\text{bleached})} \tag{5}$$

Using the results from the Table 1 and the equation (5), the integrated intensity modulation of about 50 % was achieved, which is considerable value that gives the opportunity for implementation of the chemically deposited WO_3 films in electrochromic devices such as electrochromic windows.

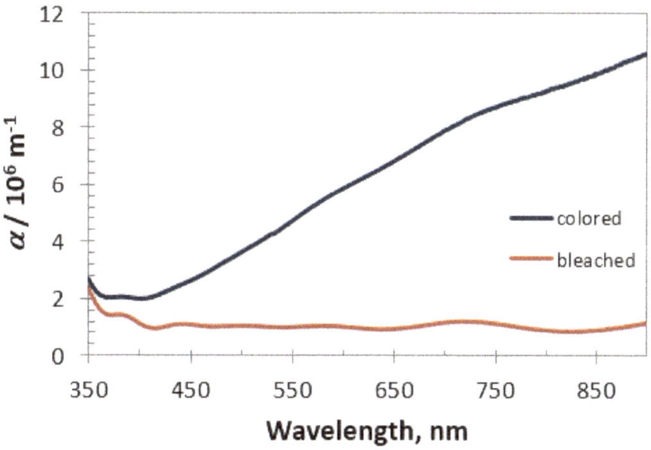

Figure 9. *Absorption coefficient spectra of the chemically deposited WO_3 film in the bleached and colored states.*

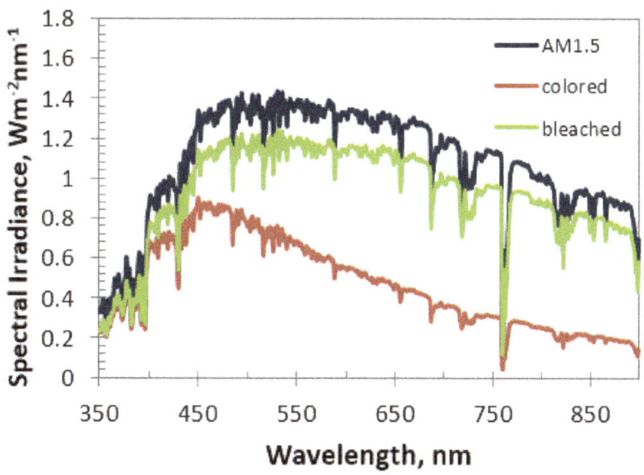

Figure 10. *Spectral intensity of the transmitted AM 1.5 solar irradiance spectrum through the WO_3 film in bleached and colored states*

Conclusions

Tungsten oxide thin films investigated in this work were deposited onto FTO coated glass substrates by chemical bath deposition method. The method is simple, economical, and has benefit of being easily realizable from the point of view of industrialization, especially on large area devices, with the required electrochromic properties. The obtained films exhibited good electrochromic properties. They were stable and exhibited excellent reversibility, with color changed from originally colorless into deep blue when negative potential was applied, and back to colorless when the potential was reversed. Transmittance difference of more than 60 % was achieved in the red region of the visible spectrum. Also, by controlling the coloring potential, intermediate states of coloration were achieved. Optical energy gaps were evaluated from the transmittance measurements for the both, bleached and colored states of the films, assuming a direct semiconductor transition mechanism. The coloration efficiency (at 700 nm) was found to be 22.11 cm^2C^{-1}, the value higher compared with those obtained for electrodeposited and sol-gel coated WO_3 thin films, but lower compared to the values obtained for sputtered WO_3 thin films. The switching times between transparent and blue states of the WO_3 thin film were found to be 9.3 s for coloring, and 1.2 for bleaching. The maximum light intensity modulation ability of the films, as the AM 1.5 spectrum is taken for an input, was calculated to be about 50 % which is considerable value which makes chemically deposited tungsten oxide thin films suitable for application in electrochromic devices.

References

[1] K. Bange, *Solar Energy Materials & Solar Cells*, **58** (1999) 1-131

[2] C.G. Granqvist, *Handbook of Inorganic Electrochromic Materials*, Elsevier, Amsterdam, Holland, 1995, p.

[3] D.R. Rosseinsky, R.J. Mortimer, *Acvanced Materials***13**(11) (2001) 783-793

[4] E. S. Lee, D. L. Di Bartolomeo,*Solar Energy Materials & Solar Cells*, **71**(4)(2002)465-491

[5] C. G. Granqvist, Solar Energy Materials & Solar Cells, **60**(2000)201-262

[6] M. C. Rao, O. M. Hussain, *Research Journal of Chemical Sciences***1** (7) (2011) 92-95

[7] P. M. S. Monk, *Critical Reviews in Solid State and Materials Sciences***24** (1999) 193-226

[8] A. J. More, R. S. Patil, D. S. Dalavi, S. S. Mali, M. G. Gang, J. H. Kim, *Materials Letters***134** (2014) 298-301

[9] C. Y. Kim, S. G. Cho, S. Park, D. K. Choi, *Journal of Ceramic Processing Research***10** (6) (2009) 851-854

[10] A. A. Joraid, *Current Applied Physics* **9** (2009) 73-79

[11] R. U. Kirss, L. Meda, *Applied Organometric Chemistry* 12 (1998) 155 – 160

[12] D. Gogova, K. Gesheva, A. Szekeres, M. Sendova-Vassileva, *physica status solidi (a)***176** (1999) 969-984

[13] C. M. Wang, C. Y. Wen, Y. C. Chen, K. S. Kao, D. L. Cheng, C. H. Peng, *Integrated Ferroelectrics: An International Journal***158** (1) (2014)62-68

[14] T. S. Yang, Z. R. Lin, M. S. Wong, *Applied Surface Science***252** (2) (2005) 2029-2937

[15] F. Zhang, H. Q. Wang, S. Wang, J. Y. Wang, Z. C. Zhong, Y. Jin, *Chinese Physics B***23** (9) (2014) 098105(1-6)

[16] A. P. Baker, S. N. B. Hodgson, M. J. Edilisinghe, *Surface and Coatings Technology***153** (2) (2002) 184-193

[17] M. Ristova, J. Velevska, M. Ristov, *Solar Energy Materials & Solar Cells* 71 (2002) 219-230

[18] T. Todorovski, M. Najdoski, J. Velevska, *International Journal of Pure and Applied Chemistry***1**(4) (2006) 549-552

[19] R. Neskovska, M. Ristova, J. Velevska, M. Ristov, *Thin Solid Films***515** (2007) 4717-4721

[20] S. Demiri, M. Najdoski, J. Velevska, *Material Research Bulletin***46** (2011) 2484-2488

[21] J. Velevska, M. Pecovska-Gjorgjevich, M. Najdoski, N. Stojanov, *Silpacorn University Science and Technology Journal***5** (1) (2011) 34-42

[22] R. S. Mane, C. D. Lokhande, *Materials Chemistry and Physics***65** (2000) 1-31

[23] J. Cheng, D. B. Fan, H. Wang, B. W. Liu, Y. C. Zhang, H. Yan, *Semiconductor Science and Technology*, **18** (2003) 676-679

[24] M. Najdoski, T. Todorovski, *Materials Chemistry and Physics***104** (2007) 483-487

[25] J. M. O-Rueda de Leon, D. R. Acosta, U. Pal, L. Castaneda, *Electrochimica Acta***56** (2011) 2599-2605

[26] R. Bhat, I. Bhaumic, S. Ganesamoorthy, A. K. Karnal, M. K. Swami, H. S. Patel, P. K. Gupta, *Physica Status Solidi* A **209** (1) (2012) 176-180

[27] F. I. Ezema, A. B. C. Eqwealor, R. U. Osuji, *Superficies Vacio***21** (2008) 6-10

[28] A. Tumuluri, K. Lakshun Naidu, K. C. James Raju, *International Journal of ChemTech Research***6** (6) (2014) 3353-3356

[29] M. Ristova, R. Neskovska, V. Mircevski, *Solar Energy Materials & Solar Cells***91** (2007) 1361-1365

[30] Y. Djaoued, S. Balaji, R. Brüning, *Journal of Nanomaterials,* **2012** (2012) Article ID 674168, 9 pages

[31] K. Srinivasa Rao, B. RajiniKanth, G. Srujana Devi, P.K. Mukhopadhyay, *Journal of Materials Science-Materials in Electronics***22** (2011) 1466-1472

[32] J. R. De Andrarde, I. Cesarino, R. Zhang, J. Kanicki, A. Pawlicka, *Molecular Crystals and Liquid Crystals***604** (2014) 71-83

[33] M. Deepa, M. Kar, S. A. Agnihotry, *Thin Solid Films*, **468** (2004) 32-42

[34] C. G. Kuo, C. Y. Chou, Y. C. Tung, J. H. Chen, *Journal of Marine Science and Technology***20** (4) (2012) 365-368

[35] C. Y. Kim, S. Park, *Asian Journal of Chemistry***25** (10) (2013) 5874-5878

[36] S. H. Lee, H. M. Cheong, C. E. Tracy, A. Mascarenhas, A. W. Czanderna, S. K. Deb, *Applied Physics Letters***75** (11) (1999) 1541-1543

[37] J. Velevska, M. Pecovska-Gjorgjevich, N. Stojanov, M. Najdoski, *International Journal of Sciences: Basic and Applied Research***25** (3) (2016) 380-392

[38] Z. Jiao, X. W. Sun, J. Wang, L. Ke, H. V. Demir, Journal of Physics D: Applied Physics **43** (2010) 285501 (6pp)

[39] Z. Jiao, J. Wang, L. Ke, X. Liu, H. V. Demir, M. F. Yang, X. W. Sun, *Electrochimica Acta* **63** (2012) 153-160

[40] ASTM GI73-03 Standard Tables for Reference Solar Irradiances: Direct Normal and Hemispherical on 37^0 Tilted Surface (2012)

9

Anticorrosive behaviour of lumefantrine hydrophobic layer on mild steel surface

Pavithra M. Krishnegowda, Venkatesha T. Venkatarangaiah ✉,
Punith Kumar M. Krishnegowda*, Anantha N. Subba Rao, Shubha H. Nataraj

Department of Chemistry, Kuvempu University, Shankaraghatta-577451, Shimoga, Karnataka, India
**Department of Materials Engineering, Indian Institute of Science, Bangalore-560012, Karnataka, India*

✉Corresponding Author: drtvvenkatesha@yahoo.co.uk

Abstract
The surface modification of mild steel was achieved by chemical treatment in lumefantrine (LF) solution. The surface morphology and wettability of modified surface was analysed by 3D profilometer and contact angle goniometer. The corrosion inhibition performance of modified mild steel surface in 1.0 M HCl solution was investigated by potentiodynamic polarization and electrochemical impedance techniques.Electrochemical measurements illustrate that the corrosion of mild steel in acidic chloride medium get substantially reduced by introducing LF film on its surface (94 % efficiency). Quantum chemical parameters were evaluated by ab initio method and they confer appropriate theoretical support to the experimental findings.

Keywords
Acid solutions; Mild steel; Polarization; Electrochemical impedance spectroscopy ;Acid corrosion

Introduction

The service life of metals can be improved by their surface modification using electroactive compounds [1]. Organic molecules containing O, N, S and polar functional groups are the extensively used electro active compounds, owing to their strong adsorption tendency on the metal surface. These electro active compounds form a thin film or molecular layers on the metal surface. Traditionally, hetero aromatic compounds have been used for mild steel (MS) protection [2-7]. However, these compounds are harmful to the environment due to their toxicity. Therefore, it becomes important to search for new, nontoxic and effective organic compounds to serve the same purpose. In this regard, drugs are the promising alternatives, referred as green corrosion inhibitors [8,9]. Drug molecules have been used for the corrosion inhibition of MS in different corrosive media and remarkable results have been reported [10-14]. In the present work, the antimalarial drug 'Lumefantrine' has been selected to protect the MS from corrosion in acidic chloride medium.The MS was chemically treated with $CHCl_3$ solution containing lumefantrine (LF) and the corrosion behaviour of modified mild steel (MMS) surface was examined by electrochemical techniques.

Experimental

Materials and sample preparation

The experiments were performed with MS specimens having composition 0.04 % C, 0.35 % Mn, 0.022 % P, 0.036 % S, and the remainder being Fe. The MS coupons of1 cm^2 area (exposed) with a 5 cm long stem isolated with araldite resin were used for electrochemical measurements. Prior to each experiment, the metal samples were abraded with series of emery papers (grade No. 220, 660 and 1200) followed by thorough wash using millipore water, acetone and then dried in hot air.

The aggressive solution of 1.0 M HCl was prepared from AR grade 37 % HCl and millipore water. The millipore water was obtained from Elix 3 Milli-pore system (Resistivity greater than 18 MΩ cm at 25 °C). Lumefantrine was obtained from Ramdev Chemicals India Pvt. Ltd., Mumbai and its structure is as shown in Figure 1. The different concentrations of $CHCl_3$ solutions of LF were prepared by dissolving desired amount of LF in $CHCl_3$. The MS coupons were dipped in 100 ml of $CHCl_3$ solutions of LF for about 1 hr to obtain MMS. After surface modification, the metal samples were dried and stored in desiccator for further corrosion studies. The MS coupon modified in 3mM LF solution was used for surface analyses.

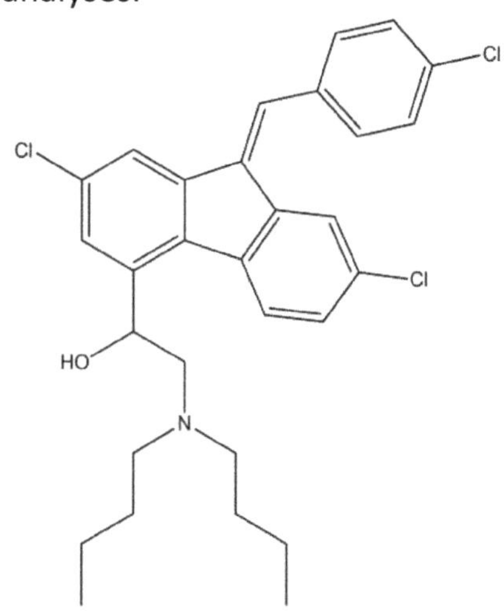

Figure 1. *Structure of lumefantrine*

Contact angle measurement

An OCA 30 from DataPhysics Instruments GmbH, Germany was used to measure the contact angles. The contact angle of water on the metal sample was measured by the sessile drop method. A commercial goniometer was used with the volumes of water droplets fixed at 2 µL and the measurements were performed at room temperature.Images of the Air/Liquid/Solid (A/L/S) systems were captured and processed using the 32-bit SCA 20 software.

Infrared spectra

An infrared reflection absorption spectroscopy (GX spectrometer, Perkin Elmer, U.S.A.) instrument outfitted with a liquid nitrogen cooled mercury cadmium telluride (MCT) detector was used to take IR spectra. Before starting the experiments the sample and detector chambers were purged with nitrogen gas. The reported IR spectra were obtained with reference to bare steel substrate over 1024 optimized scans at 4 cm^{-1} resolution by using polarized beam spectra of modified surface.

The surface morphology

The surface of MS samples was examined by Zeta-20 True Color 3D Optical Profiler (Zeta Instruments, CA, U.S.A.). After the experiment, the metal samples were kept in a vacuum desiccator and were mounted on sample holder under the objective of the Optical Profiler, and the 2D photos were taken from the 50× magnified surface.

Quantum chemical studies

Quantum chemical calculations were performed using standard HyperChem, Release 8.0 software (Hypercube, Inc. GmBH Austria, USA). Geometrical structure of LF molecule was optimized by parametric (PM3) method. The optimized structure which has minimum energy value was used as the input structure for the geometrical optimization using ab initio method (Hartree–Fock method, 6-31G** basis set). The Polak–Rieberre algorithm which is fast and accurate was used for computation. The energy parameters in the form of root mean square gradient were kept at 0.1 kcal/Å mol. 0.1 kcal/Å mol.

Electrochemical measurements

Electrochemical measurements were conducted in a conventional glass cell using CHI 660C electrochemical analyzer (CH Instruments, Austin, TX, USA). MS specimen (1 cm^2), a platinum electrode and Ag/AgCl electrode were used as working, auxiliary, and reference electrodes respectively. Prior to each electrochemical measurement, a stabilization period of 30 min was allowed to establish a steady state open circuit potential (OCP). Each experiment was carried out in triplicate and the average values of corrosion parameters are reported.

The potentiodynamic polarization (PDP) measurements were carried out over a potential range of −200 mV to +200 mV at OCP with a scan rate of 0.5 mV s^{-1}. The corrosion kinetic parameters such as corrosion potential (E_{corr}), corrosion current density (I_{corr}), and anodic (β_a), and cathodic (β_c) Tafel slopes were evaluated by the software installed in the instrument.

The percentage inhibition efficiency η_T / %was computed from I_{corr} values using the following expression;

$$\eta_T / \% = \frac{I^o_{corr} - I_{corr}}{I^o_{corr}} \times 100 \qquad (1)$$

whereI^o_{corr}andI_{corr} are the corrosion current densities without and with inhibitor, respectively.

The electrochemical impedance spectroscopy (EIS) measurements were carried at OCP in the frequency range 1 mHz to 100 kHz with 5 mV sine wave as the excitation signal. Impedance data were analyzed using ZSimp-Win 3.21 software.

The inhibition efficiency η_Z/ % was evaluated from charge transfer resistance (R_{ct}) values using the following equation

$$\eta_Z / \% = \frac{R_{ct} - R^o_{ct}}{R_{ct}} \times 100 \quad (2)$$

whereR^o_{ct} and R_{ct} are the charge transfer resistances without and with inhibitor, respectively.

Results and discussion

Morphology and contact angle measurement

The 3D profilometer images and contact angle of a water droplet (2 μL) on the naked MS surface and MMS surface are presented in Figure 2. In general, low contact angle represents the hydrophilic surface and high contact angle represents the hydrophobic surface [15-16]. In the present study, the contact angle of the naked MS is 77.15 ° (Figure 2a) and that of LF covered MS is 92.9 ° (Figure 2b). This indicate that the LF modified MS surface is more hydrophobic than the bare MS surface. Moreover, the noticeable change in contact angle value on MMS may be due to the formation of a well packed layer of LF molecules. The compact layer on MMS may be attributable to the adsorption of differently oriented LF molecules on different sites of MS surface, which affords a barrier effect against the diffusion of chloride ions [17]. On the other hand, the surface of polished MS is comparatively rough and the smooth surface of MMS may be due to the uniform coverage of LF molecules. Thus, the surface morphology and contact angle measurements substantiates the formation of LF film on the MS surface.

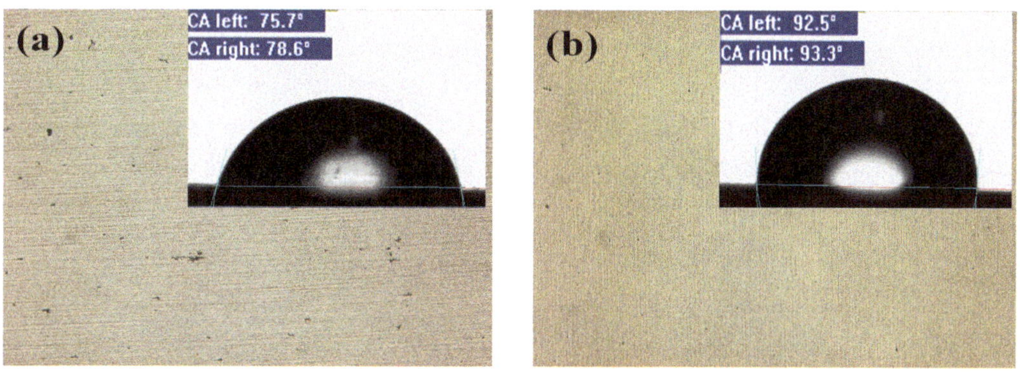

Figure 2.*Profilometer images and optical micrographs of a water dropleton (a) the naked mild steeland (b) modified mild steel.*

IR spectral analysis

The IR spectra were used to determine the presence of LF film on the modified MS surface and are depicted in Figure 3.By comparing Figure 3a and 3b,it can be observed that certain peaks have been disappeared completely and some have been shifted. Figure 3a shows the IR spectrum of LF molecule. The characteristic bandat 3490 cm^{-1} and 1459 cm^{-1} wereattributed to stretching vibration of the O-Hand aromatic C=C groups, 1151 cm^{-1}corresponds to C-N bond, 2864 cm^{-1} corresponds to aromatic and 2953 cm^{-1} corresponds to aliphatic C-H stretching. Also the band at 874 cm^{-1} is attributed to C-Cl stretching. The IR spectrum of LF film on MMS is shown in Figure 3b. In this spectrum, C-N stretching frequency is shifted to 1122 cm^{-1}, and aromatic C=C stretching frequency is shifted to 1409 cm^{-1}. Furthermore, the O-H band at 3490 cm^{-1} gets disappeared in Figure 3b. Based on these observations, it can be concluded that the adsorption of LF molecule on MS surface takes place through C=C, O-H and C-N bonds and confirms the adsorption of LF molecules on the MS surface.

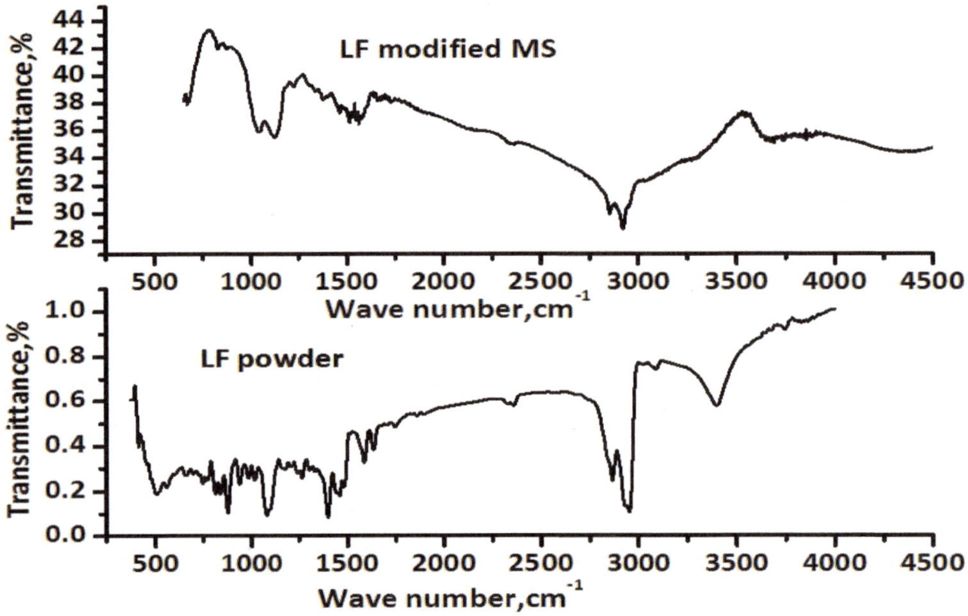

Figure 3. *IR spectra of (a) LF powder and (b) LF film on the mild steel surface.*

Quantum chemical Study

Quantum chemical calculations will provide information about the molecular structure of LF molecule responsible for the formation of film on the MS surface.The quantum chemical calculations were carried out with the help of complete geometry optimization by ab initio method using 6-31G** basis set. The geometry of LF molecule was optimized by Hartree–Fock method and the optimized structure, distribution of highest occupied molecular orbitals (HOMO) and lowest unoccupied molecular orbitals LUMO are shown in Figure 4. The calculated quantum chemical parameters are presented in Table 1.

Table1*. Quantum chemical parameters of LF*

Quantum chemical parameters	
E_{HOMO} / eV	–8.009
E_{LUMO} / eV	1.525
ΔE / eV	9.534
μ / D	4.495
Total energy, k cal mol^{-1}	–1670734.09

The spatial orbitals are distributed randomly on O, N and phenyl rings present in the molecule. Moreover, LF molecule possesshigh dipole moment (μ) and the low energy gap (ΔE) values which indicate that electron transfer from LF to the surface takes place during the process of adsorption. Conversely, the adsorption centers of organic molecules can be approximated by net atomic charges in the molecule. The net atomic charges of some atoms present in LF molecule are given in Table 2. The regions of highest electron density on organic molecule are the most plausible sites of adsorption [18,19]. Therefore, N, O and some C atoms are the active centers, which have the strongest ability to bond to the metal surface. Thus, oxygen atom, nitrogen atom and π electrons in the phenyl ring are the main active sites of adsorption on MS surface. By reviewing these findings, it can be deduced that LF forms a compact film by the process of adsorption on the surface of MS. As a result, the above experimental findings of IR spectral analysis are in good agreement with the quantum chemical results.

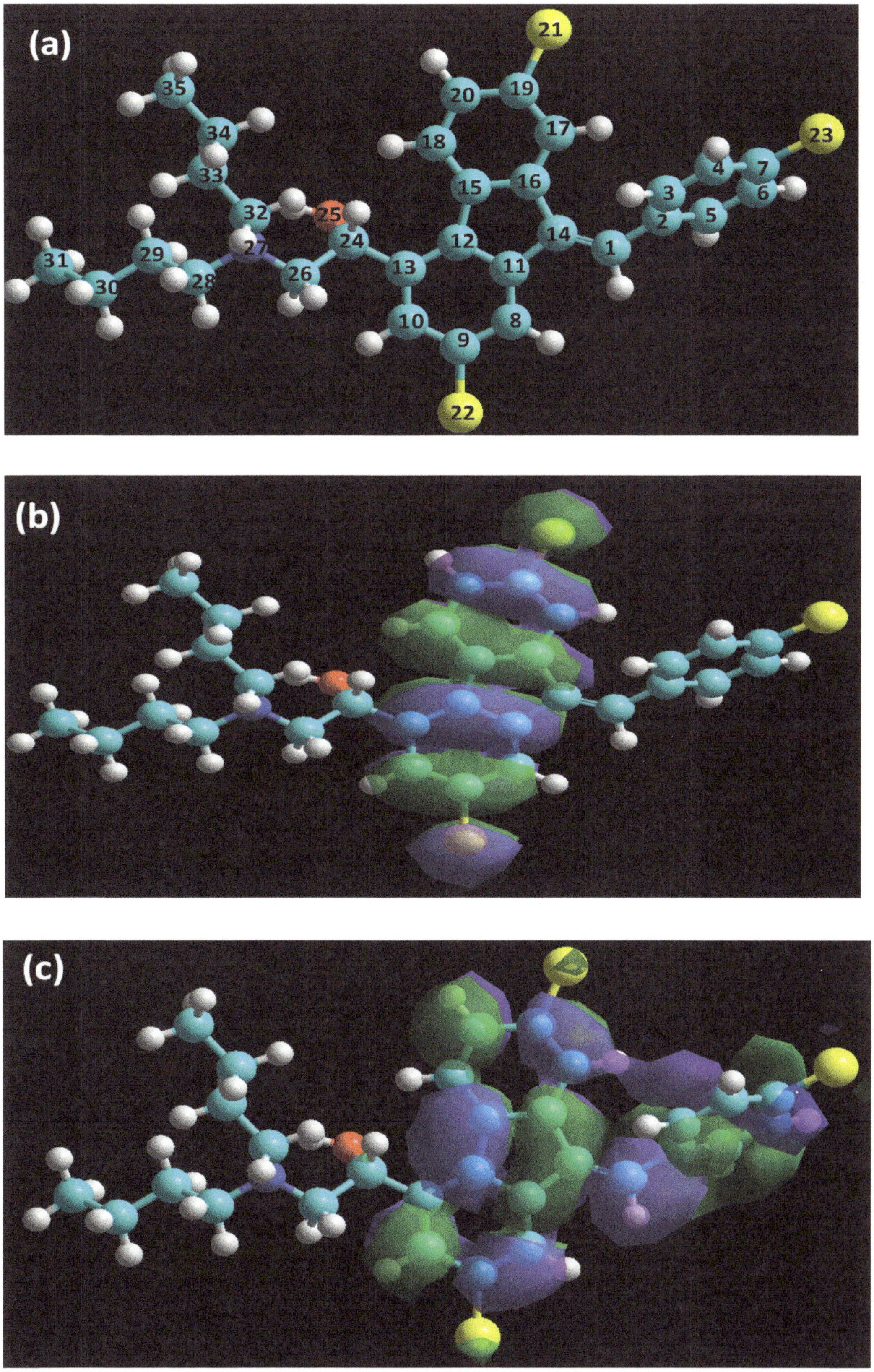

Figure 4.*(a) Optimized structure, (b) distribution ofHOMOand (c) distribution of*LUMO of LF molecule.

***Table 2.** List of atoms having highest electron density in LF molecule.*

Atom	Charge
C (2)	-0.013
C (3)	-0.026
C (5)	-0.026
C (18)	-0.025
Cl (21)	-0.116
Cl (22)	-0.116
Cl (23)	-0.116
O(25)	-0.386
N (27)	-0.301
C (29)	-0.041
C (30)	-0.055
C (31)	-0.065
C (33)	-0.041
C (34)	-0.055
C (35)	-0.065

Potentiodynamicpolarisation measurements

Electrochemical analyses were performed to investigate the kinetics of electrode processes as well as the surface characteristics of the electrochemical system. The potentiodynamic polarization curves in 1M HCl for the naked MS and MMS in different concentrations of LF solutions are given in Figure 5. The electrochemical corrosion parameters such as corrosion potential (E_{corr}), corrosion current density (I_{corr}) , anodic tafel slope (β_a), cathodic tafel slope (β_c) and inhibition efficiency (η_T/ %) are tabulated in Table 3 .

It is apparent from Figure 5 that increasing the LF concentration reduces both the cathodic and anodic current densities and the E_{corr} get slightly shifted towards the anodic direction. Moreover, the decrease in both the anodic and cathodic current densities enumerated that the LF film formed on the MMS surface suppressed both the anodic and cathodic reactions [20]. This phenomenon may be due to the adsorption of the LF molecules, which in turn decreased the attack of chloride ions on the MS surface.

***Table 3.** Polarization parameters in 1 M HCl for the bare mild steel and modified mild steel in different concentrations of LF solutions.*

c/mM	$-E_{corr}$/ mV	$-\beta_c$/ mV dec^{-1}	β_a / mV dec^{-1}	I_{corr}/ µA cm^{-2}	η_T/ %
0.00	449	106.46	84.38	257.5	
0.50	376	157.28	64.71	43.67	83
0.75	370	148.77	71.03	30.89	88
1.00	373	142.07	63.38	17.41	93
3.00	365	154.30	60.75	14.58	94
5.00	364	143.97	56.27	19.15	92

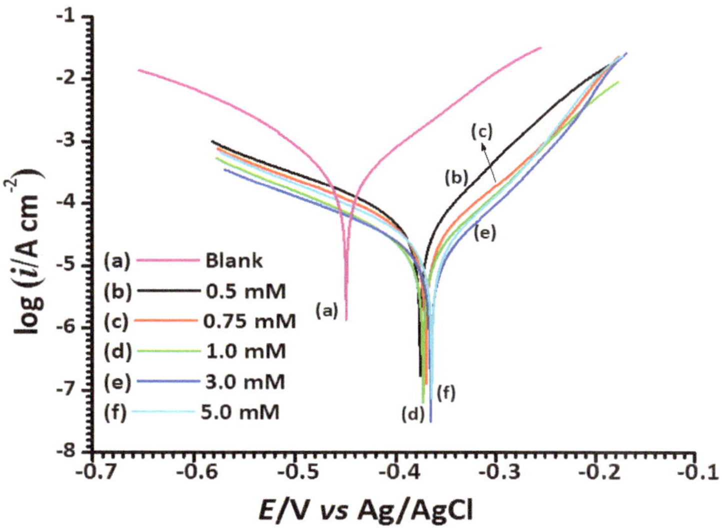

Figure 5. *Potentiodynamic polarization curves in 1 M HCl for the naked mild steel and mild steel modified inpresence of different concentrations of LF.*

It is apparent from the Table 3 that with increasing concentration of LF, I_{corr} values lowered to its minimum value at 3 mM and the η_T reached the maximum value 94 %. This implies that the film formed on the MS becomes denser with increase in LF concentration from 0.5 to 3 mM. But the η_T value decreases when the LF concentration is 5 mM. This may be due to the fact that the LF molecules get rearranged on the MS surface after the number of adsorbed molecules reaches a maximum value [21]. This makes easy for the acidic chloride ions to attack the MS surface through the interspaces. Furthermore, the values of β_a and β_c change with increase in inhibitor concentration and it indicates that the film inhibits both the anodic and cathodic reactions. By reviewing all these results it can be considered that the LF layer acts as a good protective film.

Electrochemical Impedance spectroscopy measurements

The Nyquist and Bode plots of the bare MS and LF covered MS in 1M HCl are given in Figure 6. For the naked MS, the impedance data were analyzed by the electrochemical equivalent circuit shown in the inset of Figure 6a. In this circuit, R_s represents the solution resistance, R_1 is the charge transfer resistance corresponding to the corrosion reaction at the MS/solution interface, and CPE_1 represents the constant phase element (*CPE*) as a substitute for the double-layer capacitance (C_{dl}). The impedance of *CPE* is defined as

$$Z_{CPE} = Q^{-1}(j\omega)_{-n} \quad (3)$$

where *Q* is the CPE constant, ω is the angular frequency, $j^2 = -1$ is the imaginary number and *n* is the CPE exponent which gives details about the degree of surface inhomogeneity resulting from surface roughness, inhibitor adsorption, porous layer formation etc.

For the bare MS, the Nyquist plot is the slightly depressed capacitive loop (Figure 6a) which indicates the roughness and other inhomogeneity of the MS surface [22]. The Nyquist and bode plots for MMS in different concentration of LF solutions were fitted by the equivalent circuit shown in Figure 7(a). R_s, R_1, and CPE_1 have the same meaning of elements in Figure 7a. In addition, R_2 represents the membrane resistance, which reflects the protective property of the film and CPE_2 is the capacitance of the film formed on MS [23-24]. The obtained impedance data are tabulated in Table 4.

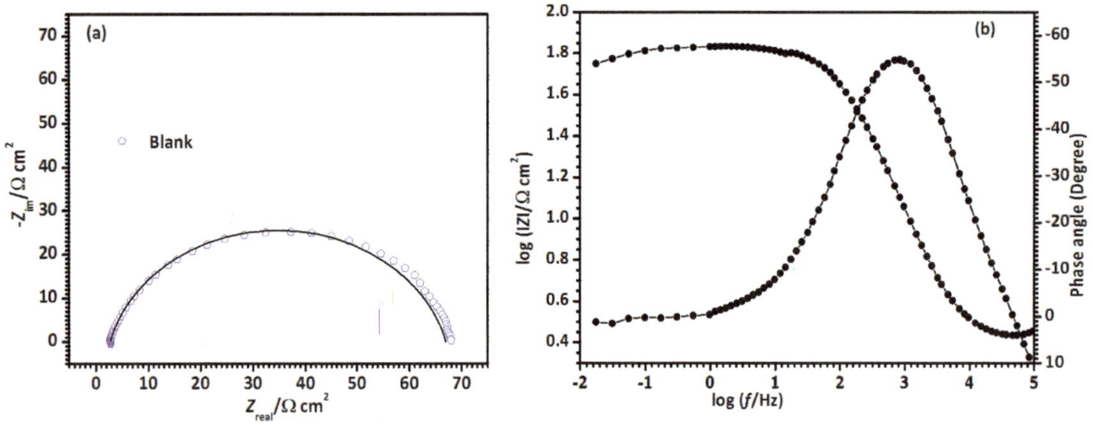

Figure 6. *(a) Nyquist plot and (b) Bode plot of naked MS in 1 M HCl*

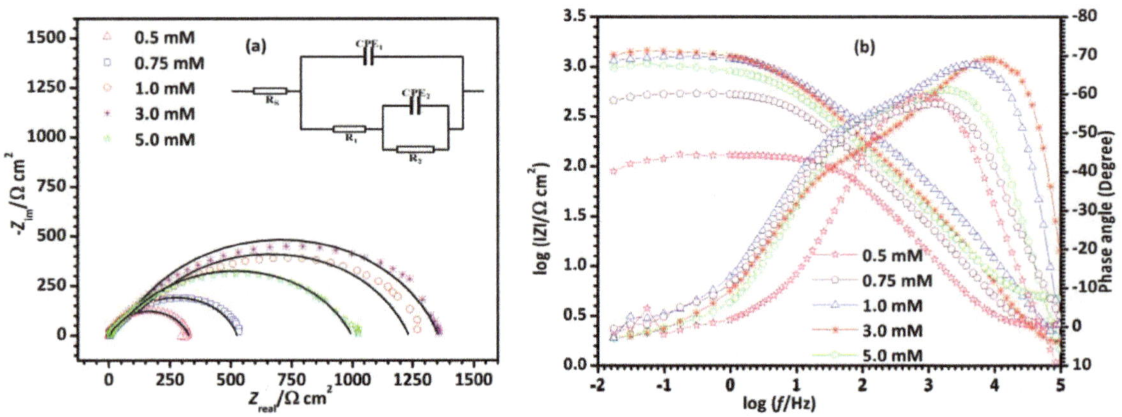

Figure 7. *EIS results in 1 M HCl for LF covered mild steel modified in differentconcentration of LF solutions: (a) Nyquist plots; (b) Bode plots*

Table 4.*Impedance parameters in 1 M HCl for the naked mild steel and mild steel modified in presence of different concentrations of LF solutions.*

c/ mM	R_1/ Ω cm^2	Q_1/ μΩ^{-1}S^n cm^{-2}	n_1	R_2/ Ω cm^2	Q_2/ μΩ^{-1}S^n cm^{-2}	n_2	η_z/ %
0.00	64.14	56.26	0.856	--	--	--	
0.50	303.5	109.4	0.733	21.1	19.38	0.999	79
0.75	499.3	51.68	0.829	24.95	17.16	0.954	87
1.00	1136	34.93	0.730	98.02	64.51	0.921	94
3.00	1292	37.84	0.815	61.75	11.20	0.946	95
5.00	950.2	59.65	0.761	41.58	33.71	0.883	93

It is apparent from Figure 7a that, the diameter of the capacitive loop increased to its maximum when the concentration of LF solution was 3 mM, implying that the compact film is formed on the surface of MS in this concentration. Bode plot shows two overlapped phase maxima, indicating the model of two time constants and the Z modulus increases with the increase of LF concentration within 3 mM, suggesting that a dense film formed on the MS exhibit better protection efficiency. It is evident that the diameter of the capacitive loop and the logarithm of Z modulus decreases when the concentration of LF is 5 mM.The inhibition efficiency calculated from R_1 also increases with increase in LF concentration, and reaches maximum (95 %) at 3 mM concentration. This may be due to the hydrophobic film of LF molecules on MS surface which inhibited the dissolution reactions of MS to a greater extent in 1 M HCl. The variation in R_1 values imply that the corrosion reaction on the MS surface was inhibited by the absorbed inhibitor film.

Surface morphology

Morphologies of metal samples were analysed by the true color images obtained from 3D profilometer. The true color images of bare MS and MMS immersed in 1M HCl is shown in Figure 8. The surface of bare MS is completely damaged with large number of pits due to the attack of aggressive chloride ions. But MMS retains smoother surface owing to the presence of LF film which inhibits the process of corrosion in acidic chloride medium. Hence it can be concluded that, the film formed by the adsorption of LF molecules behaves as a barrier and retards MS corrosion in HCl media.

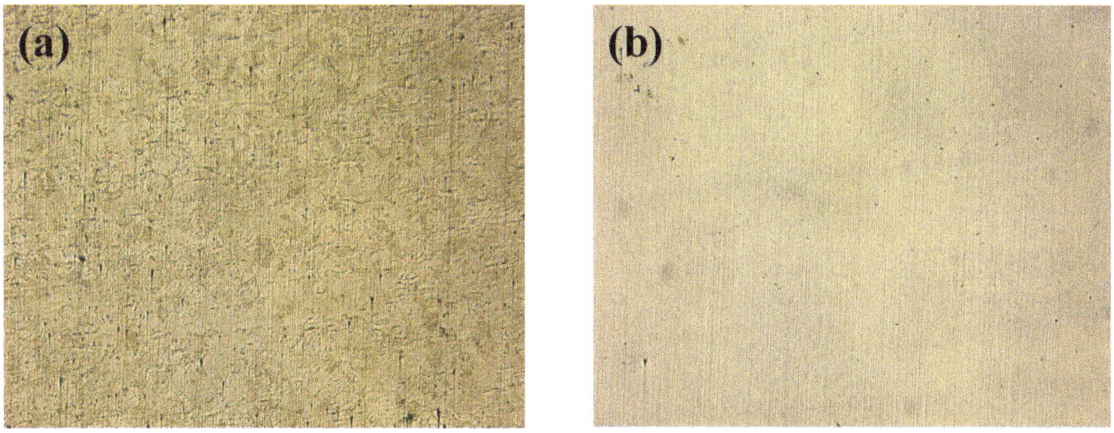

Figure 8.*Profilometer images of (a) the naked mild steel and (b) modified mild steel after immersion in 1 M HCl for 1 h.*

Conclusion

The contact angle and IR analyses indicate the presence of LF molecules on mild steel surface. The quantum chemical study reveals that the O, N and π electrons in the LF molecule are the main active sites for adsorption. The chemical treatment of MS for about 1 h leads to the formation of compact film of LF molecules and hampers acid corrosion. The LF film inhibits anodic and cathodic reactions of MS in 1 M HCl. The electrochemical studies and profilometer images confirmed the MS corrosion inhibition ability of LF molecules in acidic chloride solution.

Acknowledgement: *The authors are grateful to the authorities of Department of Chemistry, Kuvempu University, Karnataka, India for providing lab facilities. Authors also thank Department of Science and Technology, New Delhi, Govt. of India [DST: Project Sanction No. 100/IFD/1924/2008-2009 dated 2.07.2008] for providing instrumental facilities. The authors are also gratified to University Grant Commission, New Delhi, Govt. of India [UGC grant Ref. 41-231/2012(SR) dated 16.07.2012] for providing HyperChem software facility.*

References

[1] T.T.Qin, J. Li, H.Q. Luo, M. Li, N.B. Li, *Corros. Sci.***53** (2011) 1072–1078.
[2] Q.Q. Liao, Z.W. Yue, D. Yang, Z.H. Wang, Z.H. Li, H.H. Ge, Y.J. Li, *Corros. Sci.***53** (2011)1999–2005.
[3] D.Q. Zhang, L.X. Gao, G.D. Zhou, *Corros. Sci.***46**(2004) 3031–3040.
[4] N. Bellakhal, M. Dachraoui, *Mat. Chem. Phy.***85**(2004) 366–369.

[5] M. Dudukcu,G. Avcı, *Res. Chem.Intermed.***41**(2015)4861–4871.

[6] H. B. Ouici,M. Belkhouda, O. Benali , R. Salghi, L. Bammou, A. Zarrouk, B. Hammouti, *Res. Chem.Intermed.***41**(2015) 4617–4634.

[7] M. K. Pavithra , T. V. Venkatesha, M. K. Punith Kumar,K. Manjunatha, *Res. Chem.Intermed.***41**(2015) 7163–7177.

[8] A. Samide,B. Tutunaru, C. Negrilab,*Chem. Biochem. Eng. Quart.***3**(2011) 299–308.

[9] G. Gece, *Corros. Sci.***53** (2011) 3873–3898.

[10] I. Naqvi, A. R. Saleemi, S. Naveed, *Int. J. Electrochem. Sci.***6**(2011) 146 –161 .

[11] I. Ahamad, R. Prasad, M.A. Quraishi, *Corros. Sci.***52**(2010) 3033–3041.

[12] I. Ahamad, R. Prasad, M.A. Quraishi, *J. Solid State Electrochem.***14** (2010) 2095–2105 .

[13] U. M. Eduok,M. Khaled, *Res. Chem.Intermed.***41**(2015) 6309–6324 .

[14] M. K. Pavithra, T. V. Venkatesha, M. K. Punith Kumar, N. S. Anantha, *Res. Chem.Intermed.* (2015)DOI 10.1007/s11164-015-2158-3 .

[15] S. Khorsand, K. Raeissi, F. Ashrafizadeh, *Appl. Surf. Sci.***305**(2014) 498–505 .

[16] N. Wang, D. Xiong, *Appl. Surf. Sci.***305**(2014) 603–608 .

[17] H. Dadafarin, E. Konkov , S. Omanovic, *Int. J. Electrochem. Sci.***8**(2013)369–389 .

[18] P. Wang, D. Zhang , R. Qiu, Y. Wana, J. Wua, *Corros. Sci.***80**(2014) 366–373.

[19] M. K. Pavithra, T. V. Venkatesha, M. K. Punith Kumar, N. S. Anantha,*J. Electrochem. Sci. Eng.* **5(2)** (2015) 115-127.

[20] Y-C.Pan , Y.Wena, X-Y. Guo, P.Song , S. Shen , Y-P. Dub, H-F. Yang,*Corros. Sci.***73**(2013) 274–280.

[21] W.J. Guo, S.H. Chen, H.Y. Ma, *J. Serb. Chem. Soc.***71** (2006)167 –175.

[22] J. Zhang , X.L. Gong, H.H. Yu, M. Du, *Corros. Sci.***53** (2011) 3324–3330

[23] N. E. Hamdania, R. Fdila, M. Tourabi, C. Jama, F. Bentiss, *Appl. Surf. Sci.***357** (2015) 1294–1305 .

[24] M. Lebrini, F. Robert, C. Roos, *Int. J. Electrochem. Sci.***6**(2011) 847–859.

10

Simultaneous electrochemical determination of acetaminophen and metoclopramide at electrochemically pre-treated disposable graphite pencil electrode

Shreekant M. Patil, Vijay P. Pattar, Sharanappa T. Nandibewoor✉

P. G. Department of Studies in Chemistry, Karnatak University, Dharwad 580003, India

✉Corresponding author: stnandibewoor@yahoo.com

Abstract

A sensitive and economic voltammetric method was developed for the simultaneous determination of acetaminophen (AMP) and metoclopramide (MCP) using pre-treated graphite pencil electrode (PTGPE). Compared to a graphite pencil electrode, the pre-treated electrode showed an apparent shift of the oxidation potentials in the positive direction and a notable enhancement in the current responses for both AMP and MCP. Cyclic voltammetry (CV) was used to study the voltammetric behavior of the drugs, while differential pulse voltammetry (DPV) was used to determine AMP and MCP simultaneously. The dependence of the current on scan rate, pH and concentration was investigated to boost the experimental conditions for simultaneous determination. The calibration curves were obtained over the range of 0.1×10^{-7} to 1.1×10^{-7} M, the concentration of each of both the drugs was varied by keeping the other constant, and achieved lower detection limit of 3.25 nM for AMP and 1.16 nM for MCP. The developed method was found to be selective and rapid for the simultaneous determination of AMP and MCP. The proposed method was applied simultaneously in real samples and pharmaceutical samples, with satisfactory results.

Keywords

Acetaminophen, metoclopramide, voltammetry, graphite pencil electrode, analytical applications

Introduction

The development of a sensitive, simple, reliable and rapid method for the determination of analyte is of great importance. The pre-treated graphite pencil electrode (PTGPE) has been used as a biosensor in modern electroanalytical field due to its high mechanical rigidity, highly economical, good electrochemical reactivity, ease of modification, renewal of electrode and low background current [1,2]. PTGPE has good application in analysis of drugs, detection of traces of metal ions and neurotransmitters. Metoclopramide (4-amino-5-chloro-N-[2-(diethylamino)ethyl]-2-methoxybenzamide) (MCP) as shown in Scheme 1 has wide range of clinical applications in different fields such as gastroenterology, gynaecology, surgery, radiology and cardiology. It shows antiemetic and prokinetic properties in disorders of decreased gastrointestinal motility and it is a dopamine receptor antagonist which also plays very important role as active ingredient of many pharmaceutical formations related with the modification of digestive behavior. MCP hydrochloride is commonly used in prevention and relief of nausea and vomiting [3-4] but mainly used in combination with chemotherapy, where drugs such as cisplatin, and some cytotoxic agents, are highly emetic [5].

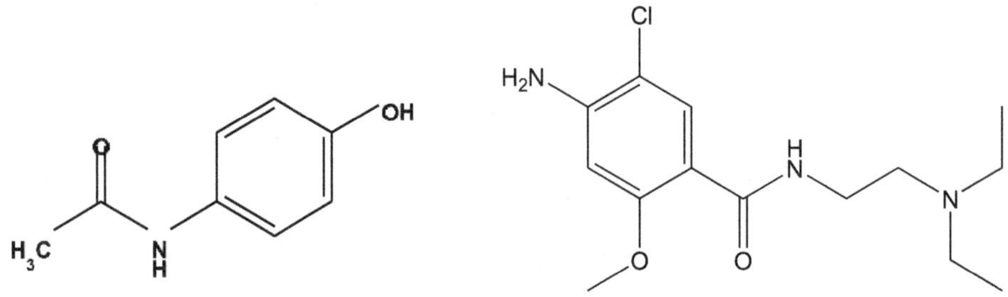

Scheme 1. *Chemical structure of AMP and MCP*

Acetaminophen (AMP) (paracetamol, N-acetyl-*p*-aminophenol) (Scheme 1) is a well-known drug which has extensive applications in pharma industries. It is a non-steroidal, antipyretic and anti-inflammatory drug [6]. It is the preferred alternative to Aspirin, specifically for patients who cannot tolerate Aspirin [7] and its use is one of the common causes of poisoning worldwide [8] and analgesic compound that has high therapeutic value. It is also used as a precursor in penicillin and as stabilizing agent for hydrogen peroxide, photographic chemical, *etc.* [9]. At the recommended dosage, there are no side effects. However, overdoses of acetaminophen cause liver and kidney damage [10] and may lead to death. It is suspected that a metabolite of acetaminophen is the actual hepatotoxic agent [11].

Acetaminophen/metoclopramide hydrochloride is an oral fixed dose combination containing the analgesic acetaminophen and the antiemetic metoclopramide hydrochloride. The combination is used to treat the migraine symptoms, both to relieve headache and to treat associated nausea and vomiting (the antiemetic). In addition to its direct antiemetic effect, MCP also stimulates gastric emptying (prokinetic), which is often delayed during migraine attacks, and accelerates the absorption of AMP. The combination of AMP and MCP has shown to enhance the analgesia relief when used to treat the arthritis pain.

As MCP is always given with AMP in any formulations, it is important to determine both the components simultaneously. Allthough there are different methods of anlysis of AMP [12] and MCP [13] separately mentioned in the literature, only a few classical methods are available [14-18] for the simultaneous determination of these componds. In the present study, the electrochemical method, a very sensitive and selective method of analysis using newly functionalised PTGPE has been used in the determination of AMP and MCP simultaneously. The proposed method was applied in real samples and pharmaceuticals.

Experimental aspects

Reagents and chemicals

The pencil-lead rods (HB 5 cm length and 0.5 mm in diameter) were purchased from a local stationary. AMP and MCP were purchased from Sigma Aldrich, India. Stock solutions of AMP and

MCP (1.0 mM) were freshly prepared in Millipore water. The phosphate buffer solutions (PBS) from pH 3.0 to 11.2 were prepared according to the method of Christian and Purdy [19]. Other reagents used were of analytical grade. All other solutions were prepared with Millipore water.

Instrumentation and analytical procedures

The voltammetric experiments were carried out on a CHI 630D electrochemical analysing system (CH instruments Inc., USA). The voltammetric experiments were carried out in a 10 ml single-compartment of three electrode glass cell with the reference electrode as silver electrode (Ag/AgCl), platinum as counter electrode and PTGPE as working electrode. pH measurements were carried with Elico LI120 pH meter (Elico Ltd., India). All experiments were performed at an ambient temperature of 298± 0.2K.

The area of the electrode was obtained by the cyclic voltametric method using 1.0 mM $K_3Fe(CN)_6$ at different scan rates. For a reversible process, the following Randles-Sevcik formula was used [20].

$$I_p = 0.4463(F^3/RT)^{1/2}n^{3/2}A_0D_0^{1/2}C_0v^{1/2} \qquad (1)$$

where I_p refers to the anodic peak current, A_0 is the surface area of the electrode, n is number of electrons transferred, v is the scan rate, D_0 is the diffusion coefficient and C_0 is the concentration of $K_3Fe(CN)_6$. For 1.0 mM $K_3Fe(CN)_6$ in 0.1M KCl electrolyte, T = 298 K, R= 8.314 J $K^{-1}mol^{-1}$, F = 96,480 Coulombs mol^{-1}, n=1, and $D_0=7.6\times10^{-6} cm^2s^{-1}$;then from the slope of the plot of I_p versus $v^{1/2}$, relation, the surface area was calculated. In our experiment the slope was 4.136×10^{-5}A $(Vs^{-1})^{1/2}$ and the area of electrode was calculated to be 0.1311 cm^2 which is three times more than that of the GPE.

Measurement procedure

Stock solutions of 1 mM of AMP and MCP were prepared by dissolving the desired amount in millipore water. Voltammograms were then recorded using voltammetric analyzer under the optimized parameters.

The parameters for DPV were, amplitude: 0.05 V; initial potential: 0.0 V; final potential: 1.4 V; increase in potential: 0.004 V; pulse width: 0.06s; sample width: quiet time: 2 s; sensitivity: 1×10^{-6} A V^{-1}.

Pre-treatment of electrode

The electrochemical treatment of GPE was performed in different supporting electrolytes by potential cycling between -2.0 V and +2.0 V with a scan rate of 50 mV s^{-1} for six scans. The investigated supporting electrolytes were each of 0.1 M H_3BO_3, $NaNO_3$, $HClO_4$, H_2SO_4, H_3PO_4, HCl, $LiClO_4$, and Na_2CO_3. The results showed that the GPE electrodes pretreated in 0.1 M HCl was the most selective and sensitive towards a AMP and MCP. So 0.1 M HCl was chosen for pre-treatment of electrodes. The prepared electrodes (PTGPE) were stored at room temperature in desiccators.

Preparation of real and pharmaceutical samples

The urine samples were collected from healthy humans and were diluted 100-fold with the phosphate buffer solution before analysis. Quantitative determination was performed by adding a standard solution of AMP and MCP to the detection system with the urine sample.

Metpar® tablets (Cipla, India) containing 500 mg of acetaminophen and 5 mg of metoclopramide were purchased from local pharmacy and were ground to a homogeneous fine powder in a mortar separately. A suitable amount of this powder was weighed and was treated with Millipore water for 20 minutes. The mixture was then filtered and solutions obtained from the filtration were diluted to 100 ml with Millipore water. Aliquots of this solution were analyzed within the calibration conditions.

Results and discussion

Cyclic voltammetry

Electrochemical response of a solution having homogeneous mixture of 1.0 µM of each AMP and MCP was estimated by cyclic voltammetry at 10 mV s^{-1} under optimized parameters using PTGPE. Anodic peaks for the oxidation of AMP and MCP were observed at 0.693 V and 1.123 V, respectively, as shown in Fig. 1. Reduction peak is absent in the reverse sweep for both compounds which clearly indicates the irreversibility of the electrode reaction. The peak intensities for PTGPE are better compared to the peak intensities obtained by using bare GPE.

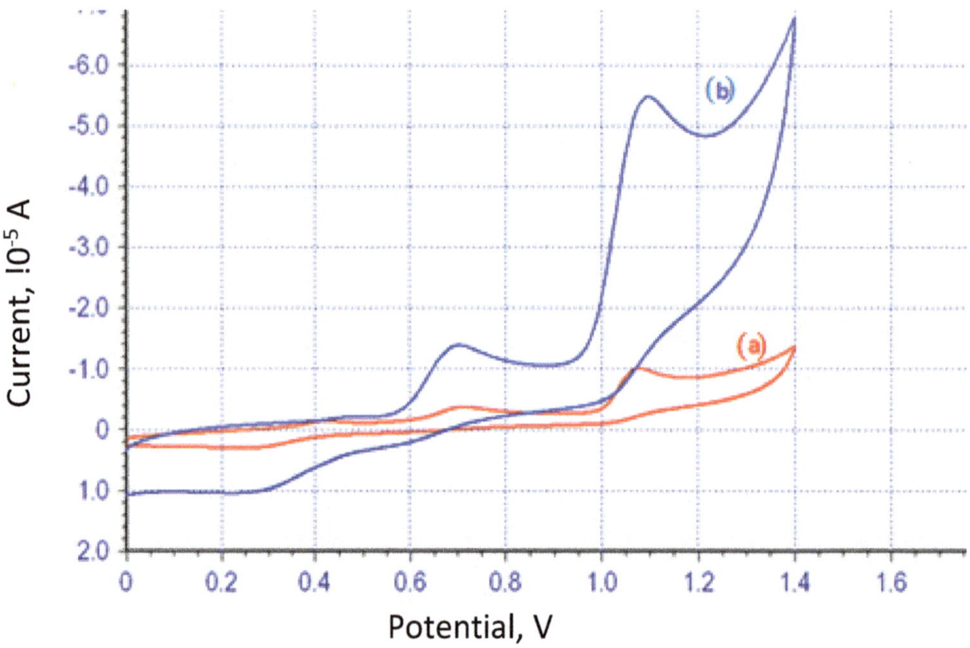

Fig. 1. *Comparison of electrochemical behavior of mixture of 1mM of AMP and 1mM of MCP at (a) bare GPE, (b) PTGPE, with scan rate 10 mV s^{-1} at pH 4.2.*

Effect of pH

The electro-oxidation of mixture of 1.0 µM each of AMP and MCP was studied over the pH range of 3.0–8.0 in phosphate buffer solution by differential pulse voltammetry which is as shown in Fig. 2A. The pH of solution influenced the peak current [21-23]. The pH dependence of the peak potential when DPV was used is shown in Fig. 2B (a and b). With an increase in pH of the solution, peak potential shifted to less positive values and was found to obey the following equations:

E_p = -0.0293 pH + 0.7214; r = 0.9821 for AMP

E_p = -0.0297 pH + 1.1483; r = 0.9615 for MCP

The best result with respect to sensitivity accompanied with sharper response and well separated peaks was obtained with pH = 4.2 (Fig. 2C), and hence it was selected for further experiments.

A

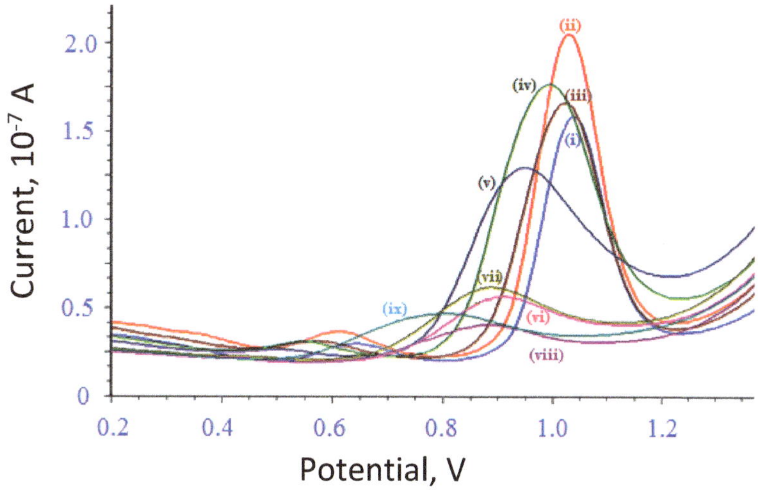

B

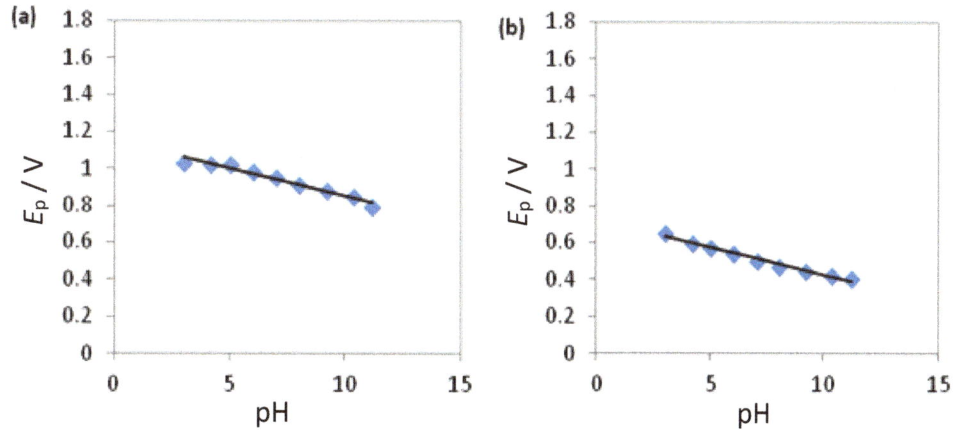

C

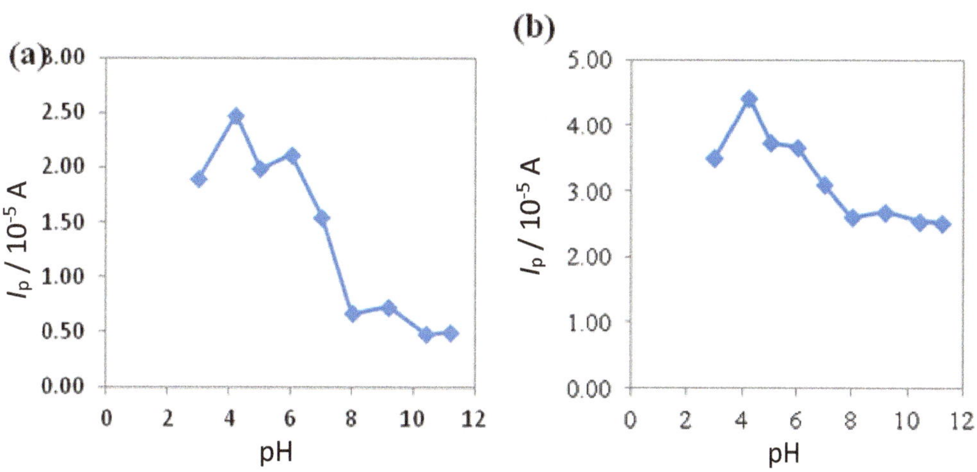

***Fig. 2**. A - dependence of pH on the oxidation of mixture of AMP and MCP at pH (i) 3.0, (ii) 4.2, (iii) 5.0, (iv) 6.0, (v) 7.0, (vi) 8.0, (vii) 9.0, (viii) 10.4, (ix) 11.2; **B** - variation of peak potential with pH for (a) 1.0 µM MCP and (b) 1.0 µM AMP; **C** - variation of peak current with pH for (a) 1.0 µM MCP and (b)1.0 µM AMP*

Effect of scan rate

The effect of scan rate on the voltammetric oxidation of a mixture of AMP and MCP was examined by cyclic voltammetry between 10 to 190 mV s^{-1} (Fig. 3A). A linear relationship was observed between log I_p and log v, (Fig. 3B) corresponding to the following equations:

A

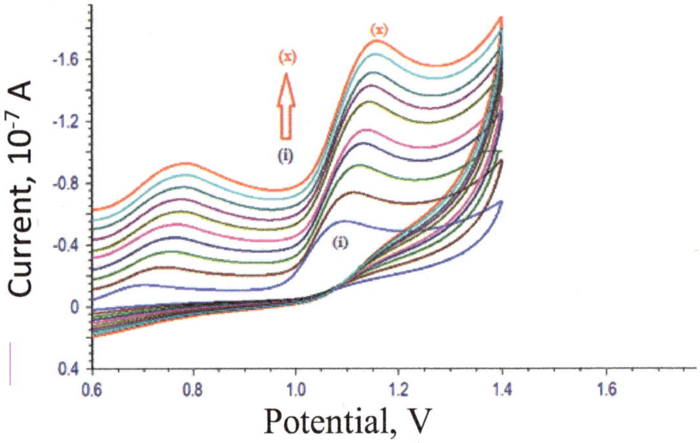

B

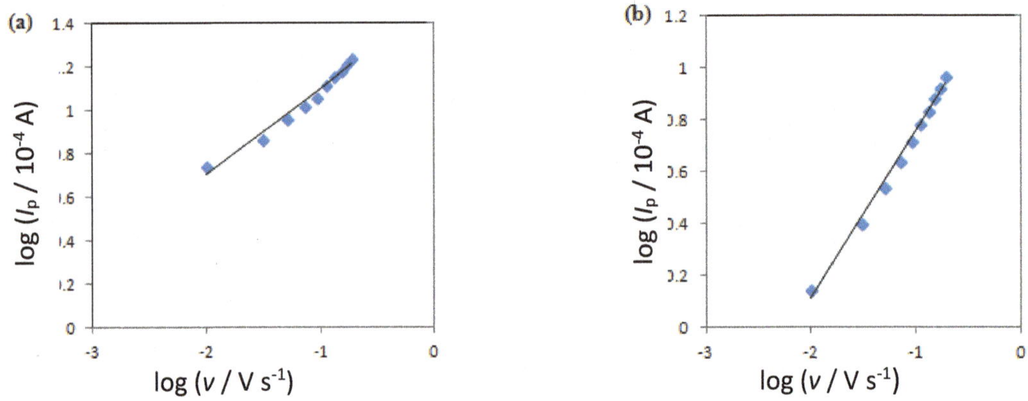

C

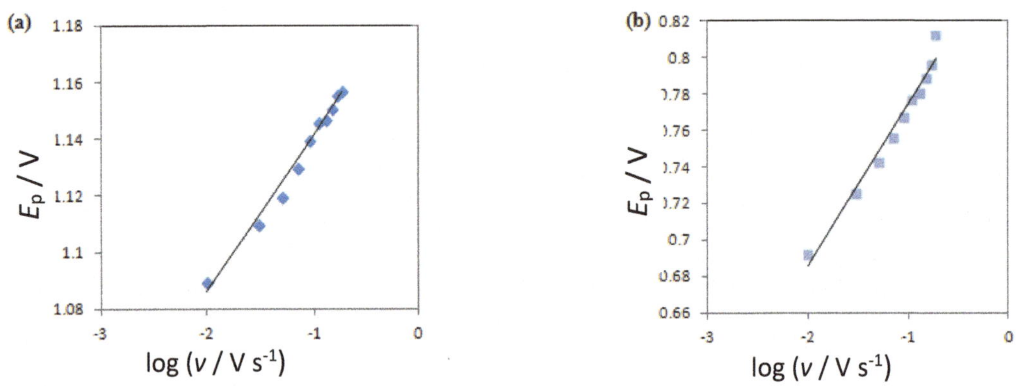

Fig. 3. A *- Effect of scan rate on the electro-oxidation of a mixture of AMP and MCP. (i) 10, (ii) 30, (iii) 50, (iv) 70, (v) 90, (vi) 110, (vii) 130, (viii) 150, (ix) 170, (x) 190 mV s-1;*
B *- observed dependence of peak current on the square root of scan rate for (a) MCP and (b) AMP;*
C *- plot of variation of peak potential with logarithm of scan rate for (a) MCP and (b) AMP.*

$\log I_p = 0.6508 \log v + 1.4111$, $r = 0.9942$ for AMP and

$\log I_p = 0.3989 \log v + 1.4981$, $r = 0.9796$ for MCP

The slopes of 0.65 V s^{-1} and 0.39 V s^{-1} were in the neighborhood of the theoretically expected value of 0.5 V s^{-1} for a purely diffusion controlled process [24] which, in turn, further confirms that the electro-oxidation of AMP and MCP were diffusion controlled. With an increase in scan rate, the peak potential shifted to a more positive value and a linear relationship was observed in the range 0.01 to 0.19 V s^{-1} as shown in Fig. 3C (a) and (b). The corresponding equations are expressed as:

$E_p = 0.0887 \log v + 0.8629$, $r = 0.9746$ for AMP and

$E_p = 0.0550 \log v + 1.1962$, $r = 0.9858$ for MCP

For an irreversible electrode process, according to Laviron [25], E_p is defined by the following equation (2).

$$E_p = E^0 + \frac{2.303RT}{\alpha nF}\log\left(\frac{RTk^0}{\alpha nF}\right) + \left(\frac{2.303RT}{\alpha nF}\right)\log v \tag{2}$$

where n is the number of electrons transferred, v is the scan rate, α is the transfer coefficient, k^0 is the standard heterogeneous rate constant of the reaction and E' is the formal redox potential. Other symbols have their usual meanings. Thus the value of αn was easily calculated from the slope, and k^0 was calculated from the intercept of E_p *vs.* log v. E^0 was obtained from the intercept of E_p vs. v plot by extrapolating to the vertical axis at $v = 0$ [24] Taking $T = 298$ K, $R = 8.314$ J K^{-1} mol^{-1} and $F = 96\,480$ C mol^{-1}, the transfer coefficient (α) and number of electrons transferred (n) were calculated for both the drugs. The α value was calculated according to the Bard and Faulkner [26] equation (3).

$$\alpha = \frac{47.7}{E_p - E_{p/2}} \text{mV} \tag{3}$$

where $E_{p/2}$ is the potential when the current is at half the peak value. From this, the value of α was calculated to be 0.29 for AMP and 0.47 for MCP. Further, the number of electrons (n) transferred in the electro-oxidation of AMP and MCP were calculated to be 2.1≈2 and 2.28≈2.0, respectively.

Plausible mechanism

Based on the experimental results, the number of electrons transferred (n) for both the drugs were calculated to be two. Hence the probable electrooxidation mechanisms for both the drugs are proposed as given in Scheme 2, which are based on earlier works [27,28].

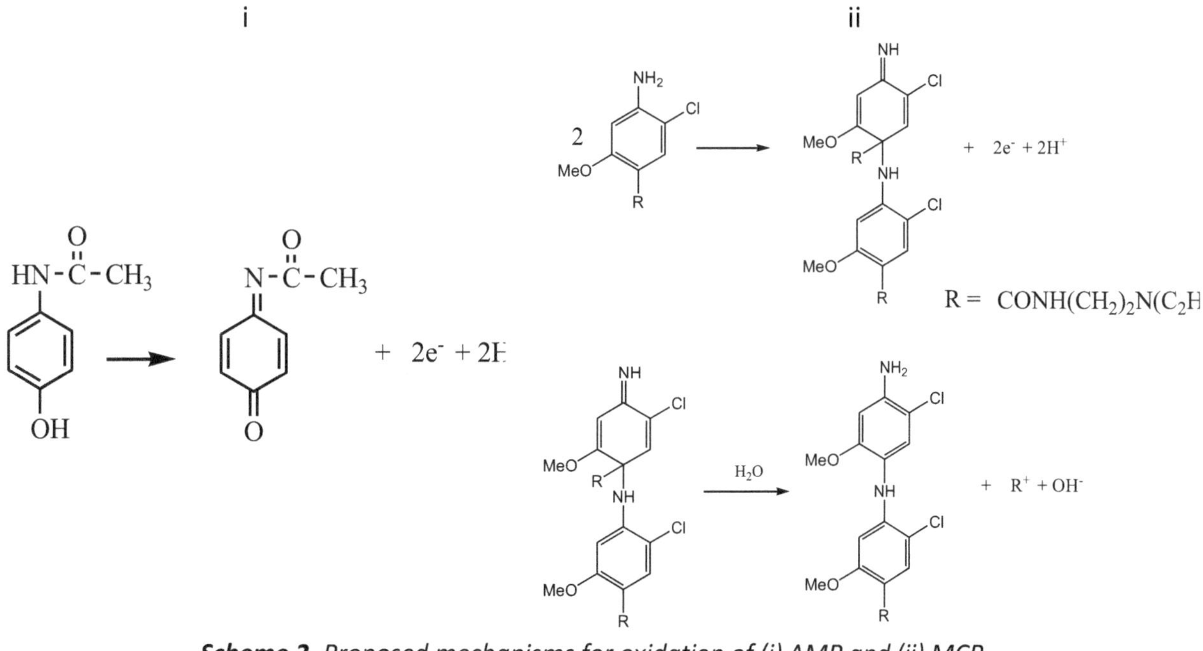

Scheme 2. *Proposed mechanisms for oxidation of (i) AMP and (ii) MCP*

Analytical application

To obtain an analytical curve for the sensor, quantitative analyses of the AMP and MCP concentrations were performed by DPV [29] at a PTGPE under the optimized experimental conditions. The oxidation of MCP occurs at high positive potential and there can be a chance of overlapping of MCP potential with the increasing concentration of drugs. Oxidation peak current was found to increase and no change in the peak current or potential of another drug indicates that these two drugs do not interact with each other. The peak current versus concentration of drug plots show a good linearity for AMP and MCP in certain concentration ranges as depicted in Fig. 4A and B. Peak current values were obtained by subtracting the background current of PBS and average of three replicate measurements were used to plot calibration curves. Linear regression equations for both the drugs arising from calibration plots are represented as:

$I_p = 0.212\ C + 0.344$, $r = 0.9703$ for AMP

$I_p = 0.928\ C + 1.257$, r=0.9767 for MCP

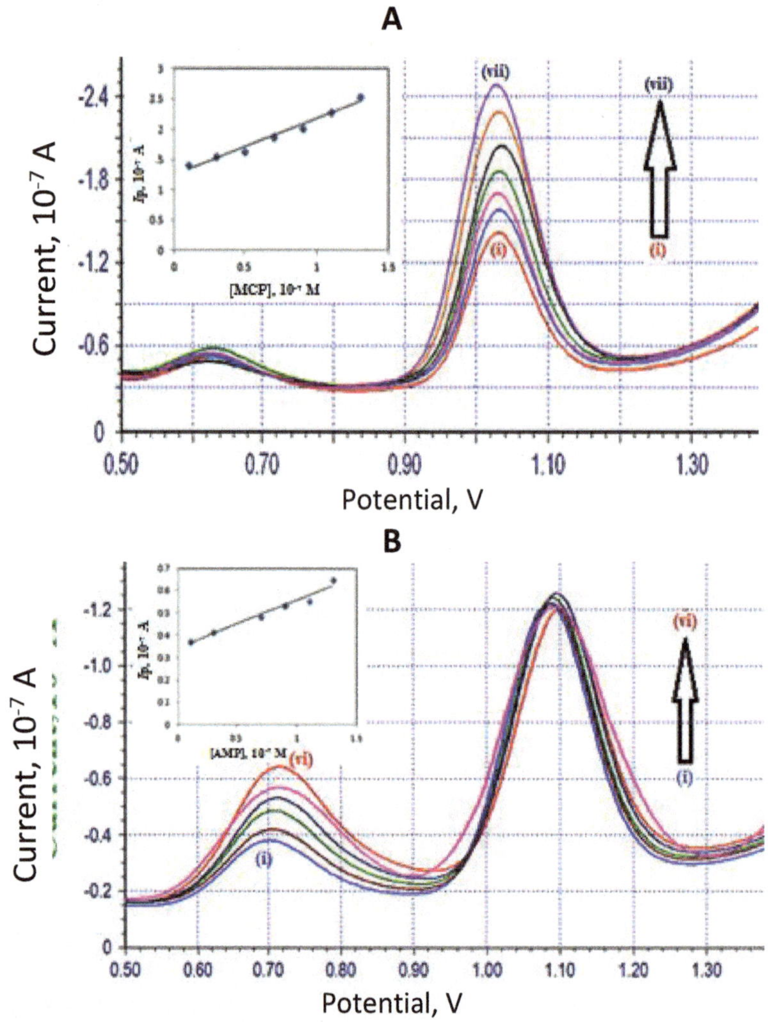

Fig. 4. *A - differential pulse voltammograms for increasing concentration of MCP at PTGPE (i) 1, (ii) 3, (iii) 5, (iv) 7, (v) 9, (vi) 11, (vii) 13, 10^{-7}M with AMP concentration 1×10^{-6}M.;* ***B*** *- differential pulse voltammograms for increasing concentration of AMP at PTGPE (i) 1, (ii) 3, (iii) 7, (iv) 9, (v) 11, (vi) 13, 10^{-7}M with MCP concentration $1\text{x}10^{-7}$M.*

The limit of detection (LOD) and limit of quantification (LOQ) were calculated by using the formulae LOD = $3s/m$ and LOQ = $10s/m$, where s is the standard deviation of peak current and m is the slope of calibration curves. LODs were found to be 3.25×10^{-9} M and 1.16×10^{-9} M for AMP and MCP respectively and the LOQs were calculated as 10.84×10^{-9} M and 3.88×10^{-9} M, respectively. Since, it is important to calculate validation parameters for any analytical method, the calibration characteristics obtained for MCP and AMP are given in Table 1.

Table 1. *The calibration characteristics for metoclopramide and acetaminophen at PTGPE.*

	MCP	AMP
Linearity range, 10^{-7} M	0.1-1.1	0.1-1.1
Slope of the calibration plot	0.928	0.212
Intercept	1.275	0.344
Correlation coefficient (r)	0.9767	0.9703
RSD of slope, %	1.21	0.82
RSD of intercept, %	0.12	0.65
Number of data points	5	5
LOD, nM	1.16	3.25
LOQ, nM	3.88	10.84
Repeatability of peak current, %	0.79	1.14
Repeatability of peak potential, %	0.51	0.35
Reproducibility of peak current, %	0.92	0.71
Reproducibility of peak potential, %	0.24	0.43

The precision of the method was calculated by repeating five experiments on the same day in standard conditions (repeatability) and over two days from the different standard solutions (reproducibility). For these studies 1.0×10^{-6} M of each of AMP and MCP standard solutions were used. From RSD values of peak potential and peak current between day reproducibility were similar to that of within a day if the temperature was kept almost unchanged which shows the excellent stability and reproducibility of PTGPE.

Effect of excipients

The effect of some common excipients used commonly in pharmaceutical preparations was examined. The tolerance limit was defined as the maximum concentration of the interfering substance that caused less than 5 % error for the determination of AMP and MCP. The effects of these excipients on the voltammetric response were obtained by analyzing sample solutions containing a fixed amount of AMP and MCP (1.0×10^{-6} M) spiked with various amounts of each excipient under the same experimental conditions. The experimental results showed that a hundred-fold excess of gum acacia, citric acid, dextrose, glucose, lactose, tartaric acid and sucrose did not interfere with the voltammetric signal of AMP and MCP. Hence, these compounds need not be extracted from these tablet additives prior to their determination in tablets.

Detection of AMP and MCP in human urine samples

The developed DPV method for the AMP and MCP determination was applied to human urine samples. The recoveries from urine were measured by spiking drug free urine with known amounts of AMP and MCP. The urine samples were diluted 100 times with the PBS before analysis without further pretreatment. A quantitative analysis was carried out by adding the standard solutions of AMP and MCP into the detection system of urine samples. The peak current increased linearly in height. The calibration plot was used for the determination of spiked AMP and MCP in urine samples. The results of four urine samples obtained are listed in Table 2. Thus, satisfactory recoveries of the analytes from the real samples were in a good agreement with the concentration ranges studied and the real ranges encountered in the urine samples when treated with drug, make the developed method applicable in clinical analysis.

Table 2. *Results of analysis of metoclopramide and acetaminophen in spiked urine samples*

MCP added, 10^{-8} M	Quantity found[(a)], 10^{-8} M	Average recovery, %	RSD[a], %
		Metoclopramide	
10	9.7	97.0	1.91
15	14.5	96.6	2.20
20	19.2	96.0	2.12
25	24.6	98.2	2.66
30	29.6	98.8	1.59
35	34.2	97.4	2.21
		Acetaminophen	
10	9.5	95.0	1.83
15	14.9	99.5	1.24
20	20.1	100.5	2.62
25	24.4	97.6	2.31
30	29.1	98.8	1.02
35	34.1	96.4	1.61

a -average of five determinations

Determination of AMP and MCP in pharmaceutical samples

This method was applied for the determination of AMP and MCP in tablets. As shown in Table 3, the content of AMP and MCP was calculated to be 495.1 mg/tablet (the nominal content is 500 mg/tablet) and 9.7 mg/tablet (the nominal content is 10 mg/tablet) respectively. To validate and obtain the accuracy and precision of the developed method, recovery studies were performed at different drug concentrations by the standard addition method. For the present study, known quantities of AMP and MCP were mixed separately with definite amounts of pre-analyzed formulations and mixtures were analyzed as before. Table 3 shows the measurement results, and the average recovery was found to be 99.02 % and 97.0 % for AMP and MCP respectively. The F and Student t tests were also calculated with confidential level of 95 % and are shown in Table 3.

Table 3. *Results of analysis of tablet containing both the analytes (acetaminophen 500 mg and metoclopramide 10 mg) in commercially available tablet samples by DPV*

	Acetaminophen	Metoclopramide
Labeled claim, mg	500.00	10.00
Quantity found, mg	495.10	9.70
Recovery[a], %	99.02	97.00
RSD[a], %	1.92	2.31
t- value at 95% confidence level	0.16	0.21
F-value at 95% confidence level	1.47	1.76

a -average of five determinations

Conclusion

The pre-treated pencil graphite electrode was applied successfully as a sensor for fast, accurate and simultaneous determination of AMP and MCP in some pharmaceutical samples. The present method is a good alternative for the analytical determination of AMP and MCP simultaneously, because it is simple, sensitive, accurate, fast, and inexpensive. The results were successfully applied in urine samples and pharmaceuticals. Furthermore, the present method could possibly be employed for pharmacokinetic studies and also in clinical and quality control laboratories.

Acknowledgement: *The author thanks the UGC, New Delhi for the award of UGC-BSR faculty fellowship to Dr. S. T. N.*

References

[1] J. I. Gowda, S.T.Nandibewoor, *Anal. Methods* **6** (2014) 5147-5154.
[2] E. Alipour, S. Gasemlou, *Anal. Methods* **4** (2012) 2962-2969.
[3] I. Parlak, R. Atilla, M. Cicek, M. Parlak, B. Erdur, M. Guryay, M. Sever, S. Karaduman, *Emerg. Med. J.* **22** (2005) 621-624.
[4] P. Norouzi, M. R. Ganjali, P. Matloobi, *Electrochem. Commun.* **7** (2005) 333-338.
[5] C. P. Page, M. J. Curtis, M. C. Sutter, M. J. A. Walker, B. B. Hoffman, *Farmacología Integrada*, Ed. Harcourt. España. 1998.
[6] L. J. Roberts, J. D. Morrow, in: *The Pharmacological Basis of Therapeutics*, 10th ed., J. G. Hardman, L. E. Limbird, A. G. Gilman (Eds.), McGraw-Hill, London, 2001.
[7] R. V. Blanke, W. J. Decker, in: *Textbook of Clinical Chemistry*, N.W. Tietz (Ed.), W.B. Saunders, Philadelphia, 1986.
[8] D. Gunnell, V. Murray, K. Hawton, *Suicide Life Threat. Behav.* **30** (2000) 313-326.
[9] W. A. Martindale, *The Extra Pharmacopoeia*, 27th ed., The Pharmaceutical Press, London, 1979.
[10] B. D. Clayton, Y. N. Stock, *Basic Pharmacology for Nurses*, Mosby Inc, Harcourt Health Sciences Company, St. Louis, Mo, USA, 2001.
[11] N. Wangfuengkanagul, O. Chailapakul, *J. Pharm. Biomed. Anal.* **28** (2002) 841-847.
[12] J. I. Gowda, D. G. Gunjiganvi, N. B. Sunagar, M. N. Bhat, S. T. Nandibewoor, *RSC Adv.* **5** (2015) 49045-49053.
[13] S. M. Patil, S. T. Nandibewoor, *Anal. Bioanal. Electrochem.* **7** (2015) 387-400.
[14] S. J.Wadher, P. R.Pathankar, M. Puranik, R.O. Ganjiwale, P. G. Yeole, *Ind. J. Pharm. Sci.***70** (2008) 393-395.
[15] A. E. El-Aleem, S. M. Khalile, A. M. Badawy, O. K. El-Naggar. *Research J. Pharma. Dosage Forms and Tech.* **5** (2013) 270-274.
[16] G. Shubhangee, K. Manish, N. Swapnil, *Int. J. Pharm. Li. Sci*, **1** (2010) 127-132.
[17] N. P. Dudhane, S. S.Vidhate. B. H.Borkar, R. T. Lohiya, M. J. Umekar, *J. Pharm. Sci. & Res.* **2** (2010) 48-52.
[18] G. Vamshikrishna, M. Neelima, M. Bhavani, G,Sreekanth, S. Shobha Rani. *IOSR J. Pharm. Bio. Sci.* **9** (201) 69-78.
[19] G. D. Christian, W. C. Purdy, *J. Electroanal. Chem.* **3** (1962) 363-367.
[20] B. Rezaei, S. Damiri, *Sens. Actuators B,* **134** (2008) 324-331.
[21] B. D. Topal, B. Bozal, B. T. Demircigil, B. Ulsu, S.A. Ozkan, *Electroanalysis.* **21** (2009) 2427-2439.
[22] J. Wang, *Electroanalytical Techniques in Clinical Chemistry and Laboratory Medicine*; VCH: New York, 1988.
[23] P. T. Kissinger, W. R. Heineman, *Laboratory Techniques in Electroanalytical Chemistry*, 2nd ed.; Marcel Dekker: New York, 1996.
[24] D. K. Gosser, *Cyclic Voltammetry: Simulation and Analysis of Reaction Mechanisms*, VCH, New York, 1993.
[25] E. Laviron, *J. Electroanal. Chem.* **101** (1979) 19-21.
[26] J. Bard, L. R. Faulkner, *Electrochemical Methods Fundamentals and Applications*, Wiley, 2nd ed, 2004,

[27] B. Habibi, M. Jahanbakhshi, M. H. Pournaghi-Azar, *Analytical Biochemistry*. **411** (2011) 167–175.

[28] S. Shahrokhian, L. Naderi and M. Ghalkhani, *Electroanalysis* **27 (**2015) 2637 – 2644.

[29] N. L. Teradal, S. N. Prashanth, J. Seetharamappa, *J. Electrochem. Sci. Eng.* **2** (2012) 67-75.

11

Ellipsometric measurement of thickness of tin oxides grown by voltammetry in phosphate solution of pH 8.7

Tiago Brandão Costa*,✉, Tania Maria Cavalcanti Nogueira*,**, Ladário da Silva*,***

**Programa de Pós-Graduação em Engenharia Metalúrgica (PPGEM), Escola de Engenharia Industrial Metalúrgica de Volta Redonda (EEIMVR), Universidade Federal Fluminense (UFF), 27255-125 Volta Redonda, RJ, Brazil*
***Departamento de Metalurgia, EEIMVR, UFF, 27255-125 Volta Redonda, RJ, Brazil*
****Departamento de Física, Instituto de Ciências Exatas (ICEx), UFF, 27213-145 Volta Redonda, RJ, Brazil*

✉Corresponding author: tiagobrandao@id.uff.br

Abstract

The voltammetry induced growth of tin oxides on tin in the buffer solution of 0.18 mol L^{-1} $Na_2H_2PO_4$ and 0.18 mol L^{-1} KH_2PO_4 (pH 8.7) has been studied. Ex-situ ellipsometric measurements were made in an order to determine thicknesses of the grown oxides. From these results the film volume per charge unit, V_f, was calculated for different charge densities of the film. This parameter was used to calculate the variable ionic resistivity of the film, ρ_f, considered by the Ohmic model for the case of voltammetric growth of oxides on metals having a previously existing continuous film. Tin oxide films grown at 2 mV s^{-1} showed to be less dense for values of charge density below 50 C m^{-2}, having V_f near 5.7×10^{-10} m^3 C^{-1}. For higher values of charge density, tin oxide films become denser, having V_f near 0.5×10^{-10} m^3 C^{-1}. The calculated values of the variable ionic resistivity of the film during voltammetric growth showed that ρ_f passes through a minimum (justifying the maximum in current densities). This behavior was also found by other authors in the cases of Zn, Nb, Ni and galvanized steel sheets.

Keywords

Ellipsometry, Tin oxide, Ohmic model, Voltammetry, Variable ionic resistivity

Introduction

Active metals cannot be obtained without an initial thin film covering the surface, as have already been observed in the cases of Cd, Fe, Ni, Pb, Sn, Zn, valve metals, *etc*. For this reason, the studies of oxide growth on these metals are found very important. The growth of passivating films on metals under voltammetric conditions gives rise to peaks or polarographic waves in current/potential plots. This phenomenon is related to the active/passive transition, or growth of a continuous film on the metal surface.

Some electrochemical processes, such as electrodeposition of metallic ions [1–3] or growth of oxide on metals [4–8], can be considered as processes presenting ohmic behaviour. The Ohmic model developed by D'Alkaine [4] describes the relation between the current density and the film overpotential during the growth of passivating film under voltammetric conditions. This model considers that the relation between the electrode potential E and overpotential at the metal/film/solution interface is given by:

$$E = \eta_{m/f} + \eta_f + E_F \quad (1)$$

In Eq. (1) $\eta_{m/f}$ is the overpotential at the metal/film interface, η_f is the overpotential across the film and E_F is the Flade potential. The overpotential at the film/solution interface is considered constant, *i.e.* independent of the current density, j.

The relation between current density (j) and overpotential at the metal/film interface ($\eta_{m/f}$) is given by the general expression:

$$j = j^0_{m/f}\left[\exp(\alpha_a nf\eta_{m/f}) - \exp\{-(\alpha_c nf\eta_{m/f})\}\right] \quad (2)$$

In Eq. (2), $j^0_{m/f}$ is the exchange current density at zero $\eta_{m/f}$, f is F/RT, α_a and α_c are anodic and cathodic transfer coefficients, while n is charge of the metal ion in the film.

If V_f and q_f are respectively the volume per charge unit and the charge density related to the growing film, then the film thickness, ℓ, is given by:

$$\ell = V_f q_f \quad (3)$$

The q_f value can be calculated as:

$$q_f = q_{volt} + q_0 = \frac{1}{v}\int_{E_i}^{E} j_a dE + q_0 \quad (4)$$

In Eq. (4) q_0 is the charge density related to the amount of film initially present at the beginning of the voltammetric growth and q_{volt} is the charge density related to the amount of film which has grown on the metal surface during the voltammetric experiment. E_i is the initial potential, E is the potential attained and j_a is anodic component of the current density, taking into account Eq. (2).

For peak or plateau, transient conditions during voltammetric growth of a film on a metal surface, the following equation should be valid for any model:

$$\eta_{f,p} = \frac{v}{j_p} q_{f,p} \quad (5)$$

In Eq. (5) $\eta_{f,p}$ is the overpotential across the film at the voltammetric peak, v is the sweep rate, j_p is the current density at the peak and $q_{f,p}$ is the peak or plateau charge density.

In the case of the Ohmic model, the relation between j and η_f, even at high fields, is given as:

$$\eta_f = V_f \rho_f q_f j \quad (6)$$

In Eq. (6), ρ_f is the average ionic specific resistivity of the film.

The Ohmic model takes into consideration that at the beginning of the scanning, the current density increases as a consequence of reduction of ionic resistivity of the film due to injection of defects in the film. Beyond the peak potential, it is possible to observe reduction of the current density, what is explained by ageing of the oxide film. This phenomenon turns the film more resistive to the passage of the current due to recombination of defects and to structural rearrangement [4]. This model, however, contains an intrinsic difficulty related to impossibility to determine ρ_f and V_f separately, using only electrochemical measurements. Despite this fact, many authors [4–8] applied the Ohmic model, considering the volume per charge unit (V_f) as the constant equal to:

$$V_f = \frac{M}{nF\delta} \tag{7}$$

In Eq. (7) M is the molar mass of the film, δ is density of the film, n is the number of electrons and F is the Faraday's constant.

The real values of V_f during the voltammetric growth could only be determined by independent measurements of the film thickness. The aim of the present work is to study the voltammetric growth of tin oxides in a solution of phosphate (pH 8.7) applying the Ohmic model. This solution was chosen in view of stability of the passive film on tin at this pH, avoiding thus loss by chemical dissolution [9]. At the same time, there are some early studies concerning the passivating species under voltammetric growing conditions in the pH range 8–9 [10–14]. The thickness of the oxides grown by voltammetry will be determined by ex-situ ellipsometric measurements in an order to verify the intrinsic difficulty explained above. The ex-situ condition was first examined as a preliminary study in order to verify stability of the film in the atmosphere.

Experimental

Electrochemical measurements

The working electrode was made of a tin disc (99.999 % of purity) with the circular area of 50 μm^2. Before the experiments, the electrode surface was polished with 600-emery paper.

The electrochemical experiments were performed using the EG&G Princeton Applied Research Model 273A potentiostat. Reagents PA and purified water (Millipore Q system) were used. The electrolyte solution was 0.18 mol L^{-1} $Na_2H_2PO_4$ + 0.18 mol L^{-1} KH_2PO_4, pH 8.7. The experiments were carried out in a conventional three-compartment electrolysis cell, using a platinum wire as the counter and $Hg/Hg_2Cl_2/KCl$ (1 M) as the reference electrode, respectively.

All current and charge densities are given in terms of the geometric surface area of the analyzed samples.

Anodic voltammetries were carried out at sweep rates of 2, 5, 10, 20, 70, 100, 200, 300 and 400 mV s^{-1}, always using the same sample. For this purpose, before each voltammetry experiment, the previously grown oxide film was reduced at constant cathodic potential equal to -1.2 V during 600 s. After this treatment, the obtained voltammograms were reproducible, indicating that surface roughness of samples was also reproducible.

Ellipsometric measurements

Ellipsometry [15] is a powerful and non-destructive technique, which allows one to access optical and dielectric properties of various materials in different forms (solids, liquids, and gases), as well as thicknesses of thin films [16–17]. It can also be used to monitor and analyze both in-situ [18–19] and ex-situ [20] experiments. Ellipsometry considers the change in the polarization state of incident light upon reflection from the studied material. Initially it accesses the ellipsometric parameters Ψ and Δ. These parameters are connected by the fundamental equation of ellipsometry [16]:

$$\frac{R_p}{R_s} = \tan(\psi)e^{i\Delta} = f(n_j, k_j, d_j) \quad (8)$$

In Eq. (8), R_p and R_s stand for the complex Fresnel reflection coefficients for p and s polarized light beams, respectively. Ψ is the amplitude ratio for the reflected and polarized components (p and s), while Δ represents the phase difference between reflected p and s polarizations, created by reflection from the material. n_j, k_j, d_j are, respectively, the refractive index, the extinction coefficient and thickness of the jth layer of the material. The ex-situ ellipsometric measurements were made using a SEMILAB spectroscopic ellipsometer, model SOPRA GES 5E, equipped with a Xe lamp, over the spectrum range of 195 – 1,000 nm. The investigation of the thickness of oxide films grown on the substrate of tin has been performed in air and at ambient temperature using the incident angle of 75°. As an indirect technique, it is necessary to model the material structure with its film layers in an order to obtain each layer film thickness. The thicknesses of the oxide films were obtained by analyzing the measured ellipsometric spectra through the Drude and Gauss model [16].

Results and discussion

As was explained before, in an order to obtain reproducibility, the voltammetric experiments were always carried out using the same sample. Before each voltammetry measurement, the oxide grown in the previous experiment was reduced at the constant cathodic potential during 600 s. This cathodic treatment was sufficient to reproduce the initial roughness of the surface, which was verified by reproducible voltammograms obtained at the same sweep rate and shown in Fig. 1.

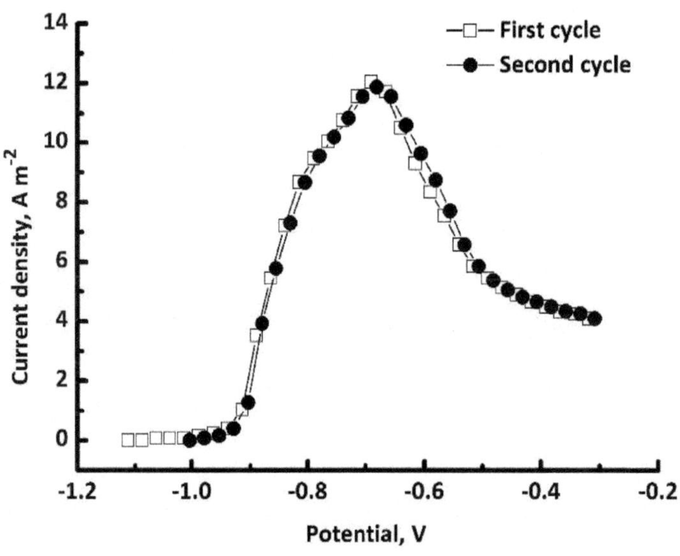

Fig. 1. *First and second cycle at 50 mV s^{-1} in the phosphate solution pH 8.7*

The anodic voltammograms for tin oxide growth at different sweep rates are shown in Fig. 2. From these results, the overpotential values in the film at the peak condition ($\eta_{f,p}$) were calculated using the Eq. (5). The value of q_0 in Eq. (4) was first considered equal to 0.0 C m^{-2}. The curve *j vs.* potential at the metal/film interface ($E_{m/f}$), also shown in this Figure, is obtained after correction of the ohmic drop through the film at the peak potential (E_p - $\eta_{f,p}$).

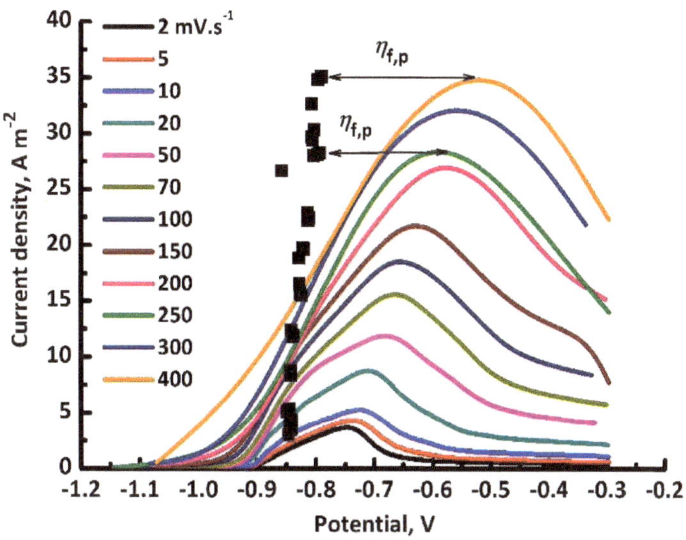

Fig. 2. *Voltammetric growths of tin oxide film, together with the plot of the calculated* j vs. ($E_p - \eta_{f,p}$) *relation at the metal/film interface, considering* $q_0 = 0.0$ C m^{-2}.

The Tafel plot of the curve *j vs.* $E_{m/f}$ in Fig. 2 is represented in Fig. 3.

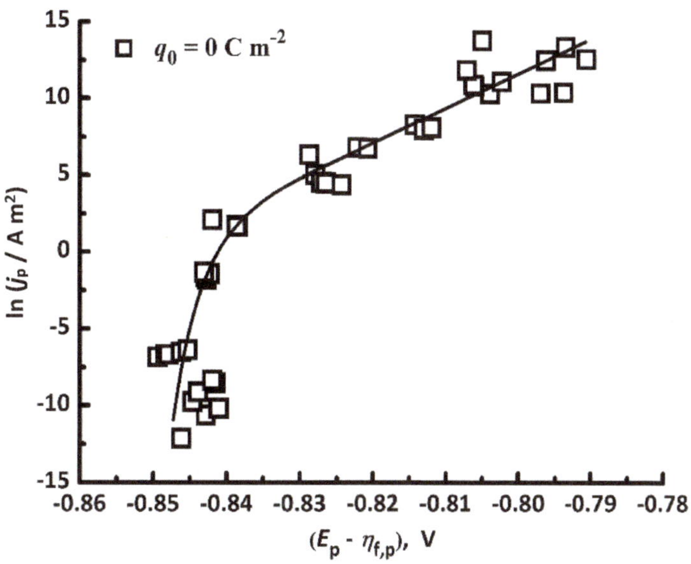

Fig. 3. *Tafel plot at the tin/oxide interface.*

In order to provide the best straight line region of the Tafel plot, the value of q_0 was taken equal to 2.5 C m^{-2}. Fig. 4 presents the obtained result.

From these results, the obtained Tafel slope (b_a) is equal to 41.995 mV dec^{-1}. By using Eq. (9) [21]

$$b_a = \frac{\alpha_{m/f} nF}{RT} \tag{9}$$

the transfer coefficients at the metal/film interface has been calculated as $\alpha_{m/f} = 1.07$.

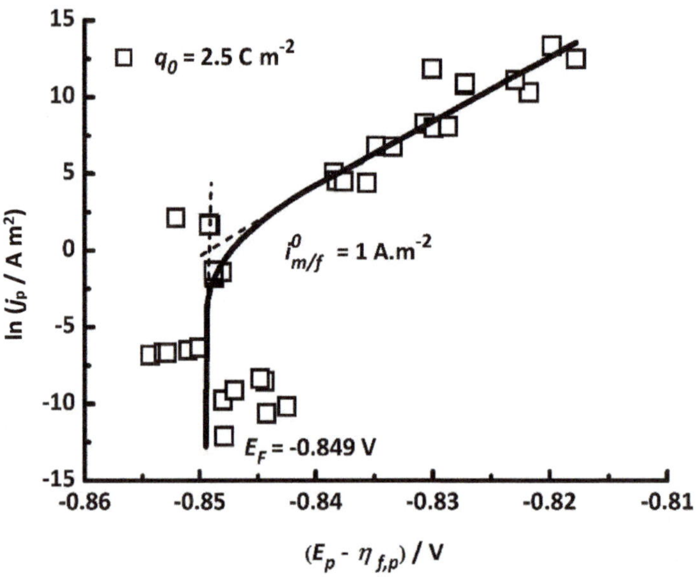

Fig. 4. *Tafel plot at the tin/oxide interface, considering* q_0 = *2.5 C m*$^{-2}$.

Fig. 5 presents the corrected curve *j vs.* $E_{m/f}$, which was obtained from the Tafel plot of Fig. 4. The anodic voltammograms are also shown in the same figure.

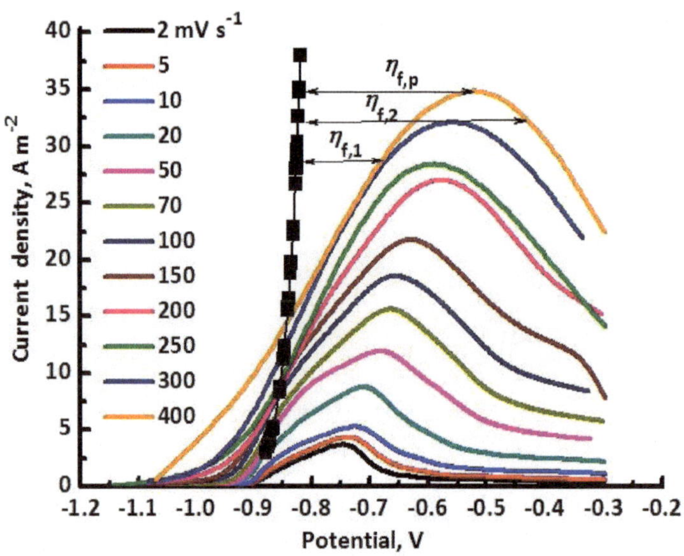

Fig. 5*. Voltammetric growths of tin oxide film, together with the plot of the calculated* j vs. *(*E – η_f*) relation at the metal/film interface, considering* q_0 = *2.5 C m*$^{-2}$ *and Figure 4.*

These results turn possible to determine the overpotential at the film (η_f) for any growing condition of the film beyond the peak condition. This can be done by calculating the difference of the potential in the voltammogram at any sweep rate and the potential in the curve *j vs.* $E_{m/f}$ for a given current density. This is shown in Fig. 5 for the voltammogram at 400 mV s^{-1}.

Thus, by calculating, η_f values for each potential of each voltammetry experiment, it becomes possible to determine the ionic specific resistivity of the film (ρ_f) and its variation with *E* and *v* using Eq. (6).

As was explained in the introduction of this work, the real values of V_f during the voltammetric growth could only be determined by an independent way of measuring the thickness of the film according to the following equation:

$$V_f = \frac{\ell A}{q_f} \tag{10}$$

In Eq. (10), ℓ is thickness and A is surface area the working electrode.

The thicknesses of oxides films grown at 2 mV s^{-1} were measured by ex-situ ellipsometry. This condition was chosen because using this sweep rate, higher thicknesses can be achieved. Fig. 6 presents the final potential (E_f) of each voltammetry experiment performed at 2 mV s^{-1}.

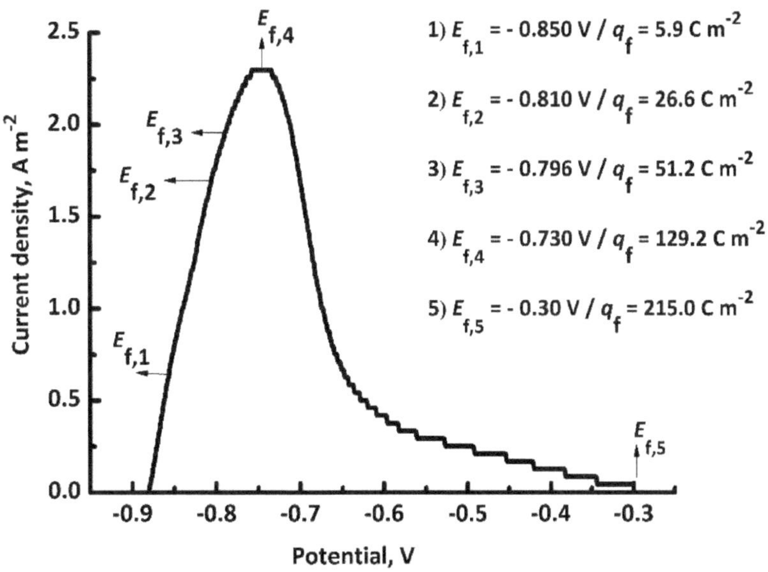

Fig. 6. *Voltammetric growth of tin oxide films at 2 mV s^{-1}, together with indication of the final potential, E_f, and corresponding film charge density values.*

As pointed out before, spectroscopic ellipsometer was used to obtain thin oxide film thicknesses for various charge densities. The proposed oxide layer structure is presented in Fig. 7. This oxide layer structure can be optically modelled using the both known refractive index (n) and extinction coefficient (k) of an available library (*nk* files) or dispersion relations for each of its layer constituents. Using the Drude and Gauss dispersion relations [16] for the film, the excellent fit was obtained between the used model and experimental data.

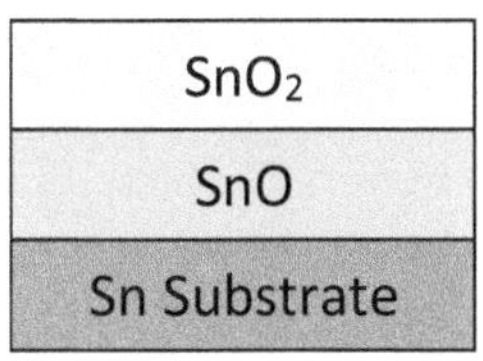

Fig. 7. *Model for the tin oxide layer structure.*

Fig. 8 exhibits typical results of fittings for measured and modelled values of tan Ψ and cos Δ for charge densities equal to 5.9 C m^{-2} and 215.0 C m^{-2}.

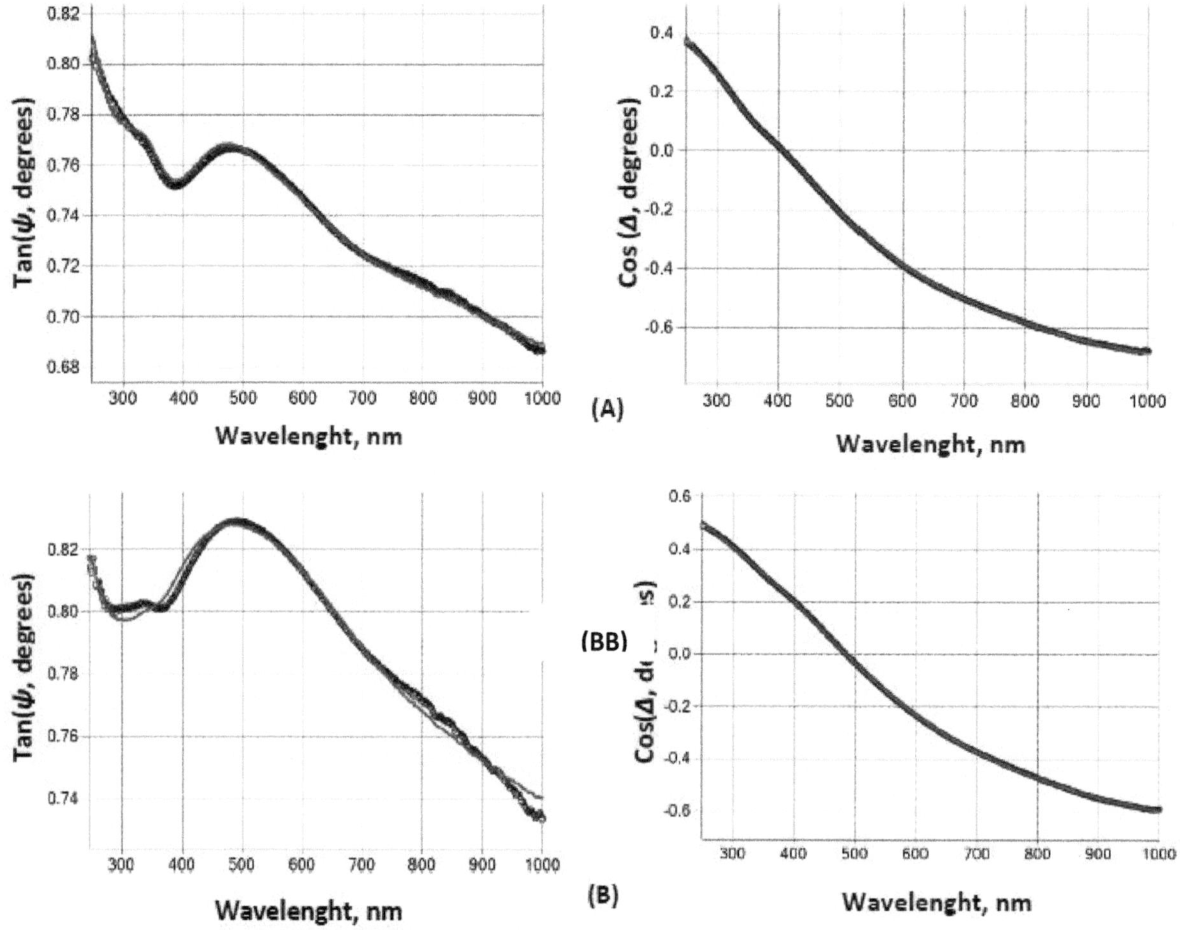

***Fig.8**. Typical plots of measured (open circles) and modelled (blue line) of tan (Ψ) and cos (Δ) for tin oxide films and charge densities equal to: (A) 5.9 C m^{-2} and (B) 215.0 C m^{-2}.*

It is clear from the graphs in Fig. 8 that excellent fits ($R^2 > 0.99$) were obtained, what make us confident to report the obtained thicknesses. The average thicknesses obtained considering eight measures at each charge density, are shown in Table 1. Except thicknesses, ℓ, in Table 1, one can also see values of the final potential, E_f, charge density of the film, q_f, and statistical results (R^2, as well as standard deviations of the average thickness values, σ). Data in Table 1 clearly show that for greater E_f and q_f, greater average film thickness is obtained.

***Table 1:** Final potentials, charge densities, thicknesses, average thicknesses and statistical results of calculations*

E_F / V	q_f / C m^{-2}	ℓ / nm								R^2	Average thicknesses, nm	σ
-0.850	5.9	3.8	3.8	3.5	3.5	3.0	3.1	3.0	3.3	**0.994**	**3.37**	**0.307**
- 0.810	26.6	4.3	3.8	4.6	4.5	3.9	4.0	3.8	3.7	**0.990**	**4.10**	**0.323**
- 0.796	51.2	4.3	4.5	4.6	4.1	3.9	4.3	4.1	4.5	**0.975**	**4.29**	**0.226**
- 0.730	129.2	9.4	9.0	9.3	8.9	9.4	9.0	9.3	9.1	**0.981**	**9.20**	**0.185**
- 0.300	215.0	10.7	11.6	10.3	11.3	10.3	11.0	12.2	10.8	**0.995**	**11.00**	**0.612**
0.100	268.1	13.7	12.5	13.2	12.9	12.4	11.2	11.8	11.5	**0.987**	**12.40**	**0.806**

In the preliminary experiment, a sample was left in the atmosphere during 24 h in order to measure the stabilized oxide film thickness and the resulting value was 2.0 nm. In Table 1, however, one can notice that the lowest average thickness for the film grown in conditions of voltammetric experiments was 3.37 nm. This suggests that here obtained results are significantly influenced by the voltammetric way of film growth.

Using the experimentally found values of average thicknesses, Eq. (10) was applied and V_f was determined for the values of charge density of voltammetry grown films at 2 mV s^{-1}. The results are shown in Table 2 and Fig. 9.

Table 2: *Film volume per charge unit obtained from the thicknesses measurements of tin oxide films for different charge densities.*

q_f / C m^{-2}	ℓ / nm	V_f / 10^{-10} m^3 C^{-1}
5.9	3.37	5.70
26.6	4.10	1.54
51.2	4.29	0.84
129.2	9.20	0.71
215.0	11.00	0.50
268.1	12.40	0.46

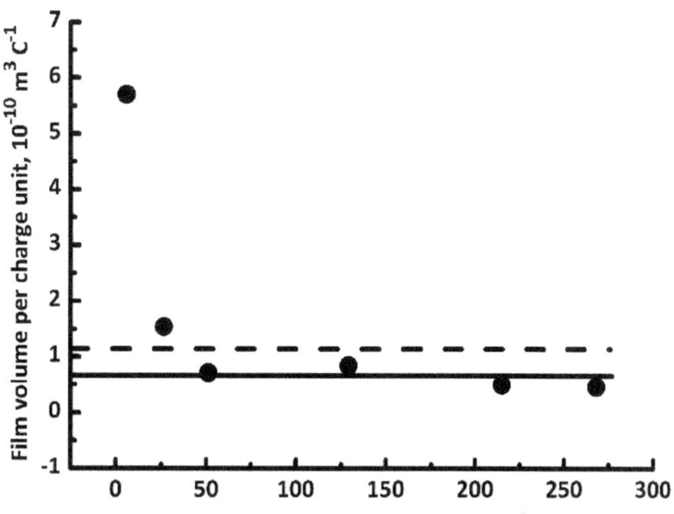

Fig. 9. *Film volume per charge unit determined for different values of charge densities. (•) Experimental V_f, (---) $V_f = 1.08\times10^{-10}$ m^3 C^{-1} (considering SnO) and (–) $V_f = 0.568\times10^{-10}$ m^3 C^{-1} (considering SnO_2)*

Data in Table 2 and Fig. 9 show that the values of V_f significantly decrease until 50 C m^{-2} of charge density of the film is achieved. This suggested that the film is less dense for lower values of charge density and becomes denser as the thickness increases [22]. Besides that, the value of V_f for charge density of 26.6 C m^{-2} is close to the value of V_f calculated by Eq. (7), considering a film of SnO (1.08×10^{-10} m^3 C^{-1}). For higher values of charge density, the value of V_f turns close to the value of V_f calculated considering a film of SnO_2 (0.568×10^{-10} m^3 C^{-1}). These results suggest that changes in the composition of the film are taking place as the potential turns more anodic. In fact, Duc and Tissot [23], studying the oxidation of tin in neutral phosphate solution, suggested that $Sn(OH)_2$ or SnO and $Sn(OH)_4$ or SnO_2 are present on the electrode surface at the beginning of polarization. At more anodic potentials only stannic species are formed. Electrochemical behavior of tin has also been studied in aqueous alkaline solutions [10–11,14,24–29]. The authors considered that primary passivation occurs due to formation of $Sn(OH)_2$ or SnO film by a dissolution-precipitation mechanism. The final passivation is characterized by a continuous film of $Sn(OH)_4$ [10]. Then, more stables species like SnO_2 or $SnO_2.H_2O$ can be formed from dehydration of this last film [9]. Some authors have suggested a duplex film [28], consisting of 5SnO $2H_2O$ and $SnO_2.H_2O$ [29]. All these results emphasized importance of ellipsometric measurements in order to determine the thickness of the films. In fact, the exact composition of the film cannot be determined electrochemically.

By introducing experimentally determined V_f values into Eq. (6) the values of ρ_f can be calculated. Fig. 10 illustrates the results obtained using the values of charge density in Table 2 for the case of the voltammetry experiment at 2 mV s^{-1}.

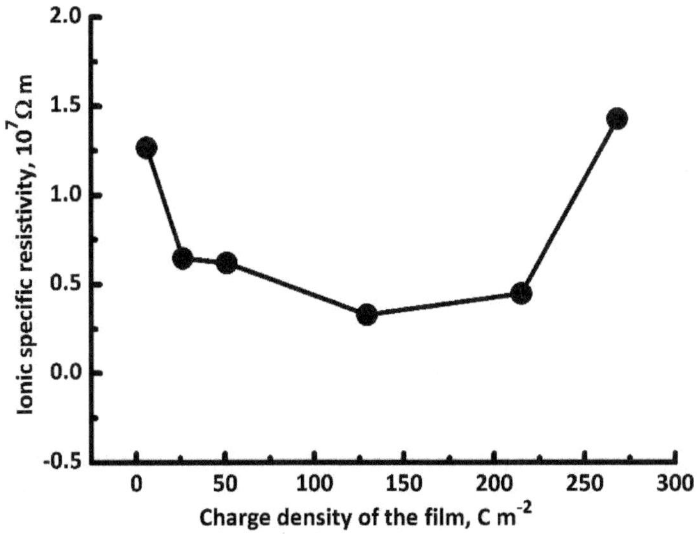

Fig. 10. *Ionic specific resistivity vs. charge density of the film for sweep rate equal to 2 mV s^{-1}.*

As can be observed in Fig. 10, ρ_f passes through a minimum (justifying the maximum in current densities). This behavior was also found by other authors for Zn, Nb, Ni and galvanized steel sheets [4–8]. According to the theory, this happens because the passage of current in the film of initial thickness (q_0) generates injection of specific defects, what resulted in decrease of ρ_f (inversely proportional to concentration of defects) [4–8]. Increase of recombining specific defects (interstitial and cationic vacancies) in the film ends up by generating the recombination reaction (interstitial cation + cationic vacancy → cationic net), and making ρ_f to increase again [4–8].

Conclusions

The ex-situ ellipsometric measurements of the thickness of tin oxides film grown by voltammetry experiments at 2 mV s^{-1} in the studied solution showed that the film is less dense for values of charge density below 50 C m^{-2} having V_f near 5.7×10^{-10} m^3 C^{-1}. For higher values of charge density, the film becomes denser having V_f near 0.5×10^{-10} m^3 C^{-1}.

From the experimentally determined values of V_f, the values of the variable ionic resistivity of the film during the voltammetric growth, ρ_f, were calculated. The behavior of ρ_f versus the charge density of the film was like that found by other authors in the cases of Zn, Nb, Ni and galvanized steel sheets.

Acknowledgements*: The author (Tiago Brandão Costa) is grateful to CAPES for his Doctor fellowship. The authors are grateful to Programa de Pós-Graduação em Engenharia Metalúrgica (PPGEM), Escola de Engenharia Industrial Metalúrgica de Volta Redonda (EEIMVR), Universidade Federal*

Fluminense (UFF) for the opportunity of realizing this work, making the Laboratory of Electrochemical available.

Notation

A electrode surface area, m^2
b_a Tafel slope
E potential, V
E_F Flade potential, V
$E_{m/f}$ potential at the metal/film interface, V
F Faraday's constant, C mol^{-1}
j current density, A m^{-2}
j_p current density at the voltammetric peak, A m^{-2}
$j^0_{m/f}$ exchange current density at zero $\eta_{m/f}$, A m^{-2}
ℓ thickness of oxide layer, m
M molar mass of the film, kg
n number of electrons
q_f charge density of the film, C m^{-2}
q_0 charge density related to the amount of film initially present, at the beginning of the voltammetric growth on the metal surface, C m^{-2}
q_{volt} charge density related to the amount of film which has grown during voltammetric experiment, C cm^{-2}
$q_{f,p}$ peak or plateau charge density, C cm^{-2}
R gas constant, 8.314 J/mol K
SE spectroscopic ellipsometry
T temperature, K
V_f volume per charge unit, m^3 C^{-1}
α_a anodic transfer coefficient
α_c cathodic transfer coefficient
δ density of the film, kg m^{-3}
$\eta_{m/f}$ overpotential at the metal/film interface
$\eta_{f.p}$ overpotential across the film at the voltammetric peak
η_f overpotential across the film
v sweep rate, V s^{-1}
ρ_f ionic specific resistivity of the film, Ω m

References

[1] A. S. Gliozzi, A. L. Alexe-Ionescu, G. Barbero, *Phys. Lett. A.* **379** (2015) 2657-2660.
[2] A. L. Alexe-Ionescu , G. Barbero , S. Bianco, G. Cicero, C.F. Pirri, *J. Electroanal. Chem.* **669** (2012) 21-27.
[3] G. Barbero, P. Batallioto, A. M. Figueiredo Neto, *J. Appl. Phys.* **101** (2007) 054102(1)-054102(5).
[4] C. V. D'Alkaine, P. C. Tulio, M. A. C. Berton, *Electrochim. Acta* **49** (2004) 1989-1997.
[5] C. V. D'Alkaine, L. M. N. Souza, F. C. Nart, *Corr. Sci.* **34** (1993) 129-149.
[6] C. V. D'Alkaine, M. N. Boucherit, *J. Electrochem. Soc.* **10** (1997) 3331-3336.
[7] C. V. D'Alkaine, M. A. Santanna, *J. Electroanal. Chem.* **457** (1998) 13-21.
[8] T. B. Costa, C. V. D'Alkaine, T. M. C. Nogueira, 67th ABM International Congress, Voltammetric growth of ZnO on galvanized steel sheets containing Sb or Pb, Rio de Janeiro, Rio de Janeiro, 2012, p. 3162.

[9] M. Pourbaix, Atlas of Electrochemical Equilibria in Aqueous Solution, Pergamon, Oxford, 1966.

[10] S. D. Kapusta, N. Hackerman, *Electrochim. Acta* **25** (1980) 1625-1639.

[11] M. Metikos-Hukovic, A. Resetic, V. Gvozdic, *Electrochim. Acta* **40** (1995) 1777-1779.

[12] T. Hurlen, *Electrochem. Acta* **39** (1994) 1359-1364.

[13] S. A. M. Refaey, *Electrochim. Acta* **41** (1996) 2545-2549.

[14] V. Brunetti, M. L. Teijelo, *J. Electroanal. Chem.* **613** (2008) 9-15.

[15] H. Fujiwara, *Spectroscopic Ellipsometry: Principles and Applications*. John Wiley & Sons Ltd, Tokyo, Japan, 2003.

[16] D. E. Aspnes, *Thin Solid Films* **571(3)** (2014) 334-344.

[17] K. Vedam, *Thin Solid Films* **313–314** (1998) 1 - 9.

[18] L. Arsov, I. Mickova, *J. Electrochem. Sci. Eng.* 5(4) (2015) 221-230.

[19] I. Arsova, Lj. Arsov, N. Hebestreit, A. Anders, W. Plieth, *J. Solid State Electrochem.* **11** (2007) 209-214.

[20] L. F. N. Guedes et al. J. Solid State Electrochem **20** (2016) 2517-2523.

[21] A. J. Bard and L. R. Faulkner, *Electrochemical Methods Fundamentals and Applications,* John Wiley & Sons, New York, United States, 2001.

[22] K. Wasa, M. Kitabatake, H. Adachi, *Thin films Materials Technology: sputtering of compound Materials*, William Andrew, New York, United States, 2004, p.28.

[23] H. Do Duc, P. Tissot, *Corros. Sci.* **19** (1979) 179-190.

[24] C. A. Gervasi, P. E. Alvarez, *Corros. Sci.* **47** (2005) 69 -78.

[25] S. D. Kapusta, N. Hackerman, *Electrochim. Acta* **25** (1980) 949-955.

[26] S. D. Kapusta, N. Hackerman, *Electrochim. Acta* **25** (1980) 1001-1006.

[27] S. D. Kapusta, N. Hackerman, *Electrochim. Acta* **25** (1982) 1886-1889.

[28] M. Metikos-Hukovic, M. Seruga, F. Ferina, Ber. Bunsenges, *Phys. Chem.* **96** (1992) 799-805.

[29] A. Ammar, S. Darwish, M. W. Khalil, S. El-Taher, *Electrochim. Acta* **33** (1988) 231-238.

12

Voltammetric growth of tin oxides in borate solution of pH 8.7

Tiago Brandão Costa[1,✉], Tania Maria Cavalcanti Nogueira[1,2], Ladário da Silva[1,3]

[1]*Programa de Pós-Graduação em Engenharia Metalúrgica (PPGEM), Escola de Engenharia Industrial Metalúrgica de Volta Redonda (EEIMVR), Universidade Federal Fluminense (UFF), 27255-125 Volta Redonda, RJ, Brazil*

[2]*Departamento de Engenharia Metalúrgica, EEIMVR, (UFF), 27255-125 Volta Redonda, RJ, Brazil*

[3]*Departamento de Física, Instituto de Ciências Exatas (ICEx), (UFF), 27213-145 Volta Redonda, RJ, Brazil*

✉Corresponding author: tiagobrandao@id.uff.br

Abstract

Voltammetry has been employed to study the growth of tin oxides in buffer solution of 0.3 mol L^{-1} H_3BO_3 + 0.15 mol L^{-1} $Na_2B_4O_7{\cdot}10H_2O$ (pH 8.7). Voltammetric data were compared with the results of tin in a phosphate solution of pH 8.7, presented in the previous work, in order to study the influence of these anions on the growth of tin oxides. The thicknesses of grown oxides were determined using ex-situ ellipsometric technique and the volume per charge unity of the film, V_f, was calculated for different charge densities of the film. The results showed that less dense films were obtained at higher sweep rates. Tin oxide films grown in phosphate solution at 2 mV s^{-1} were denser than those grown in borate solution at the same sweep rate. The kinetic parameters, determined applying the ohmic model, showed that there are no significant differences between the kinetics at the metal/film interface of tin in borate and tin in phosphate solutions. Despite these facts, the ionic specific resistivity for oxide film growth in borate solution were significantly higher than in phosphate solution. This result indicates that incorporation of anions occurs during the growth of the films.

Keywords

ellipsometry; tin passivation; ohmic model; voltammetry; variable ionic resistivity

---

Introduction

Many studies concerning the composition of tin oxides films grown by voltammetry in borate and phosphate solutions in a pH range of 5 to 9 are found in literature. The results showed that the composition of tin passivating film, in terms of Sn(II) and Sn(IV) oxides or hydroxides, depends on the growing anodic potential and different anions in solution. However, these studies do not give information about properties of tin oxides such as density and ionic resistivity [1 -11].

In our recent publication [12] we applied the ohmic model to describe the induced growth of tin oxides by voltammetry in phosphate buffer solution pH 8.7. The thicknesses of the grown oxides were measured by ex-situ ellipsometry to determine the film volume per charge unit, V_f, for different charge densities of the film. This parameter makes it possible to calculate the variable ionic resistivity of the film, ρ_f, during the voltammetric oxide growth, described by the ohmic model.

Tin oxide films grown at 2 mV s^{-1} become denser for higher values of charge density, near 50 C m^{-2}, having V_f of 0.5×10^{-10} m^3 C^{-1}. At the same time, the representation of ρ_f vs. charge density of the film presents a minimum. This behavior was also found by other authors in the cases of Zn, Nb, Ni and galvanized steel sheets [12-17].

The ohmic model developed by D'Alkaine [12-17] presented in earlier work, describes the relation between the current density and the film overpotential during the growth of passivating film under voltammetric conditions. The following equation should be valid for peak or plateau transient conditions during voltammetric growth of a film:

$$\eta_{f,p} = \frac{v}{j_p} q_{f,p} \tag{1}$$

In Eq. (1), $\eta_{f.p}$ is the overpotential across the film at the voltammetric peak, v is the sweep rate, j_p is the current density at the peak and $q_{f,p}$ is the peak or plateau charge density.

The $q_{f,p}$ values can be determined considering that :

$$q_f = q_{volt} + q_0 = \frac{1}{v}\int_{E_i}^{E} j_a \mathrm{d}E + q_0 \tag{2}$$

In Eq. (2), q_0 is the charge density related to the amount of film initially present at the beginning of the voltammetric growth and q_{volt} is the charge density related to the amount of film which has grown on the metal surface during the voltammetric experiment. E_i is the initial potential, E is the potential attained and j_a is anodic component of the current density.

The thickness of the growing film, ℓ, is given by:

$$\ell = V_f q_f \tag{3}$$

In Eq. (3), V_f and q_f are respectively the volume per charge unit and the charge density related to the growing film.

The relation between the current density, j, and η_f, even at high fields, is given by:

$$\rho_f = \frac{\eta_f}{V_f q_f j} \tag{4}$$

In Eq. (4), ρ_f is the average ionic specific resistivity of the film.

In order to calculate the values of ρ_f using Eq. (4), many authors [12-17] applied the ohmic model, considering the volume per charge unit (V_f) as the constant equal to:

$$V_f = \frac{M}{nF\delta} \tag{5}$$

In Eq. (5), M is the molar mass of the film, δ is density of the film, n is the number of electrons and F is the Faraday's constant.

In fact, the real values of V_f during the voltammetric growth could only be determined by independent measurements of the film thickness. A lot of techniques are available to access such information. Generally, it can be measured [19] or calculated [20]. Ellipsometry is an interesting technique to be considered for the thickness measurement [21]. Ellipsometry is an indirect technique, which allows one to access layer thickness in ex-situ [12,22], and in-situ [19,20,23] experiments, which is the main interest in voltammetric measurements. In an in-situ experiment it can also monitor layer growth [19,20]. As ellipsometry is an indirect technique, it is important to match the measured quantities, *i.e.*, tan Ψ and cosΔ, with their counterparts in the adopted model [21,24]. The measured quantities are functions of the various layers and substrate refraction indexes and extinction coefficients [21].

The aim of the present work is to study the voltammetric growth of tin oxides in borate solutions (pH 8.7) applying the ohmic model. The thickness of the oxides grown by voltammetry will be determined by ex-situ ellipsometric measurements in order to determine experimental values of V_f. The results will be compared with those obtained in previous work, in the case of tin in phosphate solution pH 8.7 [12], to verify the influence of these anions in the voltammetric growth of tin oxides.

Experimental

Electrochemical measurements

The working electrode was made of a tin disc (PINE, 99.99 % purity) with circular area of 0.5 cm^2. Before the experiments, the electrode surface was polished with 600-emery paper.

The electrochemical experiments were performed using the EG&G Princeton Applied Research Model 273A potentiostat. Solution of 0.3 mol L^{-1} H_3BO_3 + 0.15 mol L^{-1} $Na_2B_4O_7 \cdot 10H_2O$ (pH 8.7) was prepared from AR chemicals (Merck) and purified water (Millipore Q system). The experiments were carried out in a conventional three-compartment electrolysis cell, using a platinum wire as the counter and $Hg/Hg_2Cl_2/KCl$ (1 M) as the reference electrode, respectively. The solution was kept at room temperature (22 °C).

All current and charge densities are given in terms of the geometric surface area of the analyzed samples.

Anodic voltammetries were carried out at sweep rates of 2, 5, 10, 20, 70, 100, 150, 200, 250 and 300 mV s^{-1} always on the same surface. Before each voltammetry experiment, the previously grown oxide film was reduced at constant cathodic potential equal to -1.2 V during 600 s. After this treatment, the obtained voltammograms were reproducible, indicating that surface roughness of samples was recovered.

Ellipsometric measurements

The ex-situ ellipsometric measurements were made using a SEMILAB spectroscopic ellipsometer, model SOPRA GES 5E, equipped with a Xe lamp, over the spectrum range of 195 – 1,000 nm. The measurements were made in air and at room temperature (roughly 22 °C) with incident angle of 75°. The thicknesses of the oxide films were obtained by analyzing the measured ellipsometric spectra through the Drude and Gauss model [16].

Results and discussion

Figure 1 presents the voltammogram for tin in the borate solution. A cathodic peak (c_1) is related to the reduction of the tin oxide. In the anodic scan, two anodic peaks (a_1 and a_2) appear related to the tin oxides growth. At near 1.5 V the oxygen evolution process begins. In view of this result, the potential of – 1.2 V was chosen to reduce the previously grown oxide film before each voltammetric experiment, as explained in the Experimental section.

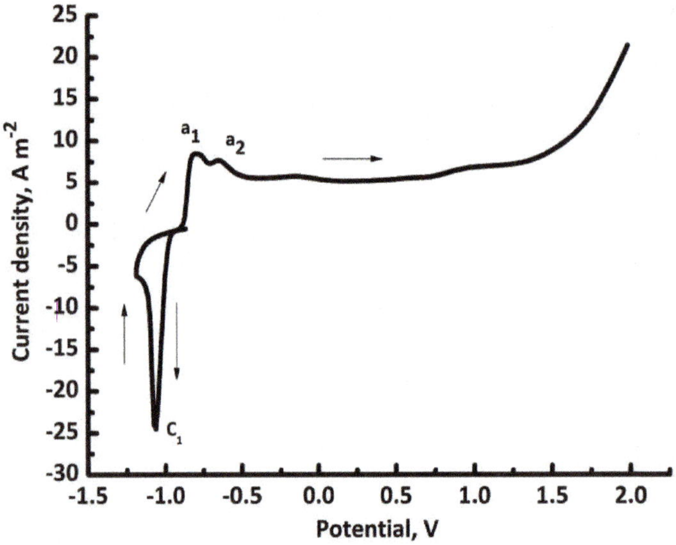

Figure 1. *Potentiodynamic curve for tin in solution borate, pH 8.7, at sweep rate 100 mV s^{-1}.*

Figure 2 presents the anodic voltammograms for tin oxide growth at different sweep rates. The overpotential values in the film at peak condition ($\eta_{f,p}$) were calculated using the Eq. (1). The values of q_{volt} and j_p were determined from the voltammograms. The value of q_0 in Eq. (2) was first considered to be equal to 0.0 C m^{-2}. The curve *j vs.* potential at the metal/film interface ($E_{m/f}$), also shown in this Figure, is obtained after correction of the ohmic drop through the film at the peak potential (E_p - $\eta_{f,p}$).

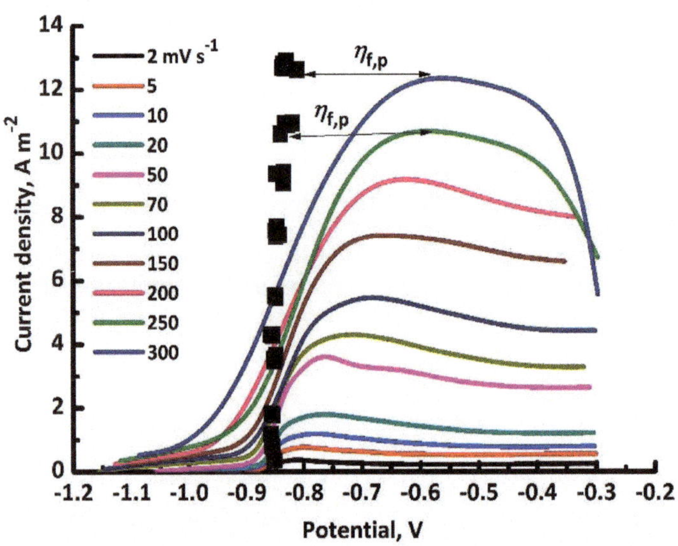

Figure 2. *Voltammetric growth of tin oxide film in borate solution at different scan rates, together with the plot of the calculated* j vs. ($E_p - \eta_{f,p}$) *relation at the metal/film interface, considering* q_0 *= 0.0 C m^{-2}* (■)$E_p - \eta_{f,p}$

The Tafel plot of the curve *j vs.* $E_{m/f}$ in Figure 2 is represented in Figure 3.

In Figure 4, the value of q_0 was taken equal to 0.8 C m^{-2}, in order to provide the best straight line region of the Tafel plot giving a Tafel slope (b_a) equal to 41.14 mV dec^{-1}.

By using Eq. (6) [25]:

$$b_a = \frac{\alpha_{m/f} nF}{RT} \quad (6)$$

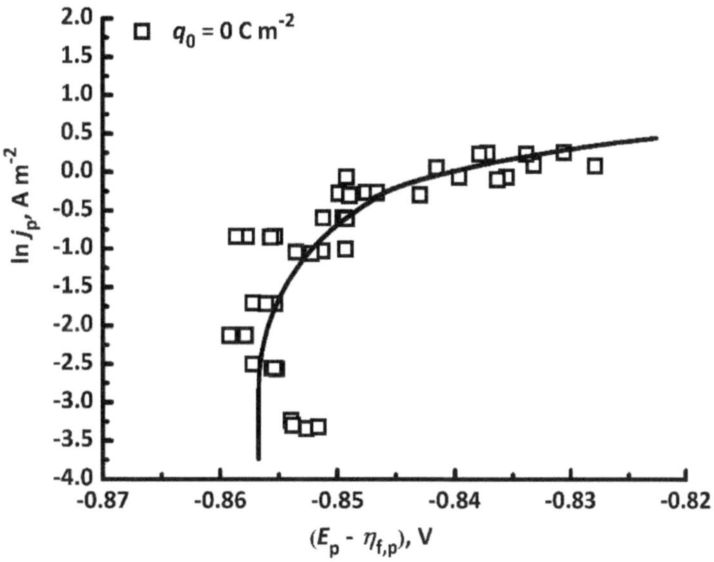

Figure 3. *Tafel plot at tin/oxide interface in borate solution.*

In Eq. (6), $\alpha_{m/f}$ is the transfer coefficient, *n* is the number of electrons, *F* is Faraday's constant, *R* is gas constant and *T* is temperature.

The product ($\alpha_{m/f}$ *n*) at the metal/film interface turned to be 1.05.

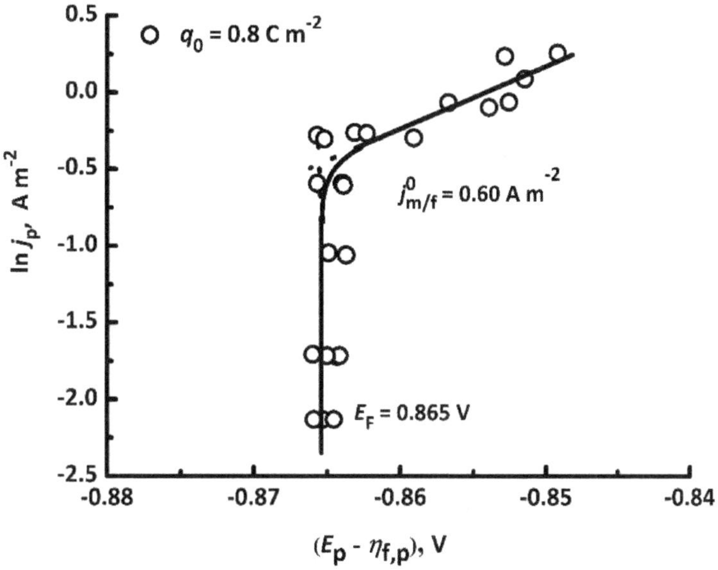

Figure 4. *Tafel plot at tin/oxide interface in borate solution, considering* q_0 *= 0.8 C m*$^{-2}$.

The value of $j^0_{m/f}$, obtained as shown in Figure 4, is equal to 0.60 A m^{-2}. These values should be compared with 1.07 and 1.0 A m^{-2}, respectively, obtained in the case of phosphate solution [12]. These results suggest that there are no significant differences between the kinetics in the two studied metal /film interfaces.

From the Tafel plot of Figure 4 the corrected curve *j vs.* $E_{m/f}$, was obtained. Figure 5 illustrates these results. The anodic voltammograms are also shown in the same Figure.

Figure 6 Present the variation of the peak charge density up to the peak potential ($q_{f,p}$) with *v*.

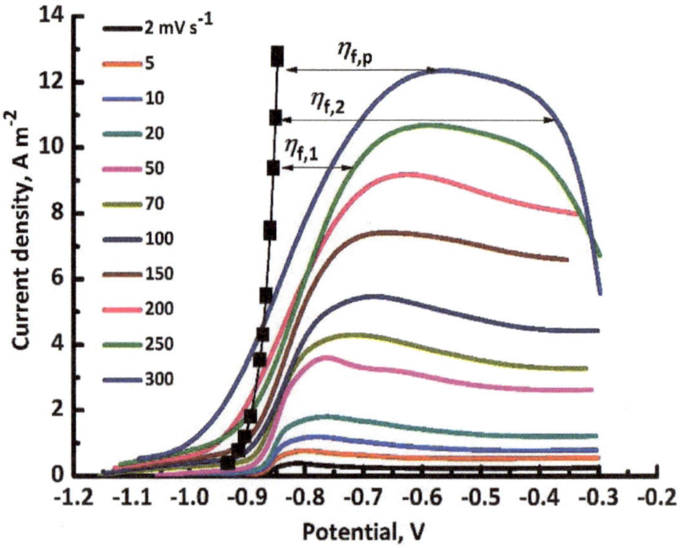

Figure 5. *Voltammetric growths of tin oxide film in borate solution at different scan rates, together with the plot of the calculated* j *vs.* ($E - \eta_{f,p}$) *relation at the metal/film interface, considering* $q_0 = 0.8$ *C m*$^{-2}$ *and Figure 4.* (■)$E_p - \eta_{f,p}$

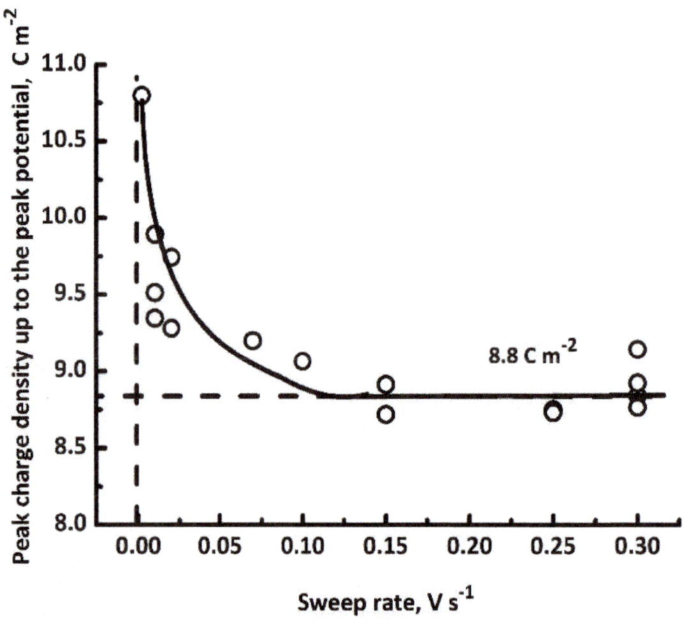

Figure 6. *Peak charge density up to the peak potential* ($q_{f,p}$) *vs. the sweep rate.*

This result shows that the charge required for film growth up to peak potential becomes constant for sweep rates higher than 0.10 V s^{-1}. According to D'Alkaine [13] this means that the parameters related to the film (aging phenomenon) become independent of the growing conditions for sweep rates higher than 0.10 V s^{-1}. In view of this fact, the variable ionic resistivities were studied for 2 and 100 mV s^{-1}.

Figure 5 can be used to determine the overpotential at the film (η_f) for any growing condition of the film beyond the peak condition, by calculating the difference of the potential in the voltammogram at any sweep rate and the potential in the curve *j vs.* $E_{m/f}$ for a given current density. This is shown in Figure 5 for the voltammograms at 250 and 300 mV s^{-1}.

By calculating η_f values for each potential, it is possible to determine the ionic specific resistivity of the film (ρ_f) and its variation with *E* and *v*, for each voltammetric curve using Eq. (4).

The real values of V_f during the voltammetric growth can be determined by an independent way of measuring the thickness of the film according to the following equation:

$$V_f = \frac{\ell A}{q_f} \quad (7)$$

In Eq. (7), ℓ is thickness and *A* is surface area the working electrode.

In the present work, the thicknesses of oxides films grown at 2 and 100 mV s^{-1} were measured by ex-situ ellipsometry. Figure 7 and Figure 8 present the final potential (E_f) of each voltammetry experiment performed at 2 and 100 mV s^{-1}, respectively.

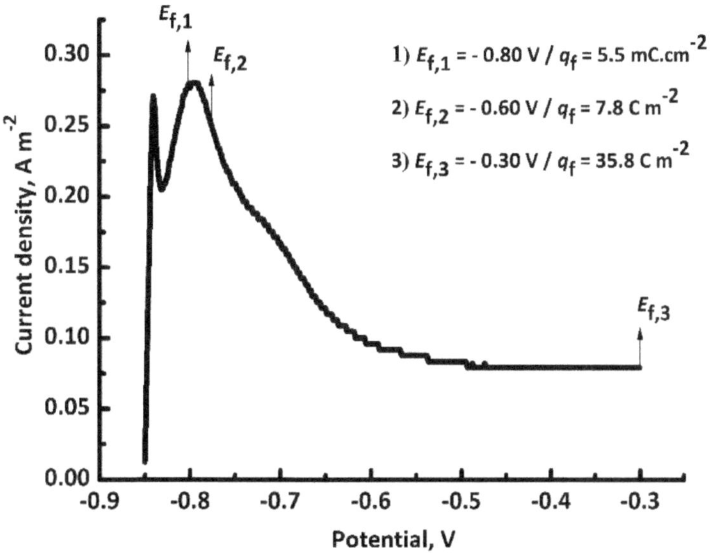

Figure 7. *Voltammetric growth of tin oxide films in borate solution, at 2 mV s^{-1}, together with indication of the final potential,* E_f, *and corresponding film charge density values.*

As pointed out before, spectroscopic ellipsometer was used to obtain thicknesses of tin oxide films grown at distinct values of charge density. Measured and modelled values of tan Ψ and cos Δ vs. wavelength presented good fitting parameters with for $R^2 > 0.97$.

The average thicknesses obtained for charge densities grown in borate solution, at 2 and at 100 mV s^{-1} are shown in Table 1 and Table 2, respectively. In Table 1 and Table 2, the values of the final potential, E_f, charge density of the film, q_f, R^2, standard deviations of the average thickness values and standard deviations (σ). Data in Table 1 clearly show that for greater E_f and q_f, greater average film thickness is obtained.

Table 1. *Final potentials at 2 mV s^{-1}, charge densities, thicknesses, average thicknesses and statistical results of calculations*

E_F / V	q_f / C m^{-2}	ℓ / nm								R^2	Average thicknesses, nm	σ
-0.80	5.5	3.1	3.3	3.5	3.3	3.3	4.2	3.4	3.3	0.971	3.43	0.311
- 0.60	7.8	5.0	5.0	4.2	5.0	4.4	5.1	5.0	4.4	0.992	4.76	0.338
- 0.30	35.8	5.6	5.6	5.8	5.8	5.8	5.5	5.7	5.9	0.991	5.71	0.127

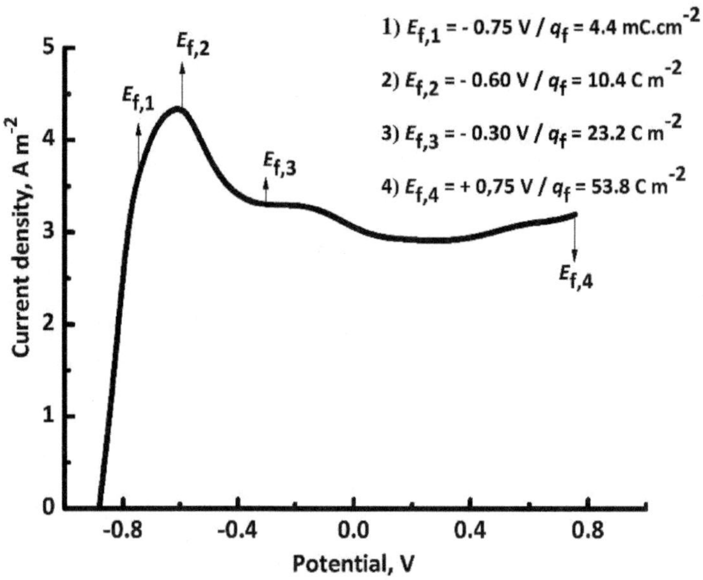

Figure 8. *Voltammetric growth of tin oxide films in borate solution, at 100 mV s^{-1}, together with indication of the final potential,* E_f*, and corresponding film charge density values.*

Table 2. *Final potentials at 100 mV s^{-1}, charge densities, thicknesses, average thicknesses and statistical results of calculations*

E_F / V	q_f / C m^{-2}	ℓ / nm								R^2	Average thicknesses, nm	σ
-0.75	4.4	4.4	4.3	4.8	4.2	4.8	4.0	4.2	4.3	0.992	4.35	0.274
- 0.60	10.4	6.4	7.3	6.6	7.3	6.6	7.3	6.3	7.5	0.983	6.90	0.455
- 0.30	23.2	7.1	7.6	7.5	7.3	7.3	7.5	7.3	7.5	0.980	7.40	0.154
0.75	53.8	9.0	8.6	9.4	8.7	9.0	8.1	9.4	8.1	0.988	8.79	0.478

Figure 9 presents the average thicknesses obtained for charge densities grown at 2 and 100 mV s^{-1}. The results for tin in phosphate solution at 2 mV s^{-1} are also presented in this same Figure.

It is possible to observe that the values of thickness of tin oxides grown at 100 mV s^{-1} in borate solution are higher than those grown at 2 mV s^{-1}. This fact shows that less dense films are obtained at higher sweep rates. In comparison with the results of tin in phosphate solution at 2 mV s^{-1}, it is interesting to notice that the films grown in this last solution are denser.

Using the experimentally found values of average thicknesses, Eq. (7) was applied and V_f was determined for the values of charge density of voltammetry grown films at 2 and 100 mV s^{-1}. The results are shown in Table 3.

Table 3. Film *volume per charge unit obtained from the thicknesses measurements of tin oxide films for different charge densities and sweep rates.*

2 mV s^{-1}			100 mV s^{-1}		
q_f / C m^{-2}	ℓ / nm	V_f / 10^{-10} m^3 C^{-1}	q_f / C m^{-2}	ℓ / nm	V_f / 10^{-10} m^3 C^{-1}
5.5	3.43	6.24	4.4	4.35	9.89
7.8	4.76	6.10	10.4	6.90	6.63
35.8	5.71	1.59	23.2	7.40	3.19
			53.8	8.79	1.63

Figure 10 presents the values of V_f vs. charge density of the film obtained in borate solution, together with the values of V_f obtained in phosphate solution [12] and the values of V_f calculated by Eq. (7), considering SnO and SnO_2.

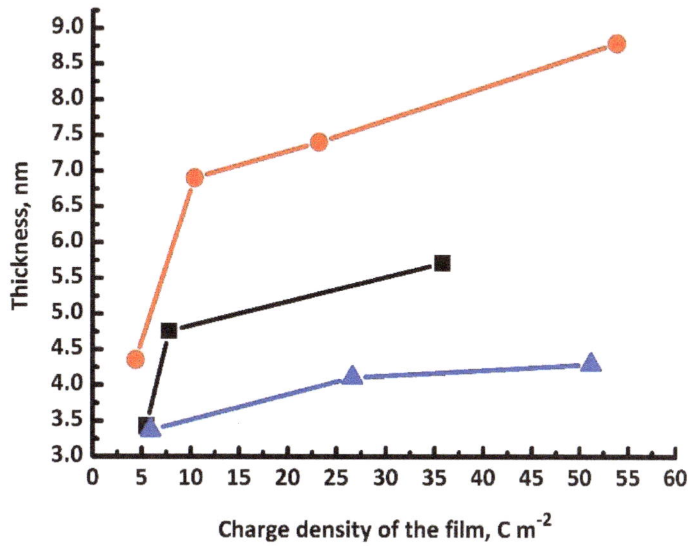

Figure 9. *Thickness* vs. *charge density of the films grown in borate solution at (■) 2, (•) 100 mV s^{-1} and in phosphate solution at (▲) 2 mV s^{-1}*

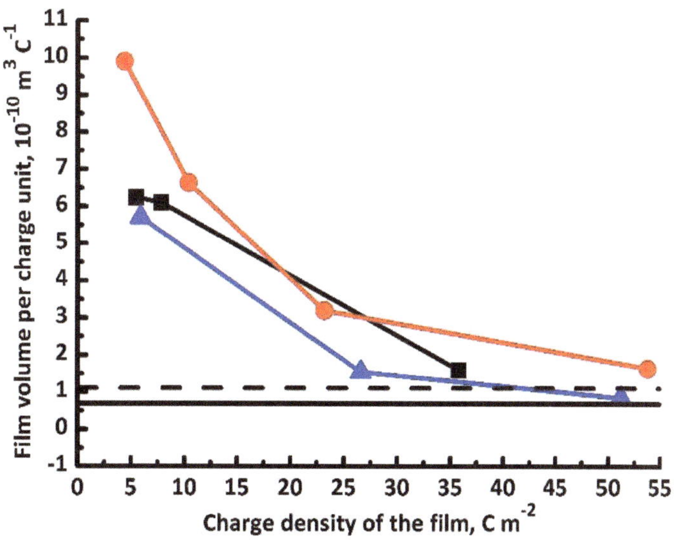

Figure 10. *Film volume per charge unit for different values of charge density. (■) Experimental V_f for tin oxide grown at 2 mV s^{-1} in borate solution, (•) Experimental V_f for tin oxide grown at 100 mV s^{-1} in borate solution, (▲)Experimental V_f for tin oxide grown at 2 mV s^{-1} in phosphate solution, (---) $V_f = 1.08\times10^{-10}$ m^3 C^{-1} (considering SnO) and (–) $V_f = 0.568\times10^{-10}$ m^3 C^{-1} (considering SnO_2).*

Data in Table 3 and Figure 10 show that, for borate and phosphate [12] solutions, the values of V_f significantly decrease until 50 C m^{-2} of charge density of the film is achieved. This result shows that the film is less dense for lower values of charge density and becomes denser as the thickness increases. This behavior suggests that changes in the composition of the film are taking place as the potential turns more anodic, approximating the values of V_f considering SnO and SnO_2. The influence of the final potential in changes in composition of the film, in terms of Sn(II) and Sn(IV), was already mentioned by other authors [1–11].

By introducing experimentally determined V_f values into Eq. (4) the values of ρ_f can be calculated. Figure 11 illustrates the results obtained using the values of charge density in Table 3 for the case of the voltammetry experiment at 2 and 100 mV s^{-1} in borate solution.

As can be observed in the case of sweep rate 2 mV s^{-1}, ρ_f passes through a minimum (justifying the maximum in current densities). This behavior was also found in previous work [1] and in voltammetric oxide film growth of Zn, Nb, Ni and galvanized steel sheets [13–17]. According to the theory this happens because the passage of current in the film of initial thickness (q_0) generates injection of specific defects, what resulted in decrease of ρ_f (inversely proportional to concentration of defects) [13–17].

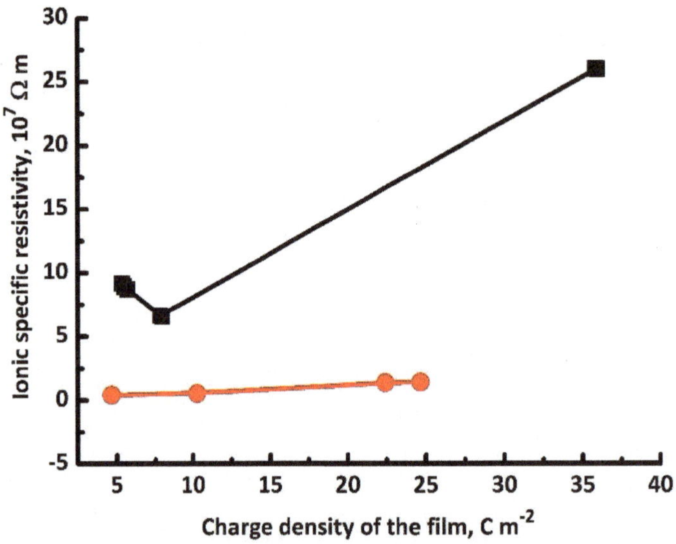

Figure 11. *Ionic specific resistivity vs. charge density of the film for sweep rate equal to 2 mV s^{-1} (■) and 100 mV s^{-1}(●) in borate solution.*

Increase of recombining specific defects (interstitial and cationic vacancies) in the film ends up by generating the recombination reaction (interstitial cation + cationic vacancy → cationic net), and making ρ_f to increase again [13–17].

In the case of the voltammetry at 100 mV s^{-1} the values of ρ_f are lower than the values of ρ_f in the voltammetry at 2 mV s^{-1}. This can be explained considering that at higher sweep rates the injection of defects occurs faster than their recombination. It is also interesting to point out that this phenomenon generates less dense films, as can be observed in Figure 9.

Figure 12 presents the values represented in Figure 11 together with the values of ρ_f *vs.* charge density of the film found in previous work in the case of tin in phosphate solution for sweep rate equal to 2 mV s^{-1}.

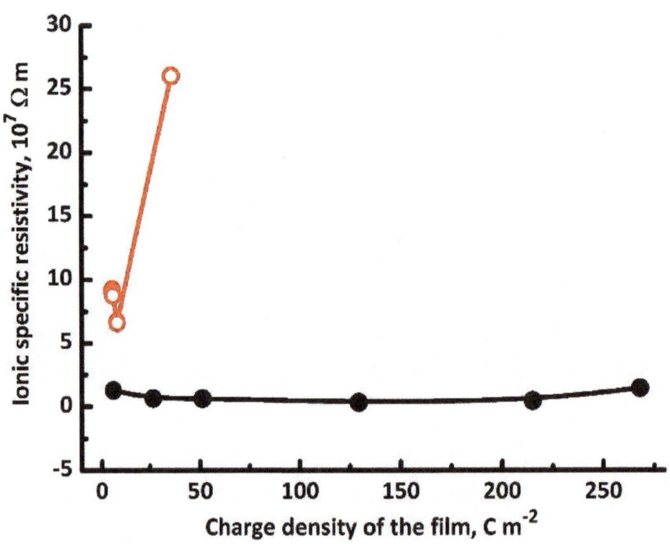

Figure 12. *Ionic specific resistivity vs. charge density of the film for sweep rate equal to 2 mV s^{-1}, for tin in borate solution (○) and for tin in phosphate solution (●).*

It is interesting to notice that the ionic specific resistivities for oxide film grown in borate solution are significantly higher than in phosphate solution. This result indicates that incorporation of anions probably occurs in the films during their growth leading to changes in ionic specific resistivities. Considering Eq. (4), it is possible to verify that this difference is a consequence of higher η_f and lower j values in the voltammetry in borate solution. Differences in film densities and V_f, in the two studied solutions, do not significantly determine the different values of ionic specific resistivities.

Conclusions

The application of the ohmic model and ellipsometric measurements in the voltammetric growth of tin oxides in borate solution pH 8.7 turned possible to verify that:

- There are no significantly differences between the kinetics in the metal/film interface of tin in borate and tin in phosphate solutions. This conclusion was supported by the obtained values of $j^0_{\mathrm{m/f}}$ and α in both cases.
- The values of V_f significantly decrease until 50 C m^{-2} of charge density of the film is achieved approximating the values of V_f considering SnO and SnO_2. This result shows that the film is less dense for lower values of charge density and becomes denser as the thickness increases.
- The ionic specific resistivities for oxide film grown in borate solution are significantly higher than in phosphate solution at 2 mV s^{-1}. This result indicates that incorporation of anions probably occurs in the films during their growth, which leads to changes in ionic specific resistiveties, contributing to different values of η_f and consequently different voltammograms.

Acknowledgements: *The author (Tiago Brandão Costa) is grateful to CAPES for his Doctor fellowship. The authors are grateful to Programa de Pós-Graduação em Engenharia Metalúrgica (PPGEM), Escola de Engenharia Industrial Metalúrgica de Volta Redonda (EEIMVR), Universidade Federal Fluminense (UFF) for the opportunity of realizing this work, making the Laboratory of Electrochemical available.*

Notation

A electrode surface area, m^2
b_a Tafel slope
E potential, V
E_F Flade potential, V
$E_{m/f}$ potential at the metal/film interface, V
F Faraday's constant, C mol^{-1}
j current density, A m^{-2}
j_p current density at the voltammetric peak, A m^{-2}
$j^0_{m/f}$ exchange current density at zero $\eta_{m/f}$, A m^{-2}
ℓ thickness of oxide layer, m
M molar mass of the film, kg
n number of electrons
q_f charge density of the film, C m^{-2}
q_0 charge density related to the amount of film initially present, at the beginning of the voltammetric growth on the metal surface, C m^{-2}
q_{volt} charge density related to the amount of film which has grown during voltammetric experiment, C cm^{-2}
$q_{f,p}$ peak or plateau charge density, C cm^{-2}
R gas constant, 8.314 J/mol K
SE spectroscopic ellipsometry
T temperature, K
V_f volume per charge unit, m^3 C^{-1}
α_a anodic transfer coefficient

α_c cathodic transfer coefficient
δ density of the film, kg m^{-3}
$\eta_{m/f}$ overpotential at the metal/film interface
$\eta_{f.p}$ overpotential across the film at the voltammetric peak
η_f overpotential across the film
v sweep rate, V s^{-1}
ρ_f ionic specific resistivity of the film, Ω m

References

[1] H. Do Duc, P. Tissot, *Corrosion Science* **19** (1979) 179-190.
[2] S. D. Kapusta, N. Hackerman, *Electrochimica Acta* **25** (1980) 1625-1639.
[3] S. D. Kapusta, N. Hackerman, *Electrochimica Acta* **25** (1980) 949-955.
[4] S. D. Kapusta, N. Hackerman, *Electrochimica Acta* **25** (1980) 1001-1006.
[5] S. D. Kapusta, N. Hackerman, *Electrochimica Acta* **25** (1982) 1886-1889.
[6] A. Ammar, S. Darwish, M. W. Khalil, S. El-Taher, *Electrochimica Acta* **33** (1988) 231-238.
[7] M. Metikos-Hukovic, M. Seruga, F. Ferina, *Ber. Bunsenges Phys. Chem* **96** (1992) 799-805.
[8] M. Metikos-Hukovic, A. Resetic, V. Gvozdic, *Electrochimica Acta* **40** (1995) 1777-1779.
[9] C. A. Gervasi, P. E. Alvarez, *Corrosion Science* **47** (2005) 69 -78.
[10] V. Brunetti, M. L. Teijelo, *Journal of Electroanalytical Chemistry* **613** (2008) 9-15.
[11] N. A. Al-Mobarak, *Chemistry and Technology of Fuels and Oils* **48** (2012) 321-330.
[12] T. B. Costa, T. M. C. Nogueira, L. Silva, *Journal of Electrochemical Science and Engineering.* **6** (2016) 303-314.
[13] C. V. D'Alkaine, P. C. Tulio, M. A. C. Berton, *Electrochimica Acta* **49** (2004) 1989-1997.
[14] C. V. D'Alkaine, L. M. N. Souza, F. C. Nart, *Corrosion Science* **34** (1993) 129-149.
[15] C. V. D'Alkaine, M. N. Boucherit, *Journal of the Electrochemical Society* **10** (1997) 3331-3336.
[16] C. V. D'Alkaine, M. A. Santanna, *Journal of Electroanalytical Chemistry* **457** (1998) 13-21.
[17] T. B. Costa, C. V. D'Alkaine, T. M. C. Nogueira, *67th ABM International Congress,* Rio de Janeiro, Brazil, 2012, p. 3162.
[18] R. Díaz, I. Díez-Pérez et al, *Journal of Brazilian Chemical Society* **14** (2003) 523-529.
[19] M. Hernández Ubeda, M.A.Pérez et al, *Journal of the Electrochemical Society* **152(1)** (2005) A37-A41.
[20] Z. Szklarska-Smialowska, W. Kozlowski, *Journal of the Electrochemical Society* **131** (1984) 499-505.
[21] H. Fujiwara, Spectroscopic Ellipsometry: Principles and Applications. John Wiley & Sons Ltd, Tokyo, Japan, 2003.
[22] L. F. N. Guedes et al, *Journal of Solid State Electrochemistry* **20** (2016) 2517-2523.
[23] W. Kozlowski, J. Flis, *Corrosion Science* **32** (1991) 861-875.
[24] D. E. Aspnes, *Thin Solid Films* **571(3)** (2014) 334-344.
[25] A. J. Bard and L. R. Faulkner, *Electrochemical Methods Fundamentals and Applications*, John Wiley & Sons, New York, United States, 2001.

13

Etoricoxib selective sensor based on uracil-5,6-diamino-2-thio hydrochloride as neutral carrier for potentiometric analysis in pharmaceutical preparations

Salwa Fares Rassi✉

Department of Chemistry, Faculty of Science, Al-Baath University, Homs, Syria

✉Corresponding Author: rassi.salwa@gmail.com

Abstract

A construction and electrochemical behavior of novel potentiometric membrane sensor responsive to the etoricoxib was described. The sensor was based on the ion-pair complex of etoricoxib (ET) with uracil-5,6-diamino-2-thio hydrochloride (UDTH) (ET-UDTH) as exchange sites in a PVC matrix with different plasticizers dioctyl phthalate (DOPH) (electrode A), dibutyl phthalate (DBPH) (electrode B), and tri-n-butyl phosphate (TBP) (electrode C). The electrodes exhibited near-Nernstian response for ET-UDTH over the concentration range 0.051-40.042 mM. The electrode offered significant advantages including long lifetime (about 2 months), excellent stability and reproducibility, good response time (10-25 s), and wide pH working range (pH 5-12). Selectivity coefficients of ET related to a number of interfering cations and some organic compounds were investigated, and there were negligible interference caused by most of the investigated species. The direct determination of 0.5-10 mM of ET showed an average recovery of 99.03-101.75 % and a mean relative standard deviation 0.40-1.88. The results were obtained by determination of ET in tablets using the proposed electrodes which were comparable favorably with those obtained by spectrophotometric method.

Keywords

Etoricoxib, Uracil-5,6-diamino-2-thio hydrochloride, PVC membrane, Potentiometric method validation.

Introduction

Etoricoxib (5-chloro-6'-methyl-3-[4-(methylsulfonyl)phenyl]-2,3'-bipyridine) represents a second-generation of COX-2 inhibitors that has been developed for the treatment of many inflammatory diseases such as rheumatoid arthritis, osteoarthritis, pain relief and acute gout, causing fewer gastrointestinal complications than conventional non-steroidal anti-inlammatory drugs (NSAIDs) [1-2]. The potential use of this COX-2 inhibitor among the general public is high due to their improved safety profile over traditional NSAIDs. Traditional NSAIDs such as aspirin and ibuprofen, inhibit both the COX-1 and COX-2 enzymes [3]. The COX-1 enzyme, which is expressed in essentially all tissues, has been found to play a role in the homeostasis of the gastrointestinal tract and kidneys, while the COX-2 enzyme is absent unless induced by an inflammatory event. Consequently, side effects of chronic use of traditional NSAIDs include gastrointestinal ulceration and bleeding [4]. Etoricoxib, a highly selective COX-2 inhibitor, was developed in order to address the safety issues associated with traditional NSAIDs, while providing pain relief from inflammatory illness such as osteo- and rheumatoid arthritis [5].

In order to provide quality control over the manufacture of any active pharmaceutical ingredient, it is essential to develop highly selective analytical techniques. Therapeutic importance of etoricoxib has prompted the development of many methods for its assay. The drug is available in tablet dosage form and is not yet official in any of the pharmacopoeias. Several methods have been reported for the analysis of etoricoxib in pharmaceutical dosage form as well as in the biological fluids and tissues, such as spectrophotometric methods [6-8], chromatographic methods HPLC [9-12], high-performance thin-layer chromatography [13], LC/MS spectrophotometry [14-18] and RP-HPLC method [6,19,20] for the estimation of etoricoxib.

Experimental

1. Reagents

All reagents used were of analytically pure grade and doubly distilled water was used throughout. High-molecular-weight poly(vinyl chloride) (PVC) was obtained from SABIC. Co., dioctyl phthalate 98.9 % (DOPH), tri-n-butyl phosphate 97 % (TBP), and di-n-butyl phthalate 99 % (DBPH) were obtained from BDH. Co., England. Tetrahydrofuran (THF) was obtained from Merck. Pure-grade etoricoxib (99.95 %) ($C_{18}H_{15}$ Cl N_2O_2S, 358.1 g mole^{-1}) was obtained from Aarti Drugs Limited (Indian). Uracil-5,6-diamino-2-thio hydrochloride (UDTH) ($C_4H_7N_4OSCl$, 194.5 g mole^{-1}) was synthesized (Figure 1) and identified in laboratory by spectroscopic means. The H-NMR and IR data (Figure 2) of UDTH are shown in Table 1.

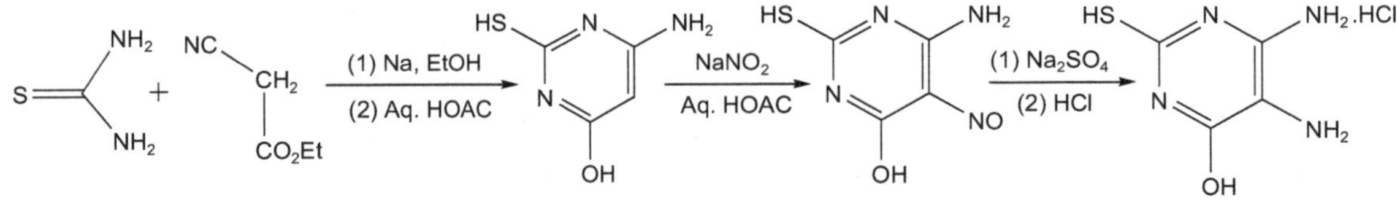

***Figure 1**. Synthesis of UDTH.*

2. Apparatus

Potentiometric and pH measurements was carried out using a digital Shott Gerate pH meter (Consort C 830, Belgium) with combined glass pH electrode. A water bath shaker (Grant instruments, Cambridge Ltd, England) was used to control the temperature of the test solutions. A saturated calomel electrode (SCE) was used as the external reference (Mettler, Switzerland) while an Ag/AgCl electrode was used as an internal reference. The electrochemical system may be represented as follows:

Ag/AgCl | inner solution | PVC membrane | test solution || KCl salt bridge || $Hg/HgCl_2$(sat.).

FT-IR 4100 (Fourier Transform Infrared Spectrometer) Jasco using KBr disk in the range 4000 - 400 cm^{-1}, Nernst glower, deuterium try glycine sulfate (DTGS), 0.1 cm^{-1}.

3. Preparation of solutions

Standard ET solutions (0.05-35 mM) were prepared in doubly distilled water. UDTH solution (10 mM) was prepared by dissolving appropriate amount of the compound in the methanol. ET and UDTH stock solutions were stored in dark bottle at refrigerator. Stock solutions of 1 M for each of LiCl, NaCl, KCl, NH_4Cl, $CaCl_2$, $MgCl_2$, $BaCl_2$, $ZnCl_2$, $MnSO_4$, $Ni(NO_3)_2$, $Co(NO_3)_2$, $Cu(NO_3)_2$, $Pb(NO_3)_2$, $FeCl_3$, $AlCl_3$, $CrCl_3$, glucose, fructose, lactose, starch, micro crystalline cellulose, carboxymethyl cellulose, polyethylene glycol, titanium dioxide, and polysorbate-80 were prepared by dissolving the appropriate amount of the compounds, and the diluted solutions from these were prepared by subsequent dilutions of the stock solutions.

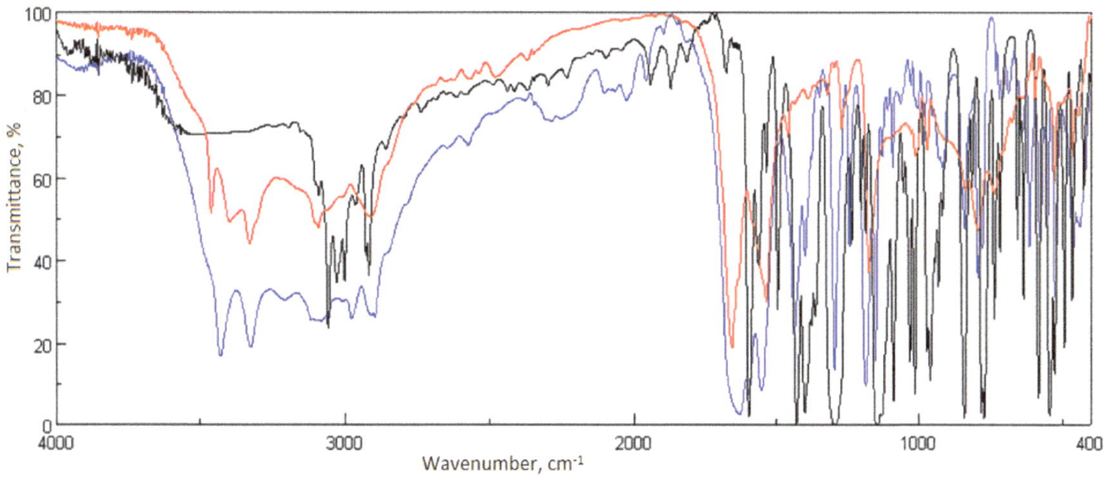

Figure 2*. IR spectrum of etoricoxib (black spectrum), UDTH (red spectrum), and ET+UDTH- complex (blue spectrum)*

4. Sample Preparation

Etoricoxib tablets were supplied by Razi Pharmaceutical Industries (Aleppo, Syria). Each tablet was labeled to contain etoricoxib 120, 90, 60 mg/tab. The homogenized powder was prepared from ten accurately weighed ET tablets. An appropriate amount of this powder was dissolved in methanol and doubly distilled water. The mixture was then filtered and made up to the mark in a 100 mL volumetric flask. Different volumes of the stock solution were taken and subjected to the direct and standard addition methods.

5. Preparation of ion-pair compound

Ion-pair (ET^+UDTH^-) was prepared by mixing equal volumes of 10^{-2} M methanolic solution of ET with methanolic solution of UDTH with stirring. Methanol was then gradually evaporated to obtain a precipitate. ^{1}H-NMR and IR data (Figure 2) of ET^+UDTH^- are shown in Table 1.

6. Construction of etoricoxib membrane electrodes

The electrodes were constructed according to the method described by Craggs *et al.* [8]. Membrane composition was studied by varying the percentages (w/w) of the ion-pair complex, PVC and DOPH (electrode A), DBPH (electrode B), or TBP (electrode C) as plasticizing solvent mediators. The optimum composition that exhibited perfect performance characteristics was reached. The membranes were prepared by dissolving the required amounts of ion pair complex, PVC and DOPH, DBPH or TBP in THF, and then the homogeneous mixture was poured into glass Petri dishes (8 cm diameter), covered with a glass plate, and allowed to evaporate overnight at room temperature. The thickness of obtained membrane was about 0.15 mm. Membranes (12 mm diameter) were cut out and then adhered to the polished end of a plastic cap attached to a glass tube using PVC-THF paste. The electrodes bodies were filled with a 1×10^{-1} M KCl solution and 1×10^{-3} M ET as the inner electrolyte, and then Ag/AgCl was inserted in it as internal reference electrode. Electrode potential was measured against the SCE as the reference electrode. Before use, the electrode membranes were conditioned by immersion in 25 mM of ET solution for 3 h.

Table 1. *H-NMR and IR data for UDTH and Complex between UDTH and ET.*

	δ / ppm			Wavenumber, cm^{-1}	
	ET^+UDTH^-	UDTH		UDTH	ET^+UDTH^-
-SH	---	12.48	-SH	2479	---
-OH	11.53	11.55	-OH	3460	3440
$-NH_3^+$	4.28	6.27	$-NH_2$	3328-3390	3250-3325
$-NH_2$	7.74	8.00	N-S	---	1186
$-CH_3$	2.76				
$-CH_3$	3.25				
9 H aromatic	7.88-8.57				

7 Selectivity of sensors

Potentiometric selectivity coefficient, $K_{A,B}^{pot}$, of an ion-selective electrode (ISE) was commonly used as quantitative expression of the ability of the electrode to respond primarily to the analyte in the presence of interfering ions. The effect of the presence of some different species on the response of ET electrodes was investigated, and the selectivity coefficient $K_{A,B}^{pot}$ of the proposed electrodes was calculated in the presence of related organic and inorganic substances using matched potential method (MPM) [9-10]. The selectivity coefficient, $K_{A,B}^{pot}$, which was measured by matched potential method was calculated according to the following equation:

$$K_{A,B}^{MPM} = \frac{(a'_A - a_A)}{a_B} \quad (1)$$

where a'_A is the known activity of primary ion, a_A is the fixed activity of primary ion, and a_B is the activity of interfering ions.

8. General procedures

The performance of the three electrodes prepared was investigated by measuring e.m.f. values of 0.05-45 mM of ET. The electrodes were calibrated by added volumes of 50 mM stock solution of ET successively in 50 ml of water to generate a total concentration ranging from 0.05 to 45 mM ET, followed by immersing the ET-electrode, together with a SCE in the solution. The potential reading was recorded after stabilization, and the e.m.f was plotted as a function of the logarithm of the ET concentration. The concentration graph was used for subsequent determinations of unknown ET concentrations. The e.m.f. measurements with the polymeric membrane electrodes were carried out with the cell assembly shown schematically in the section 2.2.

9. Potentiometric determination of ET

ET was determined potentionmetrically by the direct and standard addition methods [11-12]. In this method the proposed electrodes (A,B) (ET-UDTH) were immersed into a sample of 15 ml with an unknown concentration of a ET solution, and the equilibrium potential, E_u, was recorded. Then 1 ml of 50 mM of standard ET was added into the testing solution and the equilibrium potential, E_S, was obtained. From the potential change, $\Delta E = E_u - E_S$, we could determine the concentration of the testing sample using the equation:

$$C_X = \frac{C_S V_S}{(V_X + V_S) \times 10^{\Delta E/S} - V_X} \quad (2)$$

where C_X and V_X are concentration and volume of an unknown sample, C_S and V_S are concentration and volume of the standard, respectively. S is the slope of the calibration graph (slope of the electrode response), and ΔE is the difference in (mV) between e.m.f. after and before addition of the standard solution. Standard addition method was applied for determining ET in commercial preparations.

Results and discussion

1. Optimization of the membrane composition

The effect of the amount of ion pair in the membrane phase on the potentiometric response was investigated. The data shown in Table 2 clearly indicate that the electrode (V) with 5 wt% of ET-UDTH ion pair has performance characteristics (slope 57.00 mV decade^{-1}, at 25 °C, usable concentration range, 0.031-40.042 mM ET), and response time 25 s. For all construted electrodes, the percentage of ion-pair ranging from 1 to 8 % was found to offer better slopes and correlation coefficients. The results obtained with ion-pair for the three plasticizers are summarized in Table 2. The electrodes A and B exhibit comparable linear ranges and the lowest detection limit.

Table 2. *Optimization of the membrane ingredients.*

	Composition, % (w/w)			Slope, mV decade^{-1}	Detection limit, mM
Membrane	Ion-pair Complex	PVC	DBPH		
I	1	49.5	49.5	48.62	0.39
II	2	49.0	49.0	50.24	0.31
III	3	48.5	48.5	53.14	0.25
IV	4	48.0	48.0	55.69	0.19
V	**5**	**47.5**	**47.5**	**57.00**	**0.13**
VI	6	47.0	47.0	54.06	0.25
VII	7	46.5	46.5	52.09	0.20
VIII	8	46.0	46.0	49.51	0.34

2. Effect of the internal filling solution

The concentration of the internal solution of ET in the electrode was changed from 10 to 0.01 mM and the potential response of the electrode was measured. It was found that variation of the concentration of the internal solution did not cause any significant difference in the potential response of the electrode. ET concentration of 1 mM as internal solution was quite appropriate for proper functioning of the electrode.

3. Effect of plasticizer type on the characteristic performance of the sensor

Three plasticizers, DOPH, DBPH and TBP were evaluated. As shown in Table 3, the best performances, in terms of slopes, linear range and detection limit obtained had the following order: DBP (electrode B) > DOPH (electrode A) > TBPH (electrode C). The working characteristics for the electrodes were assessed on the basis of their calibration curves. The TBP which had a low viscosity (3.11 cSt), led to leaching of the complex from the membrane. All further studies were conducted using DBPH and DOPH as plasticizers.

Table 3. *Effect of the nature of an ion-pair and a plasticizer on characteristics of the electrodes.*

Electrode	A	B	C
Ion-pair complex, wt %	5	5	7
Plasticizer	DOPH	DBPH	TBP
Slope, mV decade^{-1}	56.26	57.00	41.23
Linear range, mM	0.051-39.810	0.031-40.042	0.653-33.140
Correlation coefficient	0.9997	0.9999	0.9699
Detection limit, mM	0.042	0.023	0.324

4. Effect of soaking

Freshly prepared electrodes must be soaked to activate the surface of the membrane to form an infinitesimally thin layer for ion-exchange process to occur [13]. This preconditioning process required different times, depending on the diffusion of ions and their equilibrium at the electrode test solution interface. Fast establishment of equilibrium was certainly a condition for a fast potential response. Thus, the performance characteristic of the ET electrode was investigated as a function of the soaking time. For this purpose the electrode was soaked in a 50 mM solution of ET, UDTH and water at room temperature. The optimum soaking time and the optimum solution was found to be 3 h for 50 mM solution of ET. During this period sensors were washed with water after each application and kept dry in air at room temperature when off-use. The results indicate that for a time of 3 h of soaking the slope remained constant at about 57.00 mV decade^{-1}, at 25 °C. *E*, mV *vs.* *t*, min plots (Fig. 3) were obtained after the electrode was soaked continuously in 50 mM ET, water and UDTH for 30-630 min.

5. Effect of pH

The influence of the pH on the potentiometric responses of membrane electrodes was investigated by following the potential variation over the pH range of 1-14. The electrode response for different ET concentration was tested at various pH values. The pH of the solution was adjusted by adding small volumes of hydrochloric acid or sodium hydroxide to the test solution. As can be seen from Figure 4, membrane electrode exhibited a negligible potential change within the pH range of 5-12. At lower pHs (pH<6) there was a decrease in potential which may be due to the interference of hydronium ion and the penetration of H_3O^+ into the membrane surface, or a gradual increase of the protonated species [14-15].

6. Effect of the temperature of the test solution

The effect of the temperature of the test solution on the potential response of the membrane was tested by following the slopes variation in the temperature range 20-65 °C (Fig. 5). The results show that within the temperature range investigated the electrodes respond practically to the ET concentration with slopes from 54.23 to 57.30 mV decade^{-1} and usable concentration range of about 0.031-40.042 mM.

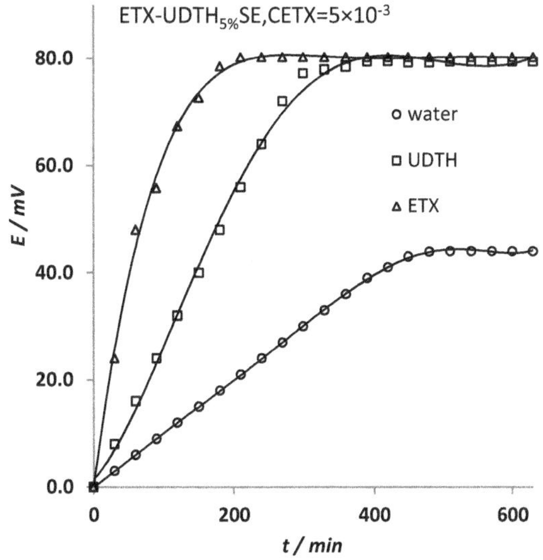

Figure 3. Effect of stock solution on response electrode

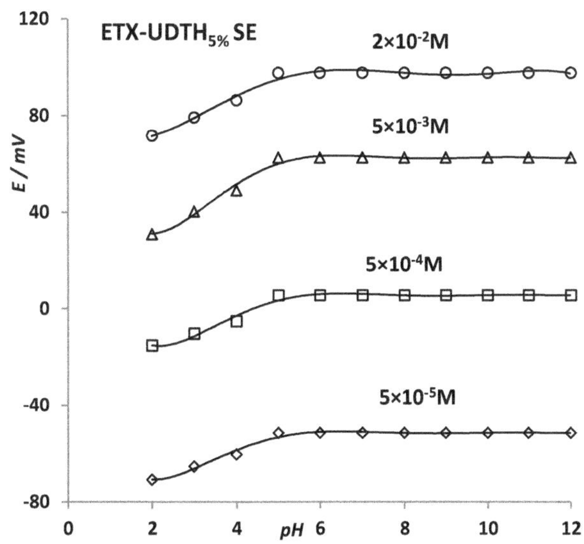

***Figure 4.** Effect of the pH on the response of the electrode*

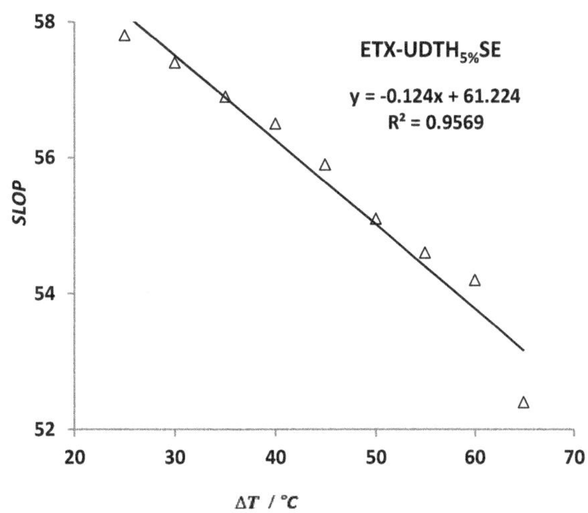

Figure 5. Effect of the temperature of the test solution on the potential response of the membrane.

7. Calibration graphs

Using the optimized membrane composition and conditions described above, the potentiometric response of the electrode was studied based on the ET concentration in the range of 0.01- 50 mM. The calibration curves for the electrodes A and B containing DOP or DBP as plasticizer gave an excellent linear response from 0.031 to 40.042 mM, as shown in Figure 6. The results given in Table 3 show the characteristics performance of the membrane electrodes. The least squares equation obtained from the calibration data was as follows:

$E/\text{ mV} = S \times \log([\text{ET,M}]) + \text{intercept}$

where E is the potential and S the slope of the electrodes.

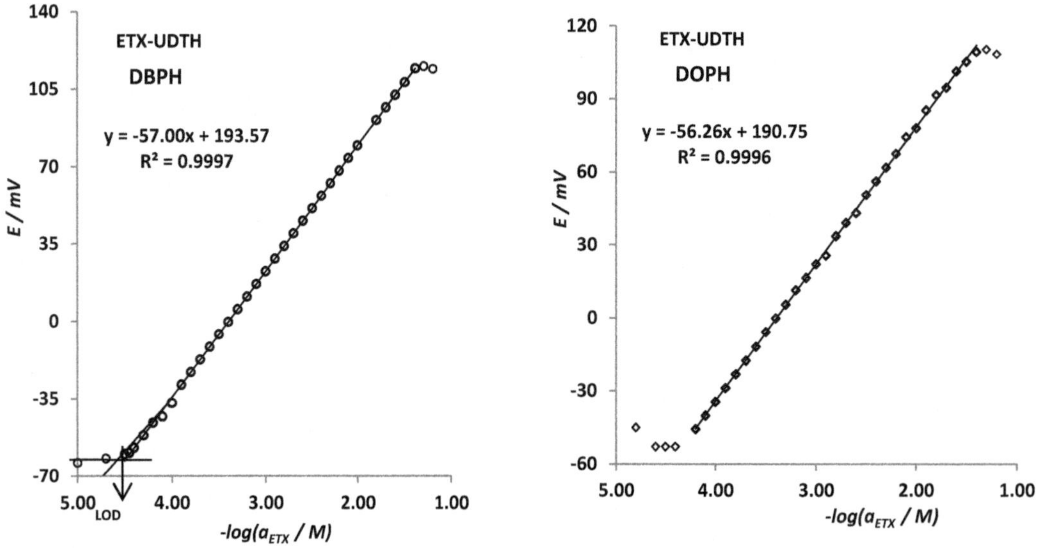

Figure 6. *Calibration graph of ET membrane electrode.*

Table 4. *Response characteristics of membrane electrodes*

Electrode number	A	B
Plasticizer	DOPH	DBPH
Parameter		
Slope, mV/decade^{-1}	56.26	57.00
Correlation cofficient	0.9996	0.9997
Linearity range, mM	0.051-39.810	0.031-40.042
Lower detection limit, mM	0.042	0.023
Response time, s	t ≤35	t ≤25
Working pH range	5.0-12.0	5.0-12.0
Temperature, °C	25	25
Life time, days	44	55

8. Response time

The time required for the electrodes to reach steady potential values, after immersing the electrode in different concentration namely from 0.05, 0.5, 5 and 25 mM of ET solution was studied. The average time was found to be short, ranging from 10 s for concentrations 25 mM and 5 mM, to about 25 s for 0.05 mM solution. Two electrodes gave the same range of response times. These values indicated the high stability of the electrodes during the measurements. A typical plot for response time is shown in Fig. 7 for the electrode based on DBP as the plasticizer.

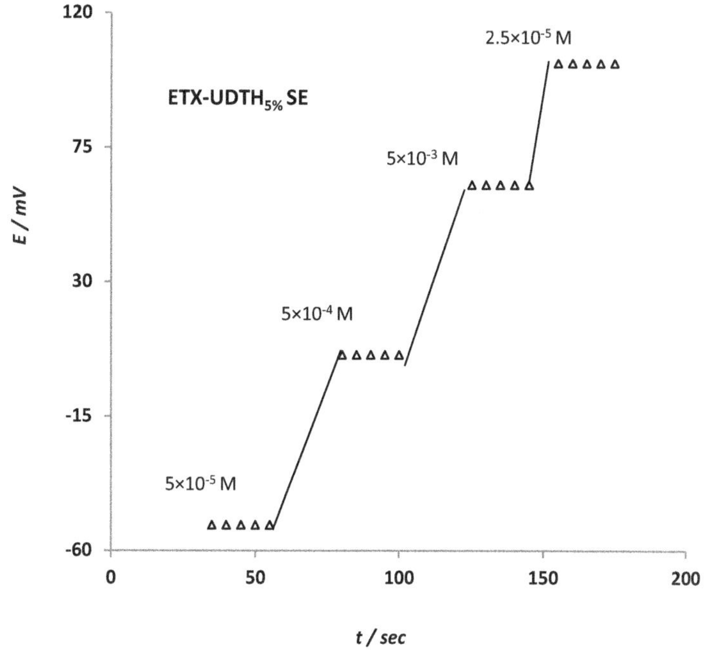

Figure 7. *Plot the response time of DBP electrode using 0.05, 0.5, 5, 25 mM etoricoxib.*

9. Lifetime

The electrode lifetime was investigated by performing the calibration curve and the periodic testing of standard solutions (0.031-40.042 mM of ET) and calculating the response slope. It was observed that the electrode B exhibited good stability in terms of slope in the linear domain of concentration and the electrodes could be used continuously for about 55 days without considerable decrease in its slope value. Electrode A exhibited good stability in terms of slopes of 56.20 to 53.4 mV decade^{-1} in the linear domain of concentration from 0.051 to 39.810 mM. This electrode could be used continuously for about 44 days. But two changes were observed, firstly, a slight gradual decrease in the slope (from 57.00 to 55.23 mV decade^{-1}) was found, and secondly an increase in the detection limit (from 0.023 to 0.2 mM) was noted. However, the electrode with DBP as plasticizers could be used for about 55 days without any considerable decrease in the slope value.

10. Selectivity of electrode

The influence of some inorganic cations such as of Li^+, Na^+, K^+, Ca^{2+}, Zn^{2+}, Ba^{2+}, Mn^{2+}, Mg^{2+}, Ni^{2+}, NH_4^+, Cu^{2+} pb^{2+}, CO^{2+}, Fe^{3+}, Al^{3+}, Cr^{3+}, and some organic compound like sugars (glucose, fructose) and excipients on the electrode response was investigated. The selectivity of the electrode was measured by applying the matched potential method (MPM). According to this method, the activity of ET was increased from a_A = 1 mM (reference solution) to a'_A=1.1 mM, and the changes in potential (ΔE) corresponding to this increase were measured. Then a solution of an interfering ion of concentration a_B is added to a new 1 mM reference solution until the same potential change (ΔE) was recorded. The selectivity factor, $K^{pot}_{A,B}$, for each interference was calculated using equation (1). The results are given in Table 5. Results revealed reasonable selectivity for ET in presence of many related substances. The selectivity coefficient obtained by this method showed that there were no significant interferences from the cations, this reflected a very high selectivity of the investigated electrode towards ET.

Table 5. *Selectivity coefficients for of the ET-UDTH responsive electrode.*

$K^{MPM}_{A,B}$	Interferent	$K^{MPM}_{A,B}$	Interferent
9.9×10^{-5}	Glucose	1.00×10^{-4}	Li^{+1}
2.0×10^{-5}	Fructose	1.25×10^{-4}	K^{+1}
2.5×10^{-5}	Lactose	9.09×10^{-5}	Na^{+1}
4.3×10^{-5}	starch	5.56×10^{-5}	$NH4^{+}$
8.0×10^{-4}	Microcrestaline cellulose	6.25×10^{-5}	Mg^{+2}
1.2×10^{-5}	Carboxymetyle cellulose	6.45×10^{-5}	Mn^{+2}
6.2×10^{-6}	polyethylene glycol	1.10×10^{-5}	Ca^{+2}
4.8×10^{-5}	titanium dioxide	7.14×10^{-5}	Ba^{+2}
8.6×10^{-6}	polysorbate 80	5.13×10^{-5}	Ni^{+2}
		5.81×10^{-5}	Cu^{+2}
		7.69×10^{-5}	Zn^{+2}
		9.52×10^{-5}	pb^{+2}
		7.81×10^{-5}	Cd^{+2}
		6.76×10^{-5}	Cr^{+3}
		1.22×10^{-5}	Fe^{+3}

Even though, the inorganic cations have different ionic sizes, mobilities and permeabilities they did not interfere with ET. The selectivity of the electrode towards neutral sugars was evaluated. The tolerance was considered as the concentration imparting a ± 0.2 mV drift in the potential reading. The results indicated that glucose, fructose, lactose and starch did not interfere. The experiments showed no interference with respect to ET response for electrodes A and B.

11. Analytical application

The ET membrane electrodes were used for the determination of ET in pharmaceutical preparations using both direct and standard-addition methods. The direct method is the simplest for obtaining quantitative results. A calibration graph was constructed and concentration of the unknown was calculated from the linear equation of the calibration curve. Direct determination of ET in tablets was carried out using the developed membrane electrodes. The results are summarized in Table 6 and 7.

Table 6. *Accuracy and precision for the determination of ET using the proposed PVC membrane sensors in pure solution*

		Direct mothed				Standard-addition method			
Electrode	Taken, mM	Found, mM	RSD, %	*R* / %	ASE	Found, mM	RSD, %	*R* / %	ASE
A	0.5	0.50	1.88	100.00	0.005	0.49	2.13	98.00	0.007
	2.5	2.49	1.80	99.60	0.026	2.44	1.97	97.60	0.034
	5	4.95	1.45	99.00	0.042	5.08	1.74	101.60	0.063
	10	9.95	0.86	99.50	0.050	9.94	1.56	99.40	0.110
B	0.5	0.49	1.52	98.00	0.004	0.51	1.46	102.00	0.005
	2.5	2.54	1.24	101.70	0.018	2.53	1.32	101.20	0.024
	5	4.98	0.81	99.60	0.023	4.98	1.01	99.60	0.036
	10	10.06	0.62	100.60	0.035	10.19	0.91	101.90	0.068

Average of four determinations

The content of drug in its formulation had good agreement with the declared amount. The standard-addition method was applied by adding a small portion (1 mL) of a 50 mM standard ET solution to 15 mL of various formulation drug concentrations (60, 90, 120) mg ET/tablet, (0.167,

0.251, 0.335) mM. The change in the potential reading (at a constant temperature of 25 °C) was recorded after each addition, and was used to calculate the concentration of ET by the equation (2). Thus, the determination of the concentration depended mainly on ΔE, hence, to obtain a noticeable ΔE we needed to prepare a higher concentration of the ET standard. Results of the standard-addition method are given in Table 7.

The determination of ET in tablet was carried out using the proposed electrode. The results were compared to those obtained using the spectrophotometric method [16]. The determination of ET in its pharmaceutical formulations Etoxia gave an average Recovery of (100.06-100.75). Mean values were obtained with a Student's t- and F-tests at 95 % confidence limits for four degrees of freedom are shown in Table 7. The data reveal that results compare favorably with those obtained by spectrophotometric. The results showed comparable accuracy (*t*-test) and precision (*F*-test).

Table 7. *Determination of ET in its pharmaceutical preparation using the proposed electrode.*

Sample	Nominal value mgET/tablet	Potentiometry		Spectrophotometry
		Direct	Standard addition	
Etoxia	60			
***R* / %±SD**[a]		100.75±0.58	100.47±0.61	99.73±0.31
***t*-Value**[b]		1.85	1.72	1.830
***F*-Value**[b]		3.50	3.87	
Etoxia	90			
***R* / %±SD**[a]		100.29±0.49	100.43±0.54	100.08±0.33
***t*-Value**[b]		1.32	2.15	0.57
***F*-Value**[b]		2.20	1.86	
Etoxia	120			
***R* / %±SD**[a]		100.06±0.23	100.46±0.55	100.46±0.49
***t*-Value**[b]		1.89	0.58	2.05
***F*-Value**[b]		1.26	4.53	

a Five independent analyses.

b Theoretical values for t- and F-values at four degree of freedom and 95% confidence limit are (t=2.776) and (F=6.26).

Conclusions

In conclusion, the developed PVC membrane sensors described in this work offer a simple, accurate, selective, and specific tool for quantitative determination of ET in some pharmaceutical formulations. The membrane sensor ET-UDTH based on DBPH, seem to be better than ET-UDTH based on DOPH with respect to calibration, slop, and accuracy. The statistical evaluations of the proposed method in comparison with spectrophotometric method indicate that the method is accurate and precise. The proposed analytical method is proved to be simple and rapid, with good accuracy.

References

[1] D. Riendeau., M.D. Percival, C. Brideau, S. Charleson, D. Dube, D. Ethier, J.-P. Falgueyret, , R.W. Friesen, R. Gordon, G. Greig, J. Guay, J. Mancini, M. Ouellet, E. Wong, L. Xu, S. Boyce, D. Visco, Y. Girard, P. Prasit, R. Zamboni, I.W. Rodger, M. Gresser, A.W. Ford-Hutchinson, R.N. Young, C.-C. Chan, *J. Pharmacol. Exp. Ther.* **296(2)** (2001) 558-566.

[2] A.D. Rodrigues, R.A. Halpin, L.A. Geer, D. Cui, E.J. Woolf, C.Z. Matthews, K.M. Gottesdiener, P.J. Larson, K.C. Lasseter, N.G.B. Agrawal, *Drug Metab. Dispos.* **31 (2)** (2003) 224-232.

[3] B. Cryer, M. Feldman, *Am. J. Med.* **104** (1998) 413-421.

[4] M.C. Allison, A.G. Howatson, C.J. Torrance, F.D. Lee, R.I. Russell, *New Engl. J. Med.* **327** (1992) 749-754.

[5] R.W. Friesen, C. Brideau, C. Chan, S. Charleson, D. Deschenes, D. Dube, D. Ethier, R. Fortin, J. Gauthier, Y. Girard, R. Gordon, G.M. Greig, D. Reindeau, C. Savoje, Z. Wang, E. Wong, D. Visco, J. Xu, R.N. Young, *Bioorg. Med. Lett.* **8** (1998) 2777-2782.

[6] R.M. Singh, Y. Kumar, D.K. Sharma, S.C. Mathur, G.N. Singh, T.A. Ansari, S. Jamil, *Indian. Drugs* **42(8)** (2005) 535-536.

[7] B.N. Suhagia, H.M. Patel, S.A. Shah, I.S. Rathod, B.P. Marolia, *Indian J. Pharm. Sci.* **67(5)** (2005) 634-637.

[8] A. Craggs, G.J. Moody, J.D.R. Thomas, *J. Chem .Edu.* **51** (1974) 541-544.

[9] Y. Umezawa, K. Umezawa, H. Sato, *Pure. Appl. Chem.* **67(3)** (1995) 507-518.

[10] Y. Umezawa, Ph.B. Hlmann, K. Umezawa, K. Tohda, Sh. Amemiya, *Pure. Appl. Chem.* **72** (2000) 1851-2082.

[11] E. Baumann, *Anal. Chem. Acta* **42** (1968) 127-132.

[12] S. TNVSS, D Aditya, T. Santosh, B. S. Sundar, IAJPR **3(4)(**2013) 3044-3056.

[13] D. G. Peters, J. M. Hayes, G. M. Hieftje, *Chemical Separations and Measurements.* W.B. Saunders, Philadelphia, 1974, pp 835.

[14] Y. M. Issa, M. M. Hassouna, F. M. Abdel-Gawad, E. M. Hussien, *J. Pharm. Biomed. Anal.* **23(2-3**) (2000) 493-502.

[15] R. I. Stefan, H. Y. Aboul-Enein, G. E. Baiulescu, *Sens. Actuators B* **37** (1996) 141-144.

[16] M. Arvand, M. Vejdani, M. Moghimi, *Desalination* **225** (2008) 176-184.

[17] K. Shah, A. Gupta, P. Mishra, *J. Chem* **6(1***)* (2009) 207-212.

[18] P.G. Sunitha, N. Deattu, R. Ravi Kumar, S. Sri Rudhra, P. Kalaimathi, B. Soundiramani, A.C.M.S **(1)** (2015) 36-38.

[19] R. Chandra, A. Sanghi, D. Kumar, K.K. Hindwan, *British Biomedical Bulletin* **2 (4)** (2014) 706-714.

[20] S. Gholve, O. Bhusnure, O. Mathew1and J. Sangshet, *Int. J. Pharm. Bio. Sci.* **7(2)** (2016) 246–253.

14

Electrocoagulation of whey acids: anode and cathode materials, electroactive area and polarization curves

Francisco Prieto-García[1], Judith Callejas-Hernández[1], Judith Prieto-Méndez[2,✉], Yolanda Marmolejo-Santillán[1]

[1]*Área Académica Química, Universidad Autónoma del Estado de Hidalgo. Carret. Pachuca-Tulancingo, km. 4.5, C. P. 42076. Mineral de la Reforma. Hidalgo. México*
[2]*Área Académica de Agronomía, Universidad Autónoma del Estado de Hidalgo. Carret. Pachuca-Tulancingo, km. 4.5, C. P. 42076. Mineral de la Reforma. Hidalgo. México*

✉Corresponding author: jud_292003@yahoo.com.mx

Abstract

Anode (Al and Fe) and cathode (graphite and Ti/RuO_2) materials have been tested for electrocoagulation (EC) and purification of the acid whey. The electroactive areas (EA) of electrodes were calculated by the double layer capacitance method. Experiments were performed by cyclic voltammetry, chronoamperometry and polarization experiments. Among cathodic materials, the Ti/RuO_2 electrode showed higher EA (2167 cm^2) than graphite (1560 cm^2). The Fe anode was found more stable than Al with greater charge transfer carried out in less time. Correlation of these results with those obtained during preliminary tests confirmed high removals (79 %) in 8 h. For the Al electrode, 24 h were required to achieve efficiency of 49 %.

Keywords

Electrocoagulation, electrode materials, acid whey, electroactive area, capacitance, Tafel slopes

Introduction

The electrodes are indispensable components of any electrochemical cell because just on their surface the processes of oxidation-reduction of the species occur together with other reactions carried out during electrolysis [1,2]. When an electrocoagulation (EC) process is applied, the electrodes must be cost-effective and easy to handle. In addition, the electrodes used as sacrificial anodes must also possess great dissolution power. The most frequent materials used for EC processes of different effluents with good removal efficiencies are aluminum (Al) and iron (Fe), but other materials such as steel [3,4] have also been tested.

In the present study, different materials acting as anodes (Al and Fe) and cathodes (graphite and Ti/RuO_2) are tested for the EC and purification of residual acid whey. The evaluation is performed with respect to the electroactive electrode area and through polarization curves.

Methodology

Determination of electroactive area (EA) of electrodes

Determination of real electroactive area (EA) was done for Al, Fe, graphite (G) and Ti/RuO_2 electrodes. The cyclic scanning voltammetry technique, exploring the BASI Epsilon EC potentiostat 2.13 USB was used. Experiments were performed in 0.1 M KCl solution acidified with HCl to maintain pH 4.7, what simulates the initial pH of the acid whey [5,6]. The electroactive area of each electrode surface was calculated by the capacitance of the double layer, C_{dl}, generated on the electrode in contact with an electrolyte [7].

A glass beaker (2 L) containing 0.1 M KCl solutions was used as the electrochemical cell in which the electrodes were immersed (Fig. 1). The electrodes were submerged to a height of 5 cm, exposing a geometric area of approximately 100 cm^2 to the solution. The dimensions of electrodes were as follows: Al (10.2 × 10 × 0.1 cm), Fe (10.1 × 10 × 0.2 cm), G (8.2 × 10 × 1.2 cm) and Ti/RuO_2 mesh (10.5 × 10 × 0.1 cm). Titanium (Ti) mesh electrode (10 × 10 × 0.1 cm) and the saturated calomel electrode (SCE) served as the counter and reference electrode, respectively.

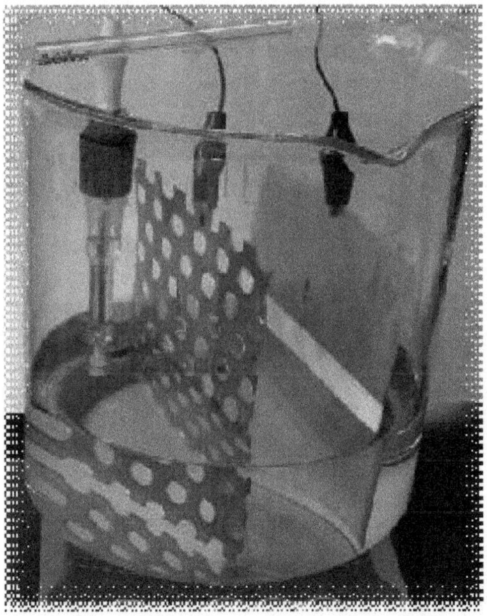

Figure 1. *Glass cell containing Ti mesh counter and SCE reference electrodes used for determination of electroactive area of working electrodes (Al, Fe, G, Ti/RuO_2)*

Cyclic voltammetry experiments were performed at different scanning rates, V = 10–100 mV s^{-1} in a potential window of ± 10 mV around the equilibrium potential (E_0). Anodic (i_a) and cathodic (i_c) currents were obtained as electrochemical responses to the potential change. The capacitive current (i_{cap}) was calculated by means of equation (1), while C_{dl} value was calculated as the slope of linear relation obtained by presenting i_{cap} *vs. V.* The EA, is obtained by means of equation (2), where C_{st} is the normal capacitance taken as C_{st} = 60 μF cm^{-2} for roughened electrodes [7].

$$i_{cap} = \frac{i_a + i_c}{2} \tag{1}$$

$$EA = \frac{C_{dl}}{C_{st}} \quad (2)$$

Measurements of polarization curves

With the polarization curves it is possible to determine the electrode of the highest oxidation capacity attained with consumption of the lowest energy. The potentioamperometric technique was explored using a sample of acid whey (30 mL) and geometrically smaller electrodes. The electrodes were submerged to a height of 2.0 cm and the exposed geometric areas were 6.96 cm^2 (Al), 8.78 cm^2 (Fe), 4.65 cm^2 (G) and 8.6 cm^2 (Ti/RuO_2). The real electrode areas were determined following the methodology described above. Three-electrode glass cell was used and shown in Figure 2a, where the working electrode is the G electrode, while Ti mesh and SCE served as the auxiliary and reference electrodes, respectively. Figure 2b shows a three-electrode cell containing 30 mL of acid whey used for chronoamperometric experiments, performed by applying different overpotentials, η, defined by the equation (3) for 45 minutes.

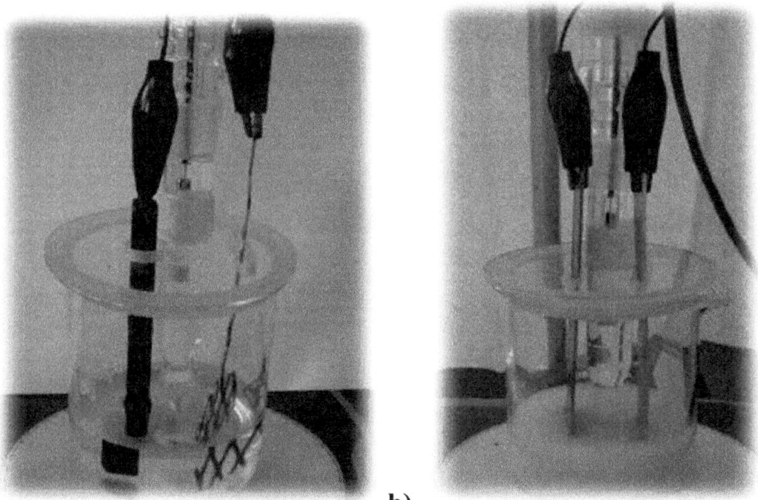

a) b)

Figure 2. *Three-electrode glass cell used for determination of electroactive area of small electrodes (G, Ti-mesh and SCE are the working, counter and reference electrodes) a). Electrolytic cell containing 30 mL of acid whey for polarization measurements (Al, Ti-plate and SCE are the working, counter and reference electrodes) b)*

Constant stirring is maintained during the test to obtain the current (i) in response. The current density (j) is calculated by the equation (4).

$$\eta = E_{applied} - E_0 \quad (3)$$

$$j = i / EA \quad (4)$$

Results and discussion

Determination of electroactive area (EA) of electrodes

Cyclic voltammograms of Al, Fe, G and Ti/RuO_2 electrodes in 0.1 M KCl recorded at different scan rates (v) are together with i *vs.* υ curves shown in Figures 3–6. Numerical data for i_{cap} determined by eq. (1) are collected in Table 1. Contrary to Figures 3 and 4 that showed ordinary capacitive current responses of G and Ti/RuO_2 electrodes, in the case of Al and Fe electrodes presented in Figures 5 and 6, the tendency is quite different. Lack from usual linear i_{cap} *vs.* v response is particularly obvious for the Fe electrode and is probably due to the dissolution of electrode in acid medium.

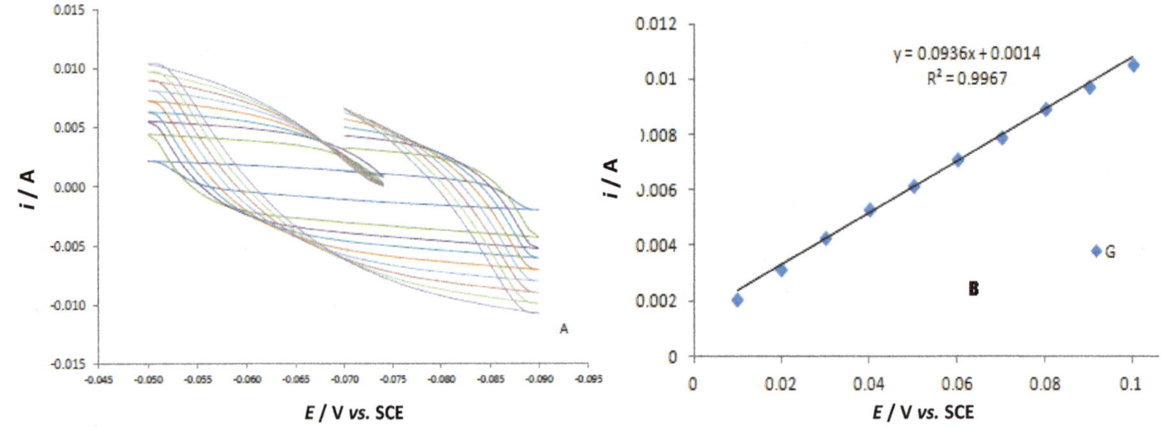

Figure 3. *Cyclic voltammograms of graphite electrode (G) in 0.1 M KCl, at different scanning rates (v = 0.01 to 0.10 V s^{-1}) (A).* i_{cap} *vs. v (B)*

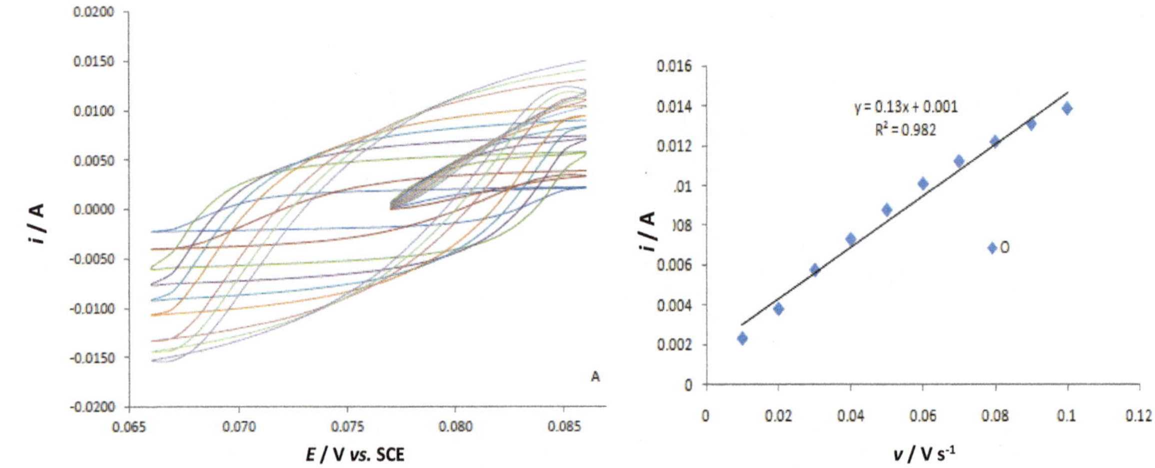

Figure 4. *Cyclic voltammograms of Ti/RuO_2 electrode in 0.1 M KCl at different scanning rates (v= 0.01 to 0.10 V s^{-1}) (A).* i_{cap} vs. *v (B)*

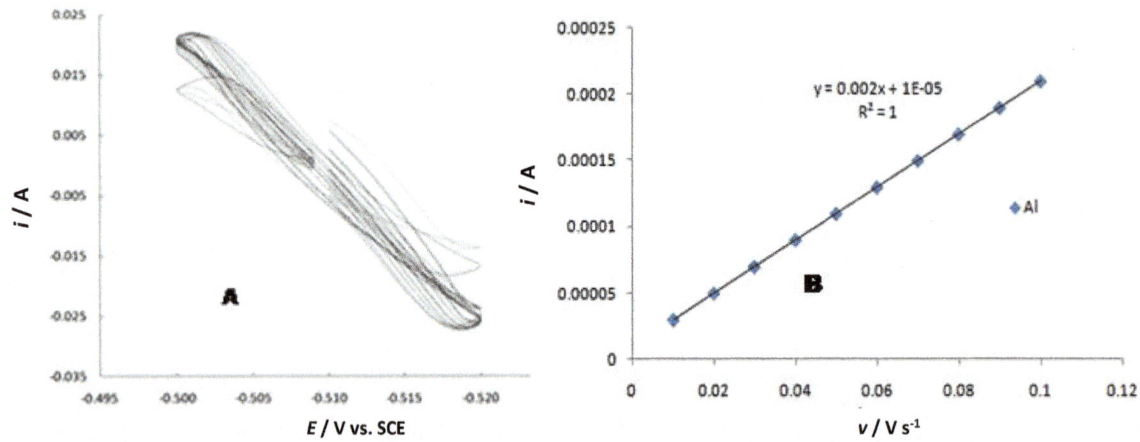

Figure 5. *Cyclic voltammograms of Al electrode in 0.1 M KCl at different scanning rates (v = 0.01 to 0.10 V s^{-1}) (A).* i_{cap} vs. v *(B)*

The electroactive area (EA) of the working electrode represents the number of active sites where the electron transfer takes place and is usually estimated as the real electrode area (Areal) defined by eq. (2). EA can increase or decrease with respect to the geometric area, what is mainly due to the modification of the electro-active electrode surface. When an EC process is carried out, the anode (Al or Fe) undergoes oxidation reaction, which leads to the release of metal ions (sacrificial anode) and concomitant dissolution process [4,8-11]. causing diminishing of the surface area. At the other hand, increase of the EA could also happen due to the porosity induced by the EC process.

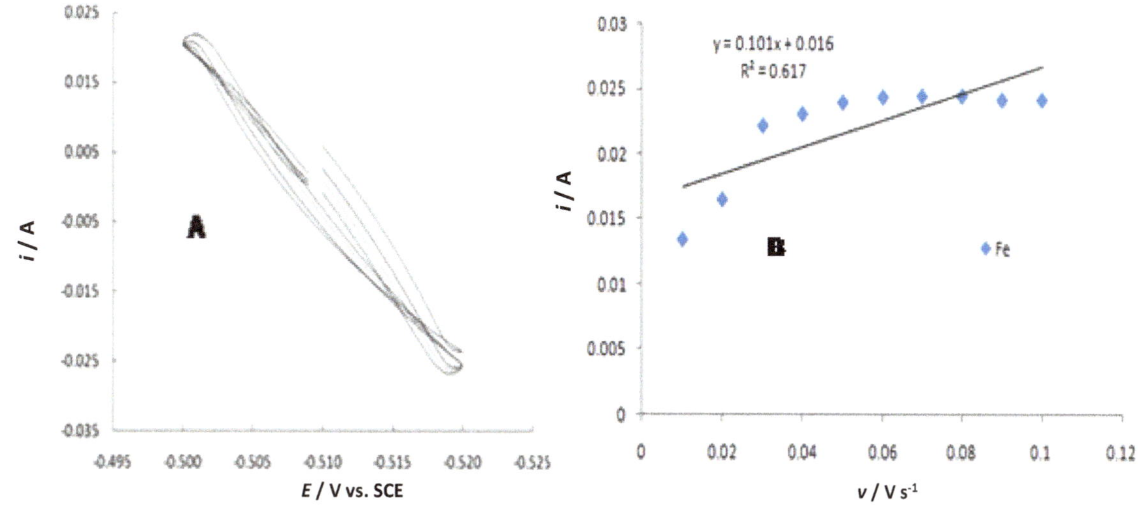

Figure 6. *Cyclic voltammograms of Fe electrode in 0.1 M KCl at different scanning rates (v = 0.01 to 0.10 V s*$^{-1}$*) (A).* i_{cap} vs. v *B)*

Table 1. *Capacitive currents (*i_{cap}*) for different EC electrodes at different scanning rates* v

V / V s^{-1}	i_{cap} / A			
	G	Ti/RuO_2	Al	Fe
0.01	0.00209	0.00229	0.00003	0.0134
0.02	0.00318	0.00378	0.00005	0.0165
0.03	0.00434	0.00573	0.00007	0.0222
0.04	0.00538	0.00728	0.00009	0.0231
0.05	0.00623	0.00874	0.00011	0.0240
0.06	0.00718	0.01007	0.00013	0.0244
0.07	0.00799	0.01121	0.00015	0.0245
0.08	0.00899	0.01218	0.00017	0.0245
0.09	0.0098	0.01308	0.00019	0.0242
0.10	0.01055	0.01385	0.00021	0.0242

Double-layer capacitance (C_{dl}) and EA values, calculated using data in Table 1 and eq. (2), are summarized in Table 2.

Table 2. *Real area (EA) and capacitance (*C_{dl}*) of different electrodes for the electrocoagulation process*

Electrodes	Equation	R^2	C_{dl} / F	EA, cm^2	GA*, cm^2	Roughness factor
G	y=0.0936x+0.0014	0.9967	0.0936	1,560	103.84	15.03
Ti/RuO_2	y=0.130x+0.0010	0.9821	0.130	2,167	107.05	20.24
Al	y=0.002x +10^{-5}	1.0000	0.0020	33.3	104.02	0.32
Fe	y=0.101x +0.016	0.617	0.1010	1,689	105.02	16.1

GA* - Geometric area

It is obvious from Table 2 that Ti/RuO_2 electrode showed the largest EA (2,167 cm^2), what is due to the RuO_2 metal oxide that generates micro, meso and macropores, and so greater surface roughness and higher capacitance (C_{dl}) values were obtained. At the same time, Al electrode showed the smallest EA (33.3 cm^2) because it is a smooth and polished metal plate that exhibits the lowest C_{dl} among all electrodes. It has generally been stated that for cathodes, greater EA is generally beneficial. Data in Table 2 show that for Ti/RuO_2 and G electrodes, EA are 2,167 cm^2 and 1,560 cm^2, respectively. This makes the Ti/RuO_2 electrode to be better cathode under the perspective of electroactive areas. Also, regarding the electroactive area, the Fe electrode having EA of 1,689 cm^2 seems better than the Al electrode having 33.3 cm^2. Although being indicative, the value of EA, however, is not a definitive factor for the choice of the most proper electrodes for EC.

Choice of better arrangement of electrodes by means of polarization curves

Aluminum electrode

Chronoamperometric, *i vs. t* responses of the Al electrode in acid way sample recorded during 45 min are presented in Figures 7 and 8. Similar behavior is observed at different potentials (E_{app}) applied between −0.650 and 1,050 mV. As E_{app} was increased, *i* was increased in the same way. However, at very high E_{app} (850 and 1,050 mV) the decomposition of the medium is already occurring, which leads to unnecessary energy consumption. Figure 9 shows *i vs. E_{app}* responses for the Al electrode (EA = 33.3 cm^2) in the acid whey sample. It is obvious from Fig. 9 that the limiting current (i_l) where diffusional control begins is reached at about 400 mV *vs.* SCE and that diffusional currents are in a range of 0.039 and 0.042 A. Despite application of higher voltages, there are no greater transformations of the species present in the medium, up to the potential suitable for secondary reactions like the electrolysis of the water. Figure 10 shows the polarization curve, log *j vs.*η, of the Al electrode. In this graph, it is observed that the limiting current is reached by diffusional control at overpotential of 200 mV. This suggests that from the results obtained previously during the EC with a removal efficiency of 79 % in 8 h, the energy required can be reduced (by decreasing η from 1.74 to 0.2 V) to obtain the same removal efficiency and possible reduction of time.

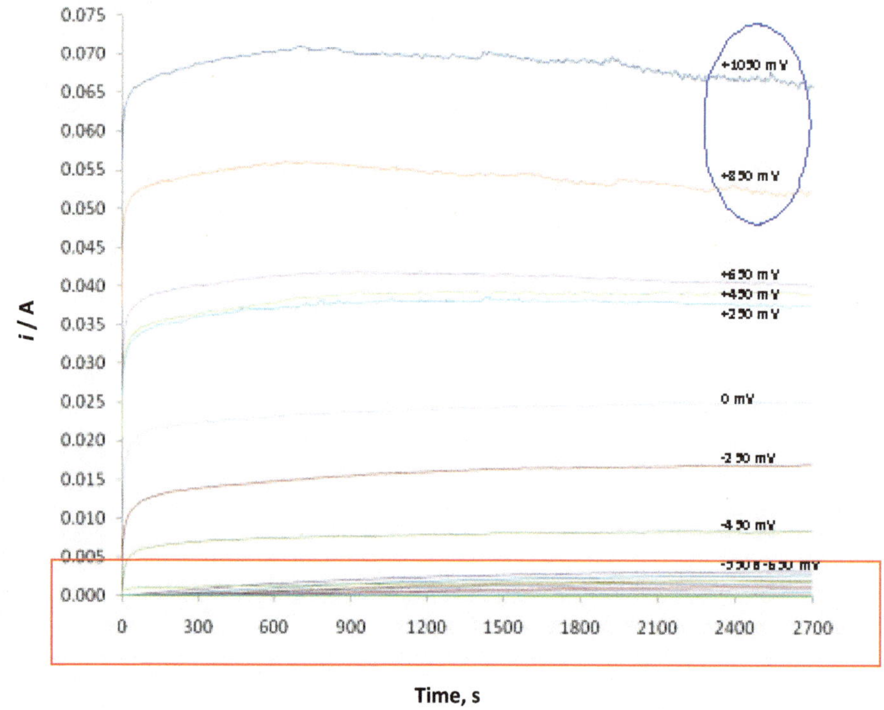

Figure 7. *Chronoamperometric* i vs. t *responses of Al electrode in acid whey sample recorded during 45 minutes at* E_{app} *between −650 mV and +1,050 mV*

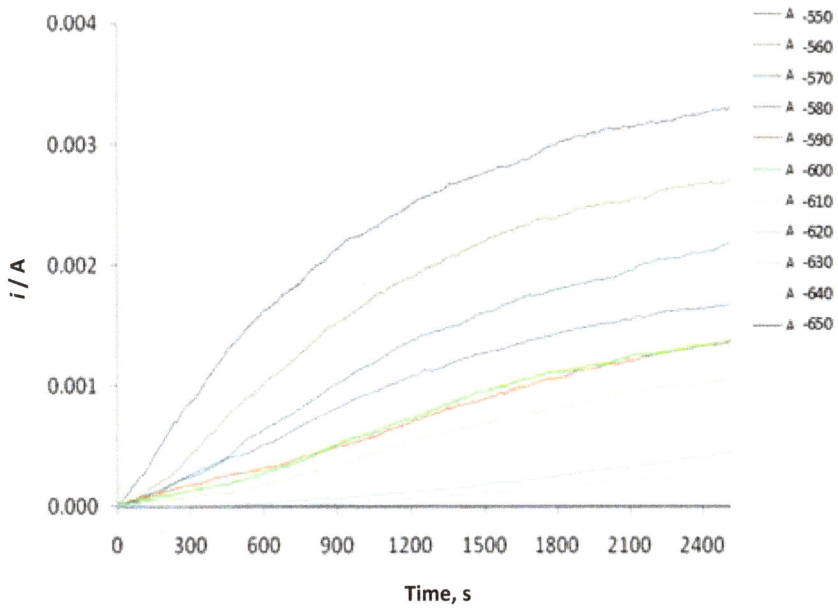

Figure 8. *Chronoamperometric* i vs. t *responses of Al electrode recorded at* E_{app} *between –650 mV and –550 mV (zoom of Fig. 7)*

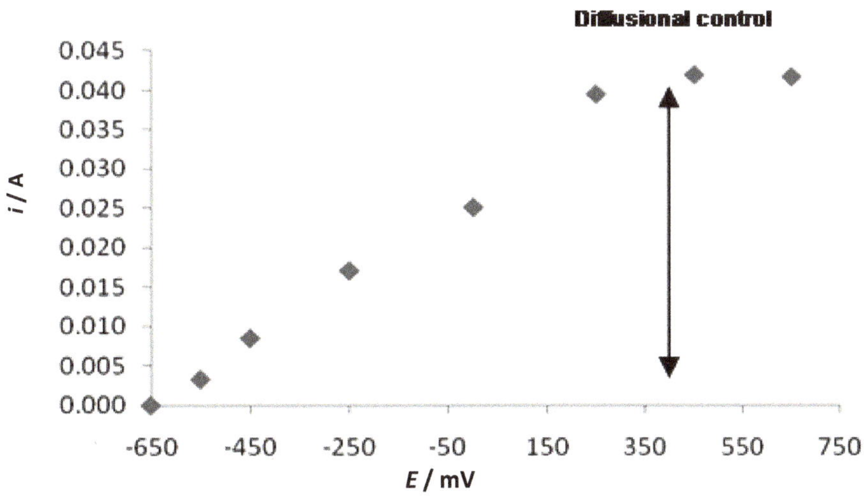

Figure 9. i vs. E_{app} *of Al electrode in acid whey sample at* E_{app} *between –650 mV and +650 mV*

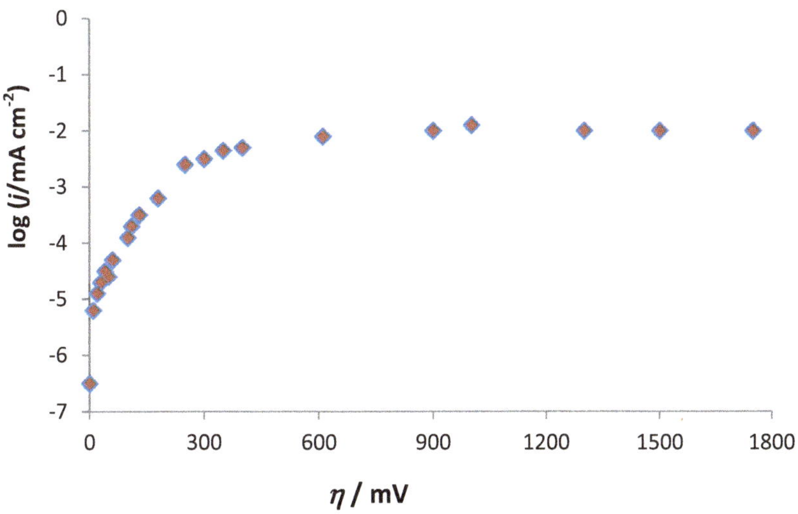

Figure 10. *log* j vs.η *dependence of Al electrode in acid whey sample*

Iron electrode

The chronoamperometric, *i vs. t* curves of the Fe electrode in acid whey sample recorded during 45 min are shown in Figures 11 and 12. Almost constant current values observed with time at each E_{app} applied between −650 and −150 mV suggest that only capacitive currents are generated at the Fe electrode in contact with the acid whey solution. When applying E_{app} between 0 to 100 mV, however, the first change in the oxidation state of Fe° to Fe^{2+} is observed with transfer of $2e^-$ coupled with the decomposition of the medium. At E_{app} between 100 to 450 mV, a second change in current is observed due to the electrochemical oxidation of Fe^{2+} to Fe^{3+}, so that the oxidation of iron is considered to occur in two stages (eqs. 5 and 6):

$$Fe^\circ \rightarrow Fe^{2+} + 2e^- \quad (5)$$

$$Fe^{2+} \rightarrow Fe^{3+} + 1e^- \quad (6)$$

Finally, by applying E_{app} higher than 450 mV the electrolysis of the medium that is decomposition of the whey begins, what is accompanied by the significant current increase with time. *i vs. E_{app}* and log *j vs. η* dependences are presented in Figures 13 and 14. Here also, two different processes which are related to iron oxidation can be observed as in the chronoamperometry experiments. With these polarization curves it is verified that in the EC system, generation of Fe^{2+} or Fe^{3+} can be favored to improve the efficiency of the system in a shorter time and lower energy cost.

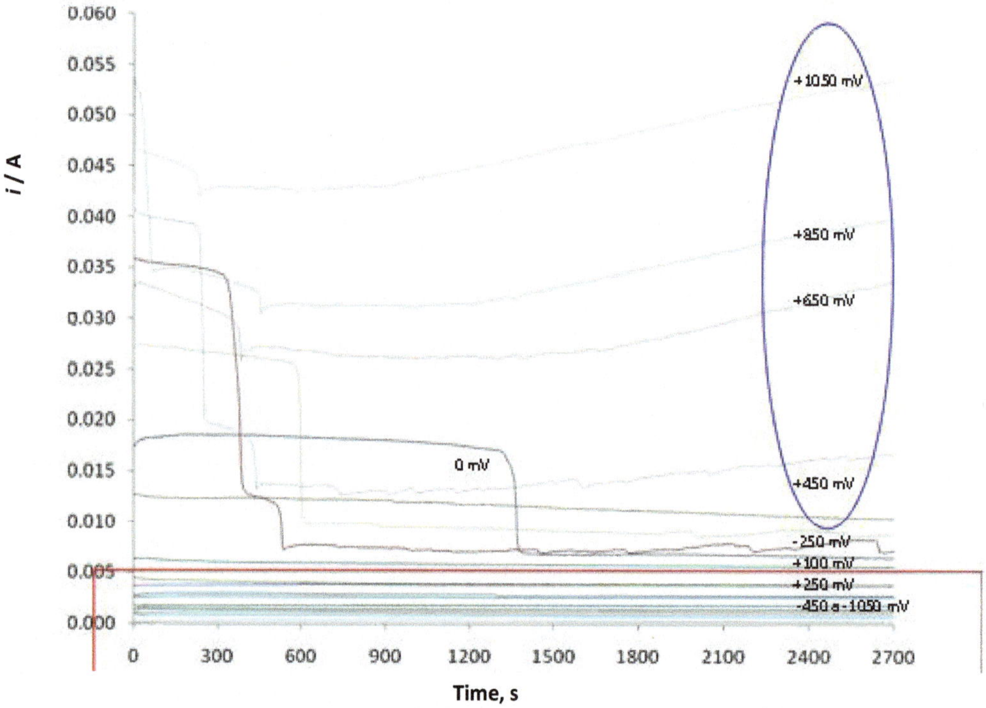

Figure 11. *Chronoamperometric* i vs. t *responses of Fe electrode in acid whey sample recorded during 45 minutes at* E_{app} *between −1050 mV and +1050 mV*

The above statement can be based on the Pourbaix diagram for the thermodynamic states of the iron in aqueous solution, where when applying potentials within the range of 0.6 to 1.6 V *vs.* SCE, the oxidation process occurs and Fe^{2+} or Fe^{3+} ions are released depending on the pH of the solution. For acid conditions, both species are present and the process of corrosion is facilitated. It should be analyzed in more detail, however, which reaction would favor the EC process for the whey solution and formation of the iron oxy-hydroxides. Thus, one would be able to establish the optimum limiting current according to the limiting reagent that controls the reaction.

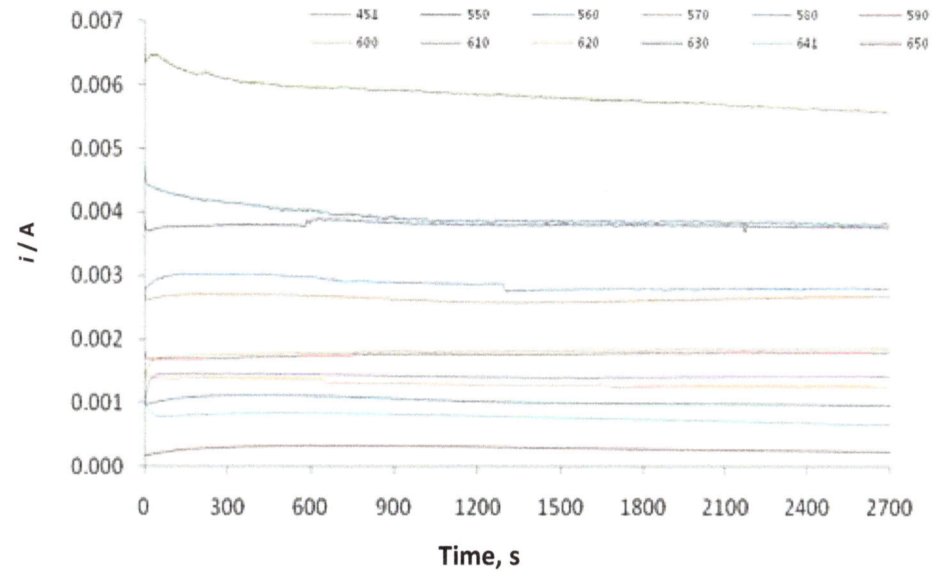

Figure 12. *Chronoamperometric* i vs. t *responses of Fe electrode at* E_{app} *between –650 mV and –450 mV (zoom of Fig. 11)*

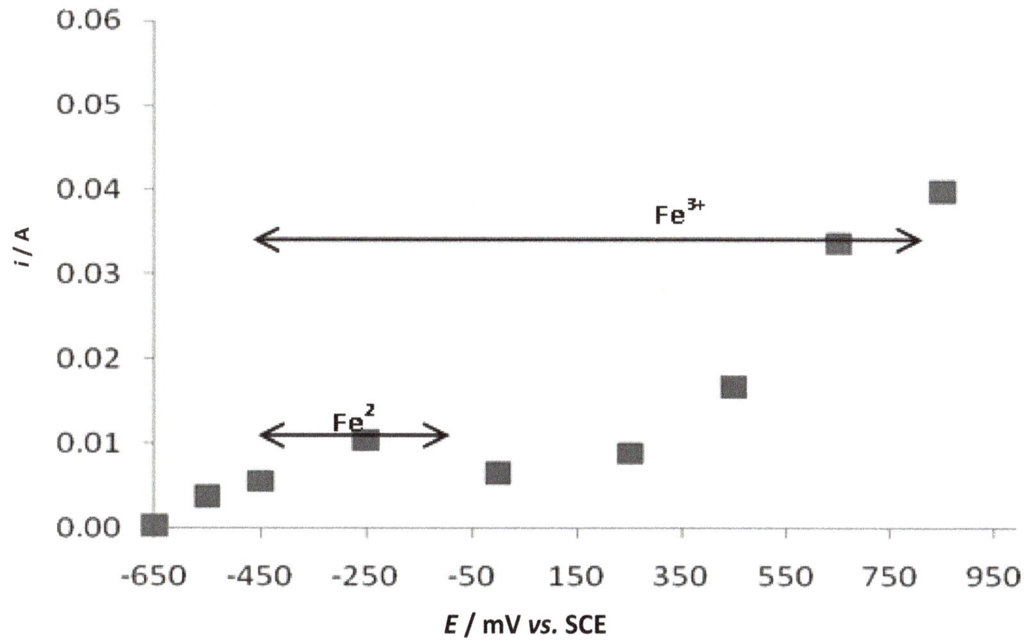

Figure 13. i vs. E_{app} *of Fe electrode in acid whey sample at* E_{app} *between –650 mV and 850 mV*

The bibliographical reports on the EC processes using iron electrodes are contradictory. Most EC studies have focused on the removal of contaminants by their co-precipitation with the solid(s) such are already formed precipitates of $Fe(OH)_2(s)$ and $Fe(OH)_3(s)$. Also, the reactions reported in the literature showed a certain ambiguity in the proposed mechanisms. Some studies [13-15] have reported that the oxidation of iron Fe^0 to Fe^{2+} occurs in an electrolytic form and produces insoluble $Fe(OH)_2(s)$ compounds by hydrolysis. Other authors [16] pointed to the formation of $Fe(OH)_2(s)$ without giving details of how it occurs. Several studies [8,17,18] reported formation of Fe^{2+} at the anode that is followed by oxidation to Fe^{3+} due to dissolved oxygen (DO) and rapid formation of $Fe(OH)_3(s)$. The third mechanism reported [7,8,10] involves the single and fast electrolytic oxidation step of Fe^0 to Fe^{3+}, followed by hydrolysis to produce $Fe(OH)_3(s)$. Finally, some studies [16,19,20] have reported formation of $Fe(OH)_3(s)$ as the final product during EC without specifying the reactions involved. According to [21], the third mechanism is questionable. These authors pointed out dependence of the kinetics of these reactions on the pH of the medium. They reported that at pH 7.5, approximately 80% of the Fe^{2+} initially formed is oxidized to Fe^{3+} in less than 10 minutes, while at pH 6.5, the same oxidation process takes about 300 minutes (5 h). The rapid increase in Fe^{2+} oxidation rate with pH increase is consistent with other reports in the literature, where based on equations (7) and (8), it was shown that increase of pH increases the oxidation rate of Fe^{2+} for about 100 times [22–24].

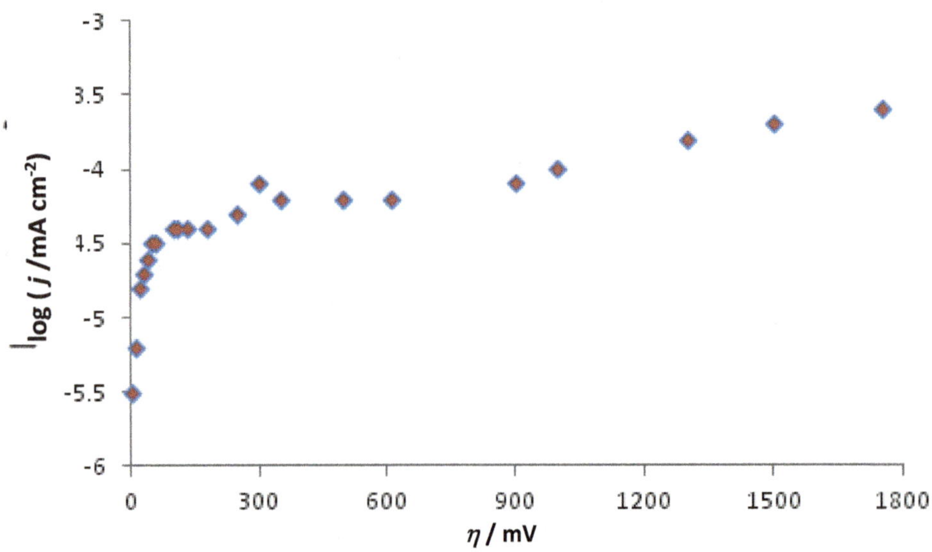

Figure 14. *Log* j vs.η *dependence of Fe electrode in acid whey sample.*

$$4Fe^{2+} + O_2 + 2H_2O + 8OH^- \leftrightarrow 4Fe(OH)_3(s) \quad (7)$$

$$d[Fe^{2+}]/dt = -k[Fe^{2+}]\, P_{O_2}\, [OH^-]^2 \quad (8)$$

In eq. (8), k is the rate constant, P_{O2} is the oxygen partial pressure, while the molar concentrations of present species are represented in the brackets.

A slight increase in the concentration of OH^- ions as observed during EC and corroborated with the increase of pH, has resulted in a greater concentration of the Fe^{2+} species forming Fe^{3+} by subsequent oxidation and $Fe(OH)_3(s)$ by precipitation. Since OH^- ions are consumed during the hydrolysis of Fe^{3+}, pH in the electrolytic cell decreases in a subsequent step that does not occur immediately. By the time elapsed between the production of Fe^{2+} and its oxidation to Fe^{3+} by the oxygen dissolved in the medium, a reaction that is much slower has been indicated.

Graphite electrode

In Figures 15 and 16, the chronoamperometric *i vs. t* responses of the G electrode recorded through 45 min are presented, showing the same behavior at different E_{app} applied. As E_{app} increases, *i* increased in the same way. At very negative E_{app} (–1050 mV), however, it can be observed that the decomposition of the medium occurs. By presenting log *j* vs. η dependence, Fig. 17 suggests that the limiting current (i_l) is occurring at overpotential of –500 mV, where diffusional control begins.

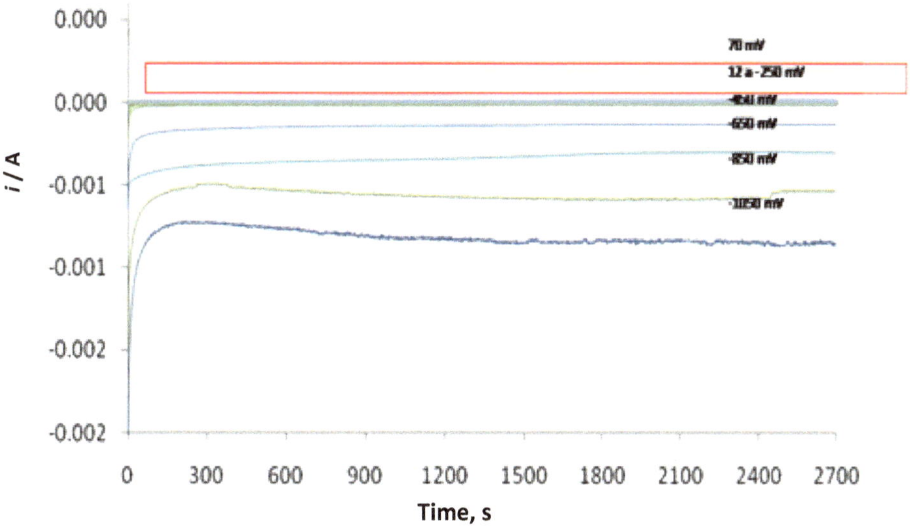

Figure 15. *Chronoamperometric* i vs. t *responses of G electrode in acid whey sample recorded during 45 min at* E_{app} *between 70 mV and –1050 mV.*

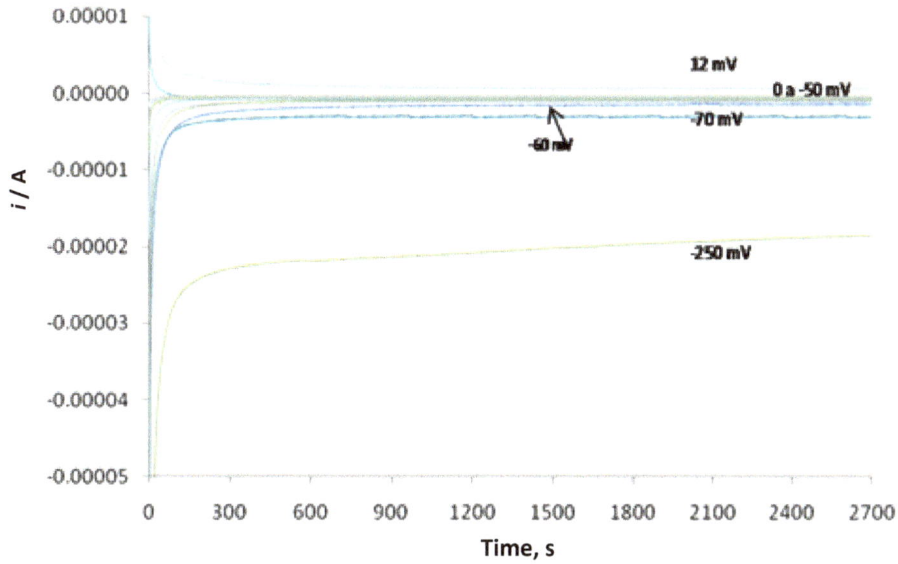

Figure 16. *Chronoamperometric* i vs. t *responses of the G electrode at* E_{app} *between 12 mV and –250 mV (zoom of Fig. 15)*

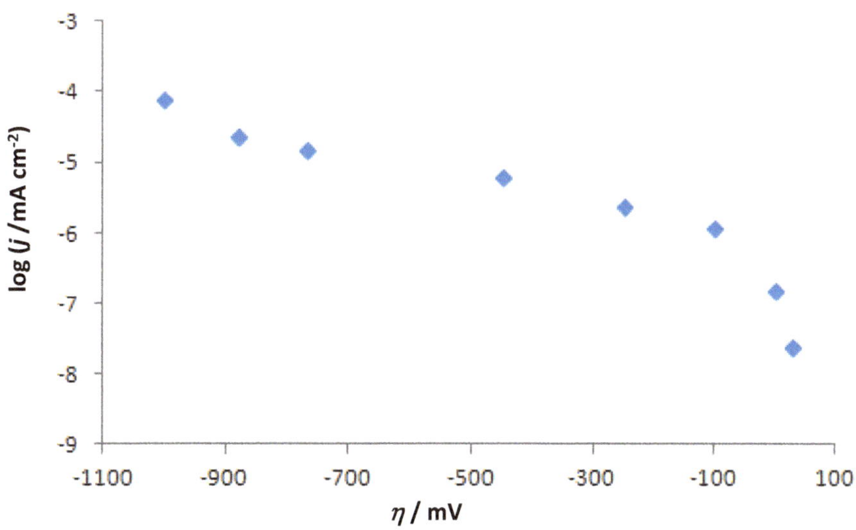

Figure 17. *log* j vs. η *dependence of G electrode in acid whey sample.*

Selection of the anodic material for the electrocoagulation process.

With the electrochemical information obtained so far, the Tafel slopes for the electrodes acting as sacrificial anodes (Al and Fe) are compared in Figure 18.

To select the best electrode, the largest slope must be taken into account. The results for Fe and Al electrodes are different, for the Al electrode, b = -54.55 mV dec^{-1} and for Fe, b = -109.98 mV dec^{-1}. In this respect, it is possible to define that the Fe electrode is considered the most effective. Figure 19 shows a better separation of the precipitate and purified liquid more transparent with the use of the Fe electrode (Figure 19 a).

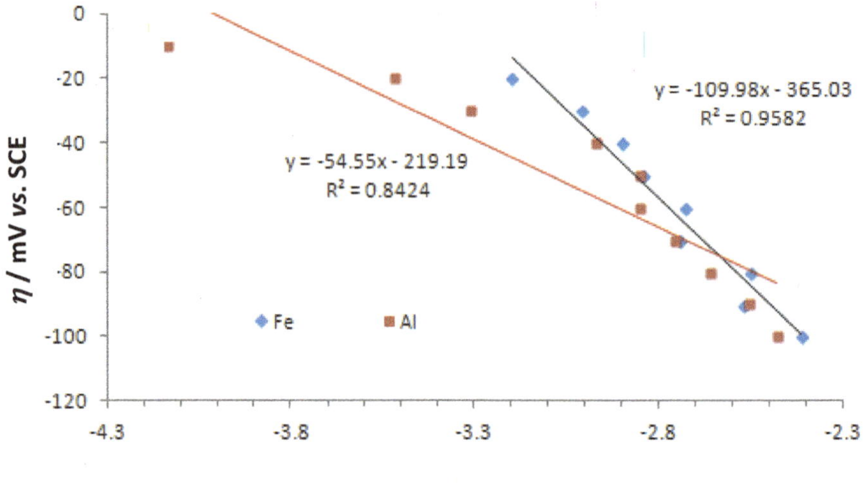

Figure 18. *Tafel slopes of Fe and Al electrodes in acid whey samples*

a)

b)

Figure 19. *Purification kinetics of acid whey with Fe (a) and Al (b) electrodes*

Conclusions

The iron electrode proved to be better anode than Al for the EC process. This is due to its stable thermodynamic behavior in the range of applied overpotentials, which translates into a greater transfer of charge in a shorter time. By correlating these results with the results obtained during the preliminary tests, it can be confirmed that the Fe electrode is better than the Al electrode. With the Fe electrode, greater removals (79 are obtained in less time (8 h)). On the other hand, with the Al electrode, the results of macroelectrolysis required more time (24 h) to achieve an efficiency of 49 %. Based on the determinations of the electroactive area, the Ti/RuO_2 electrode (2,167 cm^2) proved to be a better cathode than the graphite (1,560 cm^2).

References

[1] M. Piña, A. Martín, C. A. González, F. Prieto, A. Guevara, J. E. García. *Revista Mexicana de Ingeniería Química* **10(2)** (2011) 257-271.

[2] A. P. Restrepo, A. Arango, L. F. Garcés. *Producción + Limpia* **1(2)** (2006) 58-77.

[3] E. Ojeda, R. Hing, Y- González. *Revista CENIC Ciencias Químicas* **43(1)** (2912), 21-42.

[4] S. Tchamango, E. Nanseu-Njiki, D. Ngameni, A. Hadjiev, A Darchen. *Science of the Total Environment* **408** (2010) 947–952.

[5] P. Panesar, J. Kennedy, D. Gandhi, K. Bunko, *Food Chemistry* **105** (2007) 1-14.

[6] A. Abaigar, *ITG Ganadero* **94(8)** (2009) 13-17.

[7] N, Balasubramanian, K. Madhavan, *Chemical Engineering Technology* **24** (2001) 519–521.

[8] M. Y. Mollah, P. G. Morkovsky, A. G. Gomes, M. Kesmez, J. Parga, *Journal of Hazardous Materials* **114(1-3)** (2004) 199–210.

[9] O. T. Can, M. Bayramoglu, M. Kobya. *Industrial Engineering Chemistry Research* **42** (2003) 3391–6.

[10] M. Kobya, O. T. Can, M. Bayramoglu. *Journal of Hazardous Materials* **B1001** (2003) 63–78.

[11] H. Inan, A. Dimoglo, H. Simsek, M. Karpuzku. *Separation and Purification Technology* **36** (2004) 23–31.

[12] G. Chen, *Separation and Purification Technology* **38(1)** (2004) 11–41.

[13] V. P. Osipenko, P. I. Pogorelyi. *Metallurgist* **21(9-10)** (1977) 44–45.

[14] M. Y. Mollah, R. Schennach, J. R. Parga, D. L. Cocke. *Journal of Hazardous Materials* **84(1)** (2001) 29–41.

[15] R. Parga, D. L. Cocke, V. Valverde. *Cheical Engineering Technology* **28(5)** (2005) 605–612.

[16] D. Ghernaouta, A. Badisa, A. Kellila, A. *Desalination* **219(1-3)** (2008) 118–125.

[17] X. M. Chen, G. Chen, P. L. Yue. *Separation and Purification Technology* **19(1)** (2000) 65–76.

[18] H. K. Hansen, P. Núñez, R. Grandon. *Minerals Engineering* **19(5)** (2006) 521–524.

[19] D- Mills, *Journal American Water Works Association* **92(6)** (2000) 34–43.

[20] P. R. Kumar, S. Chaudhari, K. Khilar, S. P. Mahajan. *Chemosphere* **55(9)** (2004) 1245–1252.

[21] D. Lakshmanan, D. Clifford, and S. Gautam. *Environmental Science Technology* **43** (2009) 3853–3859.

[22] P. C. Singer, W. Stumm. *Journal American Water Works Association* **62** (1970) 198–202.

[23] W. Stumm, S. Morgan, *Journal. Aquatic Chemistry*; 1981. Wiley-Interscience, New York.

[24] M. Bond, S. Kratsis and S. Mitchell. *Journal Analyst* **122** (1997) 1147-1152.

15

Electrocoagulation removal of anthraquinone dye Alizarin Red S from aqueous solution using aluminum electrodes: kinetics, isothermal and thermodynamics studies

Abideen Idowu Adeogun*,**,✉, Ramesh Babu Balakrishnan**

**Department of Chemistry, Federal University of Agriculture, Abeokuta, Nigeria*
***Electrochemical Pollution Control Division, CSIR - Central Electrochemical Research Institute. Karaikudi 630006, India*

✉Corresponding Author: abuaisha2k3@yahoo.com

Abstract
Electrocoagulation (EC) was used for the removal of anthraquinone dye, Alizarin Red S (ARS) from aqueous solution. The process was carried out in a batch electrochemical cell with Al electrodes in a monopolar connection. The effects of some important parameters such as current density, pH, temperature and initial dye concentration, on the process were investigated. Equilibrium was attained after 10 minutes at 30 °C. Pseudo-first order, pseudo-second order, Elovic, and Avrami kinetic models were used to test the experimental data in order to elucidate the kinetics of the electrocoagulation process; pseudo-first order and Avrami models best fitted the data. Experimental data were analyzed using six isotherm models: Langmuir, Freudlinch, Redlich–Peterson, Temkin, Dubinin–Radushkevich and Sips isotherms and it was found that the data fitted well with Dubinin–Radushkevich and Sips isotherm model. The study showed that the process depended on the current density, temperature, pH and initial dye concentration. The calculated thermodynamics parameters (ΔG^o, ΔH^o and ΔS^o) indicated that the process is spontaneous and endothermic in nature.

Keywords
Batch; Electrochemical cell; Mnopolar connection; Pseudo first order; Equilibrium; Spontaneous.

Introduction

The advanced in technologies for industrialization and urbanization has substantially enhanced the degradation of the environment especially the aquatic environment, through the discharge of industrial wastewaters and domestic wastes [1,2]. The residual dyes from different sources such as: textile, paper and pulp, dye and dye intermediates, pharmaceutical, tannery, and kraft bleaching industries are considered as organic coloured pollutants [3–6]. These industries utilize large quantities of a number of dyes with residues leading to the large amount of coloured wastewaters, toxic and even carcinogenic, posing serious hazard to aquatic living organisms. Most dyes used in industries are stable to light, heat and oxidation, they are not biologically degradable and are also resistant to aerobic digestion and even when they does, they produce toxic and hazardous products [7,8].

Anthraquinone dye Alizarin Red S (ARS) is available as sodium salt of 1,2-dihydroxy-9,10-anthraquinonesulfonic acid (Fig. 1) and it is referred to as Mordant Red 3, C.I. no. 58005. Alizarin, a natural dye obtained from madder (*Rubiatinctorum, L. Rubiaceae*) by sulphonation, is a water-soluble and widely used anthraquinone dye in textile and as a stain in clinical study of synovial fluid to assess basic calcium phosphate crystals [9]. ARS is a durable pollutant when released to aquatic ecosystems. It cannot be completely degraded by general physicochemical and biological processes because of the complex structures of the aromatic rings that afford high physicochemical, thermal, and optical stability [7,10]. Therefore, most treatments for such dye-laden effluents are largely inadequate; however, removal of this dye from industrial wastewaters is a crucial process, from both economic and environmental points of view [11].

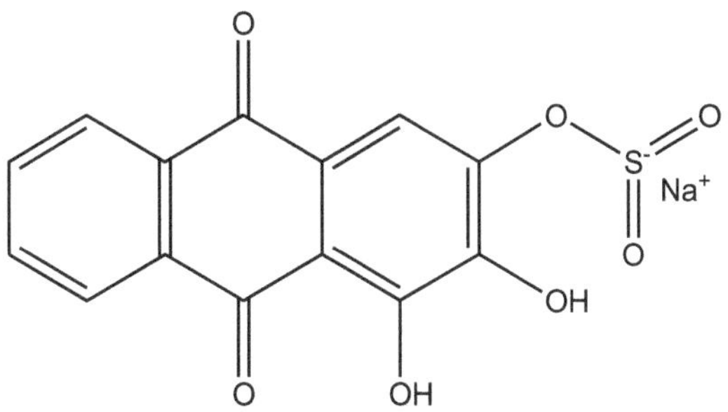

Figure 1. *Structure of Alizarin Red S (ARS)*

Electrochemical techniques which include electrooxidation, electrochemical reduction, electrocoagulation, electroflotation [12], have been developed for the treatment of organic pollutants in waste water with higher efficiency than any other biological, physical and chemical process [13,15]. Electrocoagulation has been known for some time as a process capable of fractionating a number of organic substances in a rather efficient manner. The coagulants are generated in-situ by electrooxidation of the anode. The mostly used anode materials are iron and aluminum because of their availability and relatively low cost. Electrocoagulation is accomplished in a three step processes as follows: (i) electrolytic reactions at surface of electrodes, (ii) formation of coagulants in aqueous phase and (iii) adsorption of soluble or colloidal pollutants onto coagulants and removal of them using sedimentation or flotation of flocs when hydrogen gas bubbles were produced at the cathode [16].

Several methods have been used for the removal of ARS from wastewaters, these include adsorption [17-19], electrochemical degradation [20-23] and electro-Fenton process [24] using various electrodes. However, to the best of our knowledge, kinetics and isothermal studies have not been elucidated in any of these studies.

In this study, ARS was removed from aqueous solution in a mono-polar electrochemical cell using aluminum electrodes. The effect of current density, initial dye concentration, electrolyte concentration, pH and temperature were studied. Adsorption kinetics of electrocoagulants was analyzed with pseudo-first order, pseudo-second order, Elovic, and Avrami kinetic models. The diffusion mechanism was analyzed with intraparticulate diffusion model while the equilibrium adsorption behaviour was analyzed by fitting the equilibrium data with six isotherm models. Thermodynamic parameters such as free energy (ΔG), enthalpy (ΔH) and entropy (ΔS) were also determined to understand the spontaneity of the electrocoagulation process.

The following reactions were envisaged at the electrodes:

Anodic reaction:

$$Al \rightarrow Al^{3+} + 2e^{-} \tag{1}$$

Cathodic reaction:

$$3H_2O + 3e^{-} \rightarrow 3OH- + 3/2H_2\ 3H_2O \tag{2}$$

Overall reaction:

$$Al^{3+} + 3OH^{-} \rightarrow Al(OH)_3 \tag{3}$$

Experimental

Dye solution preparation

Alizarin Red S (sodium alizarin sulphonate (Fig. 1), CI 58005, Product No. – 13005) was a product of British Drug House, Poole, England. 1000 mg L^{-1} aqueous solution of ARS was prepared with de-ionized water as the stock solution and was further diluted with de-ionized water to obtain the working standard solutions. The pH of the solution was adjusted when necessary with aliquots of 1.0 mol L^{-1} of HCl and NaOH before the commencement of the experiment. The conductivity of the solution was maintained with NaCl solution as the electrolyte.

Experimental apparatus and procedures

Figure 2 depicts the electrocoagulation cell which consists of a 0.6 L glass cell fitted with a polycarbonate cell cover with slots to introduce the electrodes, thermometer and electrolyte. The aluminum electrodes of dimensions 4.5×7×0.3 cm with inter electrode distance of 2 cm were fully immersed in the 0.5 L solution of the dye. A regulated direct current (DC) was supplied from a rectifier (0 – 2 A, 0 – 35 V; Applab 7711 multi-output). The temperature of the electrolyte was controlled to the desired value with a variation of ±1 °C by adjusting the temperature knob on the IKA RCT Basic magnetic hotplate stirrer and was allowed to equilibritate before the commencement of the experiment.

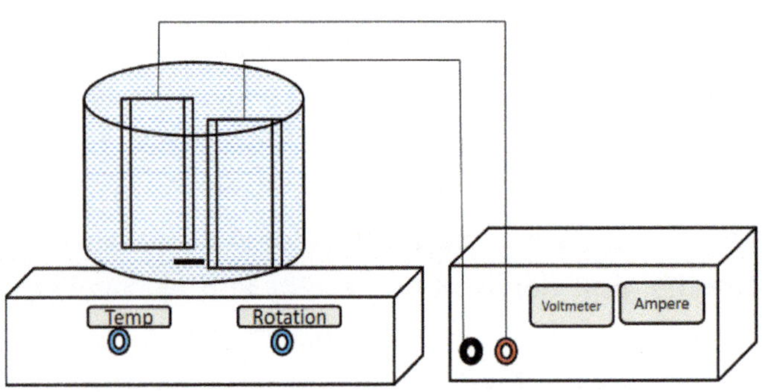

Figure 2. *Laboratory scale electrocoagulation cell*

Analytical procedure

The concentration of the dye in solution was estimated using spectrophotometer (UV-VIS –NIR VARIAN 500 Scan CARY). To compare the dye obtained from the coagulation with the original ARS dye, the flock is filtered and the residue dried at 105 °C for 30 minutes, then the FTIR of the two samples (ARS and residue) were obtained using FTIR spectrophotometer (TENSOR 27 Bruker Optik GmbH, Germany). Non-linear regression analysis method using a program written on Micro Math Scientist software (Salt Lake City, Utah) was used to obtain the least square fit for all the models.

Equilibrium studies

The effects of current density, initial dye concentration, electrolyte concentration, pH and temperature on the electrocoagulation removal of ARS were studied. Sample solutions were withdrawn at intervals to determine the residual dye concentration by using UV-VIS-NIR spectrophotometer. The amount of dye coagulated at equilibrium, Q_e / mg g^{-1}, was calculated using equation 4 below:

$$Q_e = \frac{(C_o - C_e)V}{W} \tag{4}$$

where C_o / mg L^{-1} is the initial concentration and C_e /mg L^{-1} is the concentration of the dye at equilibrium in the liquid-phase. V is the volume of the solution, L while W is the mass of the coagulant which can be estimated from Faraday Law according to the equation 5:

$$W = \frac{MIt}{nF} \tag{5}$$

M is the molar mass (g mol^{-1}) of the elements, I is the current (A), t is the electrocoagulation time in seconds, n is the number of electrons involved and F is Faraday's constant (96,485 C mol^{-1}).

The percentage dye removal as colour was estimated by:

$$\text{Colour removal, \%} = \frac{(Abs_0 - Abs_e)}{Abs_0} \times 100 \tag{6}$$

where Abs_0, is the blank absorbance and Abs_e is the absorbance at equilibrium.

Adsorption isotherms

The equilibrium data from this study were described with the six adsorption isotherm models. These are models by Langmuir [25], Freudlinch [26], Temkin [27], Dubinin and Radushkevich [28], Sip [29] and Redlich and Peterson [30]. The acceptability and suitability of the isotherm equation to the equilibrium data were based on the values of the correlation coefficients, R^2 estimated from linear regression of the least square fit statistic on Micro Math Scientist software.

Electrocoagulation kinetics studies

Since the amount of coagulant can be estimated for a given time, the pollutant removal can be modelled using an adsorption phenomenon. The procedures for the kinetics studies were basically identical to those of equilibrium tests. The aqueous samples were taken at preset time intervals, and the concentrations of the dye were similarly determined. The amount of dye removed at time t, Q_t (mg g^{-1}), was calculated using Equation 7:

$$Q_t = \frac{(C_o - C_t)V}{W} \tag{7}$$

where C_o / mg L^{-1} is the initial concentration and C_t / mg L^{-1} is the concentration of the dye at time t in the liquid-phase. V is the volume of the solution (L), and W is the mass of $Al(OH)_3$ calculated as stated in eqn. 5 above. In order to investigate the mechanisms of the adsorption process, pseudo-first order, pseudo-second order, Avrami, and Elovich models respectively were applied to describe the kinetics of adsorption of ARS to $Al(OH)_3$ generated during the electrocoagulation process. Since the diffusion mechanism cannot be obtained from the kinetic model, the intra-particulate diffusion model [8] was also tested. A model is adjudged best-fit and selected based on statistical parameters.

Statistical test for the kinetics data

The acceptability and hence the best fit of the kinetic data were based on the square of the correlation coefficients R^2 and the percentage error function which measures the differences (SSE, %) in the amount of the dye concentration coagulated at equilibrium predicted by the models, (Q_{cal}) and the actual, (*i.e.* Q_{exp}) measured experimentally. The validity of each model was determined by the sum of error squares (SSE, %) given by:

$$\text{SSE, \%} = \sqrt{\frac{((Q_{exp} - Q_{cal})/Q_{exp})^2}{N-1}} \times 100 \tag{8}$$

N is the number of data points. The higher is the value of R^2 and the lower the value of SSE; the better fitted the data.

Thermodynamics of electrocoagulation process

Arrhenius equation is applied to estimate the activation energy of adsorption according to the relationship:

$$\ln k = \ln A - \frac{E_a}{RT} \tag{9}$$

where k is the rate constant obtained from the kinetic model, E_a is the Arrhenius activation energy of adsorption, (kJ/mol), A is the Arrhenius factor, R is the universal gas constant (8.314 J mol^{-1} K^{-1}) and T is the absolute temperature. The thermodynamics parameters i.e. ΔG°, ΔH° and ΔS° were estimated using the following relation:

$$\Delta G^\circ = -RT \ln K_d \tag{10}$$

$$\ln K_d = \frac{\Delta S^\circ}{R} - \frac{\Delta H^\circ}{RT} \tag{11}$$

The equilibrium constant, $K_{d,}$ is obtained from the value of Q_e/C_e at different temperature equilibrium study. Van't Hoff plot of ln K_d against the reciprocal of temperature (1/T), should give a straight line with intercept as $\Delta S_o/R$ and slope as $\Delta H_o/R$.

Results and discussion

Batch equilibrium studies

Effect of current density

The amount of coagulants generated is related to the time and current density [31]. Current density determines the coagulant production rate, and adjusts the rate and size of the bubble production, and hence affects the growth of flocs [32,33]. The effect current density on the removal of ARS, were carried at the current density varied between j of 7.15 to 35.77 A m^{-2}. Figure 3 shows the plot of current density versus the percentage colour removal by the electrocoagulation process. From the figure it is glaring that as the current density is increased the rate of colour removal also increased from 96 to 99.31 %. Increasing current density results in a corresponding increase in the production of coagulant in the solution leading to high efficiency. The optimum current density of 35.77 A m^{-2} was used throughout the study.

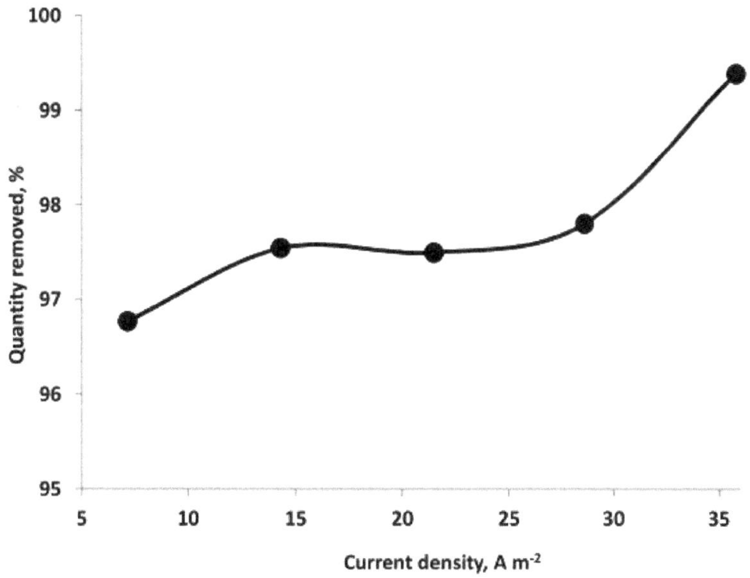

Figure 3. *Effect of current density on colour removal ([ARS] = 50 mg L^{-1}, T = 30 °C, pH 7.0 and [NaCl] = 2 g L^{-1}).*

Effect of pH on electrocoagulation process

pH is an important parameter influencing the performance of the EC process [34], it affects the chemistry of both the coagulants, dye molecules and that of electrochemical process in the solution. The effect pH on the removal of ARS was studied at initial pH varied between 3 and 11. The pH was adjusted with 0.1 M NaOH or 0.1 M HCl. The color removal percentages for dye solutions with various initial pH values were shown in Fig. 4.

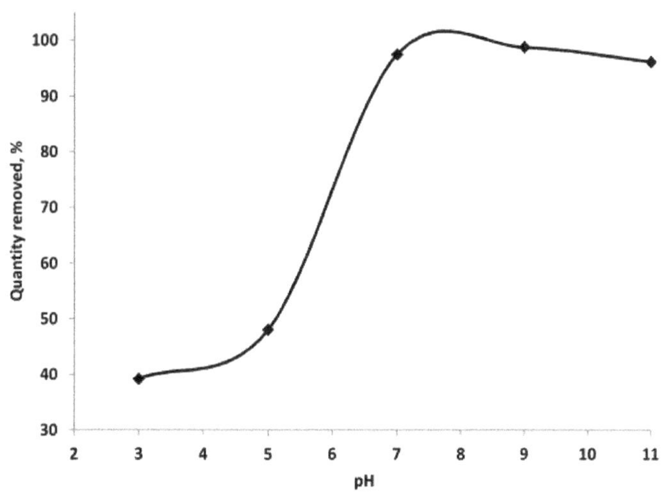

Figure 4. *Effect of initial pH on colour removal ([ARS] = 50 mg L^{-1}, T = 30 °C, [NaCl] = 2 g L^{-1}, j = 21.5 A m^{-2})*

The colour removal efficiency are optimum at the at pH range of 6.5 and 7.5 with about 97.5 % colour removal. The decrease in removal efficiency at more acidic and alkaline pH had been attributed to amphoteric behaviour of $Al(OH)_3$ which leads to soluble Al^{3+} cations (at acidic pH) and formation of monomeric anions (at alkaline pH). These ions transform finally into solid $Al(OH)_3$ according to complex precipitation kinetics thereby affecting the removal efficiency [35,36].

Effect of electrolyte concentration

Solution conductivity influences the current efficiency, cell voltage and consumption of electrical energy in electrolytic cells. The use of NaCl to increase solution conductivity is also accompanied by the production of chloride ions that reduces the effects of other anions, such as bicarbonate and sulphate which may lead to the precipitation of Ca^{2+} leading to the high ohmic resistance of the electrochemical cell [17]. The tests of electrolyte concentration effects were performed in solutions containing a constant concentration of ARS 50 mg L^{-1}, at current density of 21.5 A m^{-2} and pH 7, while the concentrations of NaCl were varied between 1 to 5 g L^{-1}. Figure 5 shows that colour removal efficiency percentage increases from 86 to 97.6 % as the electrolyte concentration rises up to 2 g L^{-1}. A further increase in electrolyte concentrations beyond these values does not significantly affect the removal efficiency of the dye from the solution. The results also suggest that high colour removal percentage with low cell voltages and low energy consumption can be obtained at NaCl concentration of 2 g L^{-1}.

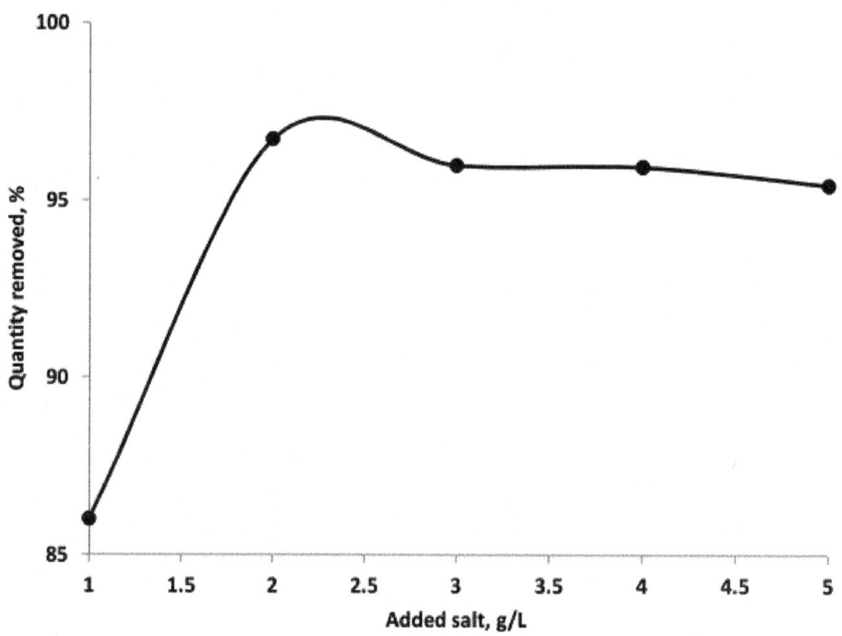

Figure. 5. *Effect of electrolyte concentrations on colour removal ([ARS] = 50 mg L^{-1}, T = 30 °C, pH = 7, j = 21.5 A m^{-2})*

Effect of initial dye concentrations

The effect of initial dye concentration on the electrocoagulation removal of ARS is shown in Figure 6 for dye concentrations increasing from 25 to 100 mg L^{-1} at constant temperature of 30 °C, current density 21.5 A m^{-2}, pH 7.0 and electrolyte concentration maintained with 2 g L^{-1} NaCl. The process showed rapid removal in the first 10 minutes for all the concentrations studied. The efficiency of the process increases from 138 to 604.8 mg g^{-1} as the initial concentration increases from 25 to 100 mg L^{-1}. As there is no significant difference in the amount coagulated after 20 minutes of the process, a steady-state approximation was assumed and a quasi-equilibrium situation was reached. The electrocoagulation curves were single, smooth, and continuous, leading to saturation. This is an indication of possible monolayer coverage on the surface of electrochemically generated coagulant [18,37].

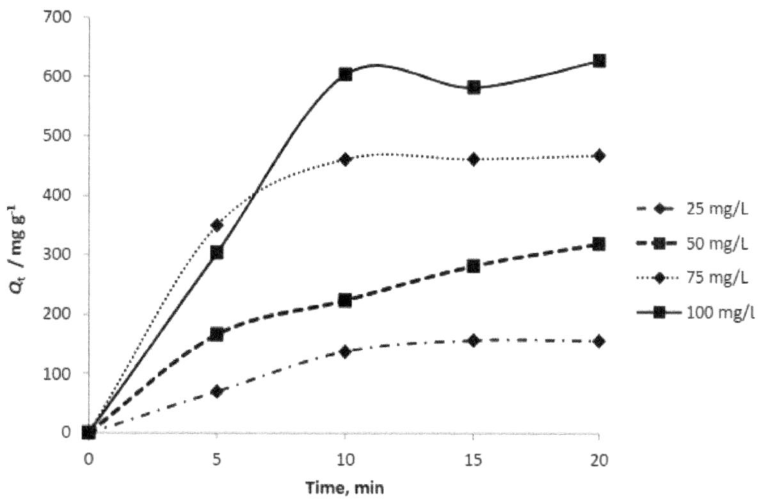

Figure 6. *Effect of initial concentration on the electrocoagulation removal of ARS (T = 30 °C, pH = 7, j = 21.5 A m^{-2} [NaCl] = 2 g L^{-1})*

Adsorption study

Adsorption isotherms

The adsorption data obtained at different initial dye concentrations were fitted into six different isotherm models. The equations representing these models and the parameters are summarized in Table 1, the details of which have been explained elsewhere [38].

Table 1. *Isotherm models used for the study of the EC removal of ARS*

Isotherm name	Isotherm model	Parameters
Langmuir	$Q_e = \frac{Q_{max} b C_e}{1 + b C_e}$	Q_{max} and b
Freudlinch	$Q_e = K_F C^{1/n}$	K_F and n,
Temkin	$Q_e = \frac{RT}{B_T} \ln(A_T C_e)$	A_T and B_T
Dubinin–Radushkevich	$Q_e = Q_s exp - \frac{\left(RT \ln\left(1 + \frac{1}{C_e}\right)\right)^2}{2E^2}$	Q_s and E
Redlich–Peterson	$Q_e = \frac{Q_o C_e}{(1 + K_R C_e^{g})}$	Q_o, K_R and g
Sip	$Q_e = \frac{Q_s (K_s C_e)^{\beta_s}}{(1 + (K_s C_e)^{\beta_s})}$	Q_s, K_s and β_s

The adsorption data fitted well with all the isotherms as shown in Figure 7, however, Dubinin--Radushkevich and Sip's isotherms have the highest R^2 (Table 2). The Q_m value of 785.31 mg g^{-1} obtained for the Langmuir isotherm model when compared with other isotherm Q values shows that the Dubinin–Radushkevich and Sip's isotherm values are lower than the Langmuir isotherm (Table 2). The $1/n$ value of > 1 obtained for Freudlinch isotherm is also indication that the adsorption is favorable.

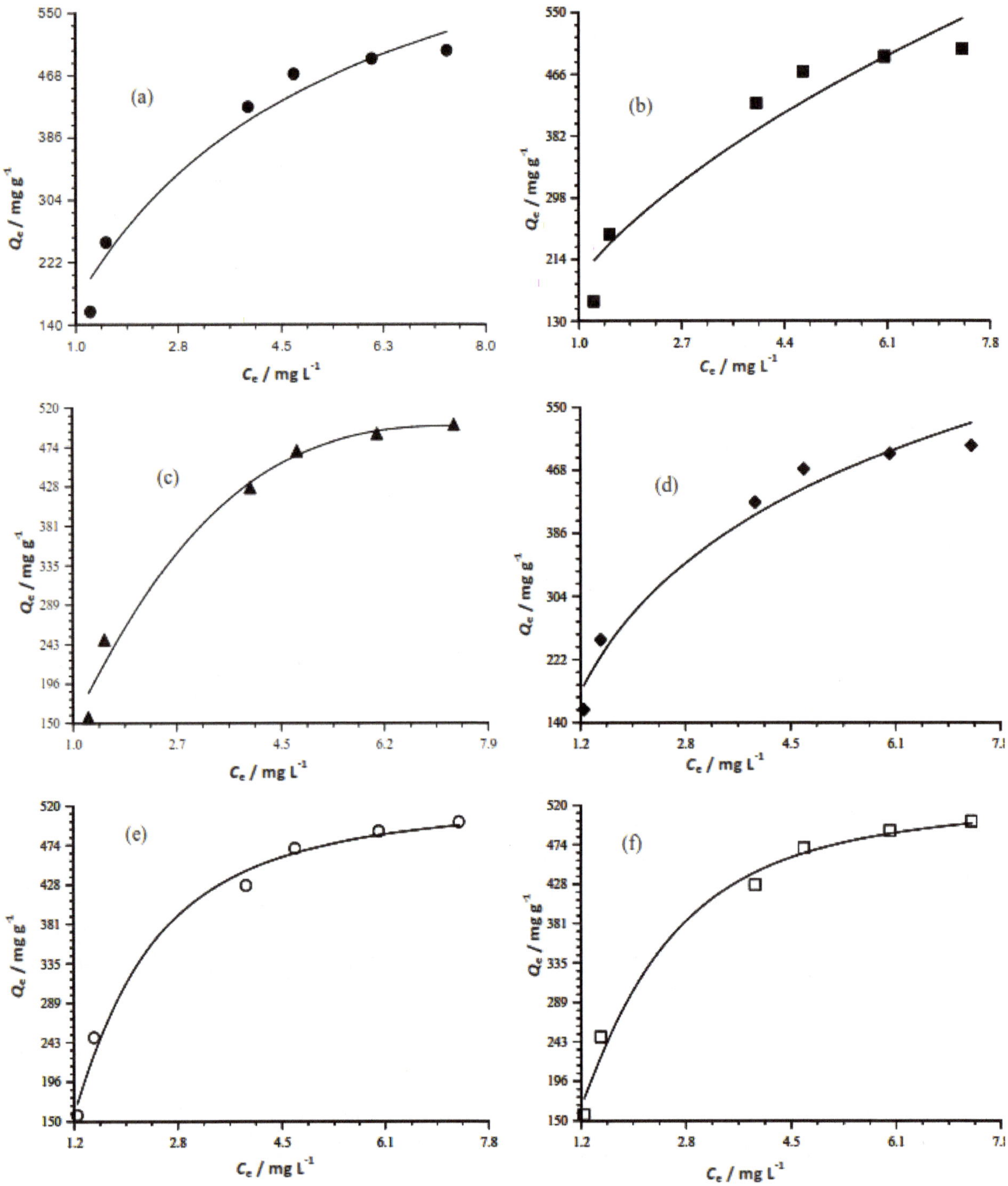

Figure 7. *Least square plots of the (a) Langmuir, (b) Freundlich, (c) Redlich–Peterson, (d) Temkin (e) Dubinin–Radushkevich and (f) Sips isotherms for electrocoagulation of Alizarin S dye at 30 °C.*

Table 2. *Isotherm parameters for electrocoagulation removal of Alizarin S*

Isotherms	Parameters	Values
Langmuir	Q_{max} / mg g^{-1}	785.31
	b / L mg^{-1}	0.27
	R_L	0.055
	R^2	0.996
Freudlinch	K_F / mg g^{-1} (mg L^{-1})$^{-1/n}$	188.85
	n	0.53
	R^2	0.992
Tempkin	A_T / L g^{-1}	13.06
	B_T / J mol^{-1} g mg^{-1}	2.12
	R^2	0.997
Dubinin–Radushkevich	Q_s / mg g^{-1}	653.10
	E / kJ mol^{-1}	5.65x10^{-7}
	R^2	0.941
Redlich–Peterson	Q_0 / mg g^{-1}	4372.23
	K_R /(mg^{-1} g)$^{1/g}$	0.036
	g	1.80
	R^2	0.998
Sips	Q_s / mg g^{-1}	524.63
	K_s / (mg^{-1} L)$^{1/\beta_s}$	0.57
	β_s	2.05
	R^2	0.999

Electrocoagulation kinetics

The plots of four different kinetic models used to explain the adsorption data are shown in Figure 8 (a - d). The equations for these models are shown in Table 3, details of which have explained elsewhere [38]. As shown in the figure, pseudo-first-order kinetic models fitted well with experimental data when compared with other models the R^2 value of the model also confirmed this (Table 4). The rate constant from all the models increases with initial dye concentration up to 75 mg L^{-1} before decreasing at 100 mg L^{-1}. This shows that at higher initial concentration the electrostatic interaction decreases at the site, thereby lowering the adsorption rate. The behaviour of Elovich constant shows that the process of adsorption is more than one mechanism.

Table 3: Kinetic models for the EC process

Isotherm name	Isotherm model	Parameters
Pseudo First Order	$Q_t = Q_e(1-e^{(-k_1t)})$	Q_e and k_1
Pseudo Secon Order	$Q_t = \dfrac{Q_e^2 k_2 t}{(1+Q_e k_2 t)}$	Q_e and k_2
Elovich	$Q_t = \dfrac{1}{\beta}\ln(\alpha\beta t)$	α and β
Avramin	$Q_t = Q_e(1-e^{-k_{av}t^{n_{av}}})$	k_{av}, and n_{av}
Intraparticulate model	$Q_t = K_{id}t^{0.5} + C_i$	K_{id} and C_i

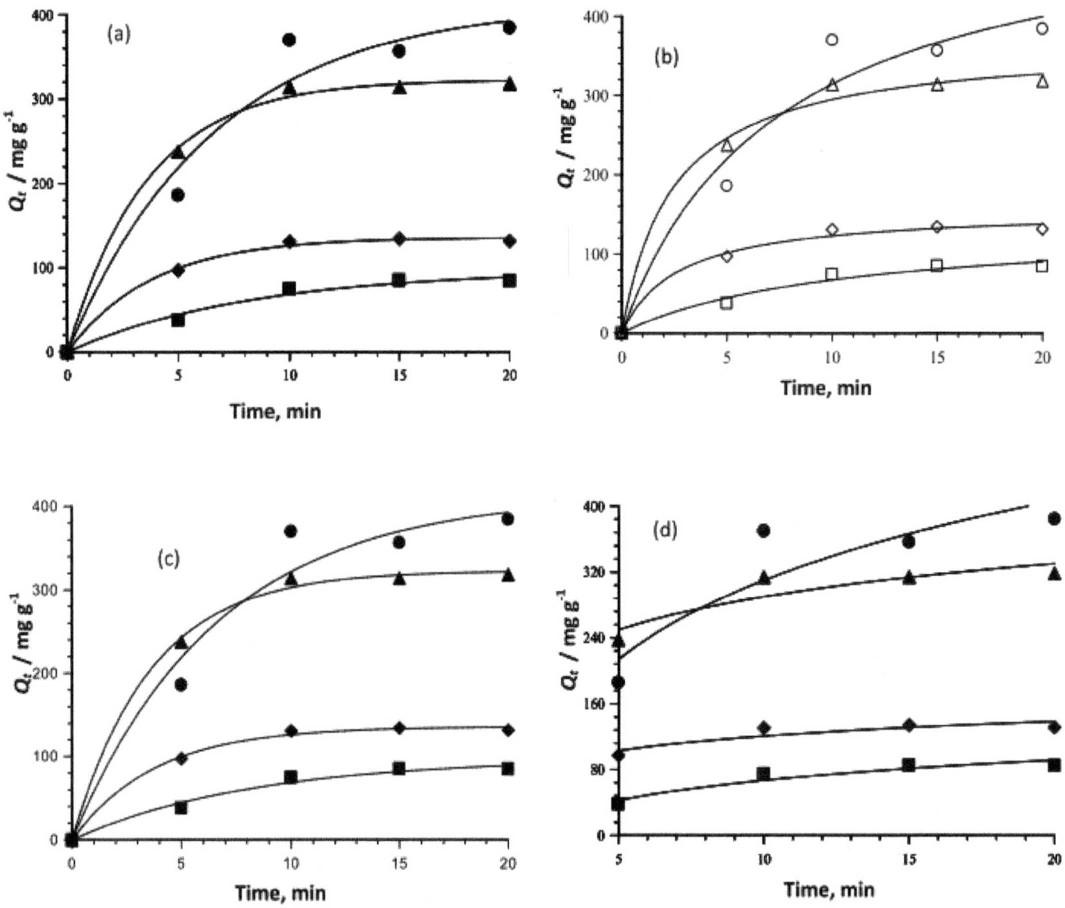

Figure 8. *Least square plots of (a) pseudo-first-order, (b) pseudo-second-order, (c) Avramin and (d) Elovich kinetic models for Alizarin S dye electrocoagulation at 30 °C.*

Table 4. *Comparison of kinetic model constant values for Alizarin Red S removal by electrocoagulation*

	C_0 / mg L^{-1}	25.00	50.00	75.00	100.00
	$Q_{e\,exp}$ / mg g^{-1}	74.52	130.75	313.80	369.92
First order	$Q_{e\,cal}$ / mg g^{-1}	97.30	135.75	322.81	413.23
	k_1 / min^{-1}	0.12	0.26	0.28	0.15
	R^2	0.995	0.999	0.997	0.992
	SSE, %	6.11	0.76	0.57	2.34
Second order	$Q_{e\,cal}$ / mg g^{-1}	136.04	155.88	367.39	548.90
	k_2 10^{-3} / g mg^{-1} min^{-1}	0.7204	2.3817	1.1079	0.2444
	R^2	0.993	0.998	0.998	0.989
	SSE, %	16.51	3.84	3.42	9.68
Avramin	$Q_{e\,cal}$ / mg g^{-1}	97.30	135.75	322.81	413.23
	k_{av} / min^{-1}	0.35	0.51	0.53	0.39
	n_{av}	0.35	0.51	0.53	0.39
	SSE, %	6.11	0.76	0.57	2.34
	R^2	0.995	0.999	0.999	0.992
Elovich	β / g mg^{-1}	0.03	0.04	0.02	0.01
	α / mg (g min)$^{-1}$	23.52	267.45	840.16	129.21
	R^2	0.993	0.997	0.998	0.989
	SSE, %	13.69	20.91	33.55	13.01

Adsorption mechanism

The mechanism of adsorption was investigated by subjecting the data to intra-particulate diffusion model. The plots are shown in Figure 9. The plots are not linear over the whole time range but rather they exhibit multi-linearity revealing the existence of two successive adsorption steps. The first stage is faster than the second, and it is attributed to the external surface adsorption referred to as the boundary layer diffusion. Thereafter, the second linear part is attributed to the intra-particle diffusion stage; this stage is the rate controlling step. Table 5 shows the intra-particle model constants for the electrocoagulation removal of ARS dye. The K_{di} values were found to increase from first stage of adsorption toward the second stage. The increase in dye concentration results in an increase in the driving force thereby increasing the dye diffusion rate.

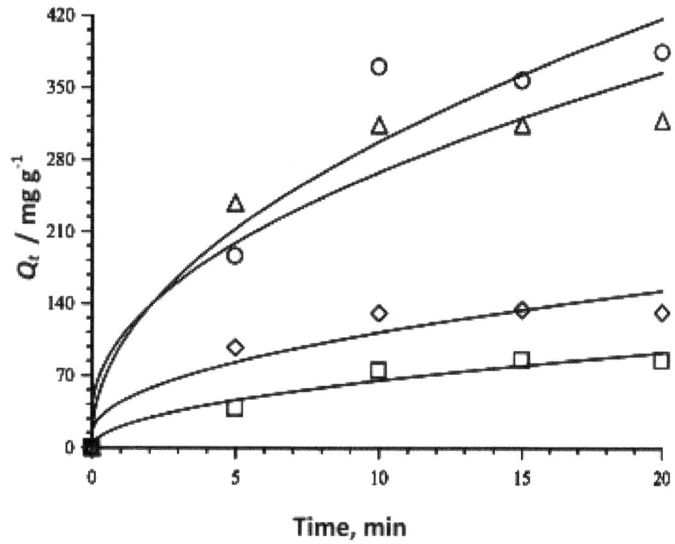

Figure 9. *Intraparticulate diffusion model plots for Alizarin S dye electrocoagulation at 30 °C.*

Table 5. *Intra-particle diffusion model for the electrocoagulation removal of Alizarin S*

C_o / mg L^{-1}	25	50	75	100
K_{d1} / mg g^{-1} min$^{-0.5}$	22.34	41.72	100.53	110.72
K_{d2} / mg g^{-1} min$^{-0.5}$	7.92	0.68	3.39	10.23
C_1	-2.73	0.85	2.91	-14.00
C_2	50.97	129.52	302.38	331.09
R_1^2	0.980	0.999	0.999	0.979
R_2^2	0.999	0.999	0.999	0.999

Thermodynamic parameters

Figure 10 shows that the rate constants vary with temperature according to Equation 9. The activation energy (3.55 kJ mol^{-1}) was obtained from the slope of the fitted equation. The free energy change, ΔG is obtained from Equations 10 and 11 according to the van't Hoff linear plots of ln K_d versus $1/T$ plot in Figure 10 (a and b). The thermodynamic parameters are presented in Table 6. It was found that the negative value of ΔG indicates the spontaneous nature of adsorption. Positive value of enthalpy change indicates that the adsorption process is endothermic in nature, and the negative value of change in internal energy (ΔG) shows the spontaneous adsorption of ARS on the coagulant. Positive values of entropy change show the increased randomness of the solution interface during the adsorption process (Table 6).

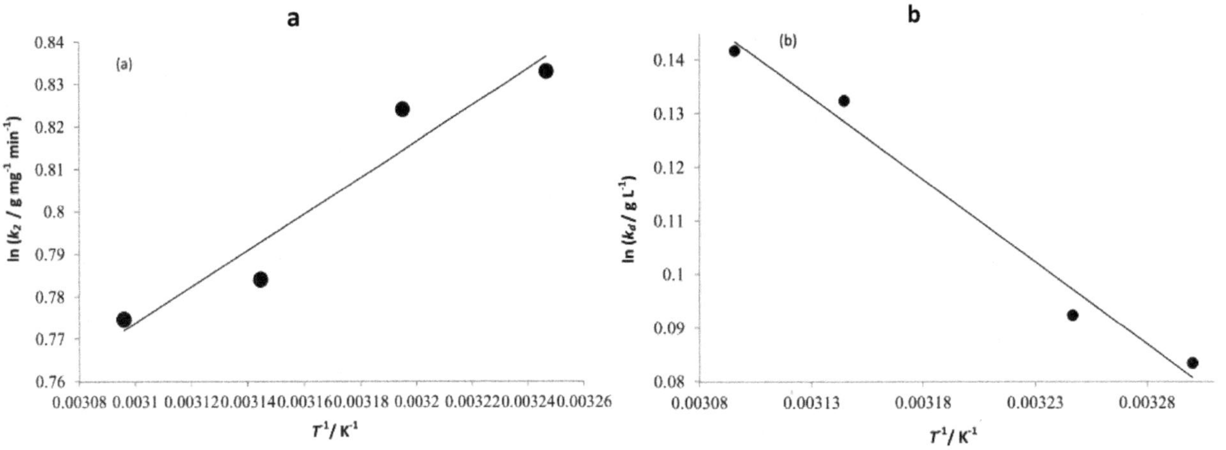

Figure 10. *Thermodynamic plot (a) ln k_2 and 1/**T** and (b) ln K_d and 1/T for the removal of Alizarin S dye by electrocoagulation.*

Table 6. *Thermodynamic parameters for the removal of Alizarin S*

T / K	K	ΔG / kJ mol^{-1}	E_a / kJ mol^{-1}	ΔS / bJ mol^{-1}	ΔH / kJ mol^{-1}	R^2
303	1.0870	-5.301	3.55	9.083	2.549	0.981
308	1.0967	-5.346				
318	1.1415	-5.437				
323	1.1522	-5.483				

FT-IR Studies of the dye solution before and after electrocoagulation

Figure 11 presents the FT-IR spectrum of the dye solution before and after the process. Before the electrocoagulation the spectrum shows several peaks; sharp and strong peak at 3440.8 cm^{-1} could be assigned to –OH stretching on the dye molecule while, that at 2821 cm^{-1} is due to –CH–. Peaks at 1593 and 1350 cm^{-1} are due to the aromatic C=C stretching. After electrocoagulation, the extra structure noted such as that at 3840 cm^{-1} may be assigned to the (O–H) stretching vibration in the $Al(OH)_3$ structures.

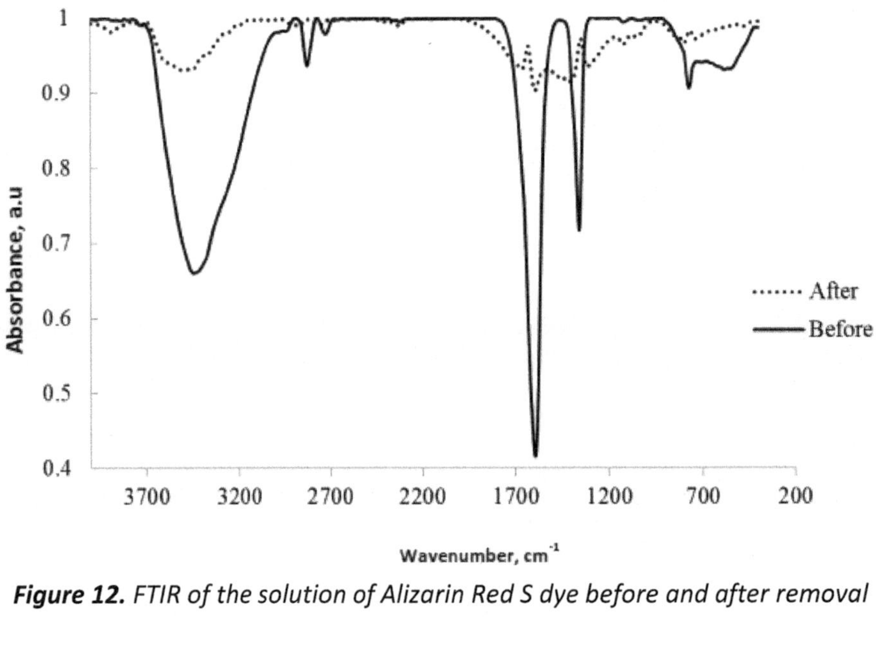

Figure 12. *FTIR of the solution of Alizarin Red S dye before and after removal*

Conclusions

This study revealed the feasibility of the use of electrocoagulation techniques for removing Alizarin Red S from its aqueous solution in a process. The process depends on numerous factors such as: current density, solution pH, temperature, initial dye concentration and contact time. The percentage removal of the dye increased with pH up to pH 7, also contact time and current density increase influence the removal positively. Equilibrium data fitted very well in the Langmuir isotherm equation, confirming the monolayer adsorption capacity of 785.31 mg g^{-1} at 303 K. The kinetics of the process is best explained using a pseudo first order kinetics model, with higher R^2 (Table 4). Intra-particle diffusion was not the sole rate controlling factor. The thermodynamics parameters obtained indicates that the process is spontaneous endothermic nature of the process. Therefore, the present findings suggested a better performance of electrocoagulation with Al electrode as an inexpensive method for the removal of ARS aqueous solutions.

***Acknowledgements**: The financial support in the form of grants from CSIR, for twelve months TWAS-CSIR Postdoctoral Fellowship, FR number: 3240275035, awarded to Abideen Idowu Adeogun that enables this work to be carried out. Also he is thankful to the authority of the Federal University of Agriculture, Abeokuta, Nigeria for granting the study leave to honour the fellowship.*

References

[1] G. Crini, *Bioresources Technology* **97** (2006) 1061-1085.
[2] S. Rajgopalan, *Pollution Management in Industries* Trivedy RK (Eds.) Environmental Publications, Karad., India (1995) p. 44.
[3] T. Routh, *Journal of Environmental Protection* **20(2)** (1998) 115-123.
[4] D.W.Kolpin, E.T. Furlong, M.T. Meyer, E.M. Thurman S.D. Zaugg, L.B. Barber, H.T. Buxton, *Environmental Science and Technology* **36(6)** (1999–2000) 1202-1211.
[5] M. Ali, T.R. Sreekrishnan, *Advance in Environmental Research* **5** (2001) 175-196.
[6] Q.Y. Sun, L.Z. Yang, *Water Research* **37** (2003) 1535-1544.
[7] R.A. Shawabkeh, M.F. Tutunji, *Applied Clay Science* **24(1–2)** (2003) 111-120.
[8] G.H. Lin, D.J. Brusick, *Journal of Applied Toxicology* **12(4)** (1992) 267-274.
[9] P. Zucca, C. Vinci, F. Sollai, A. Rescigno, E. Sanjust, *Journal of Molecular Catalysis A: Chemical* **288** (2008) 97-102.
[10] P.A. Carneiro, M.E. Osugi, C.S. Fugivar, N. Boralle, M. Furlan, M.V. Zanoni, *Chemosphere* **59(3)** (2005) 431-439.
[11] M. Panizza, P.A. Michaud, G. Cerisola, C. Comninellis, *Electrochemistry Communications* **3** (2001) 336-339.
[12] Z.X. Wang, X.C. Xu, Z. Gong, F.Y. Yang, *Journal of Hazardous Materials* **78(1)** (2012) 235-236.
[13] A. Aleboyeh, N. Daneshvar, M. B. Kasiri, *Chemical Engineering Process* **47** (2008) 827-832.
[14] F. Ghanbari, M. Moradi, A. Eslami, M.M. Emamjomeh, *Environmental Process* **1** (2014) 447-457.
[15] M.O. Nkiko, A.I. Adeogun, N.A.A. Babarinde, O. J. Sharaibi, *Journal of Water Reuse and Desalination* **3(3)** (2013) 239-248.
[16] B. Merzouk, M. Yakoubi, I. Zongo, J.P. Leclerc, G. Paternotte, S. Pontvianne, F. Lapicque *Desalination* **275(1–3)** (2011) 181-186.
[17] J. Zolgharnein, M. Bagtash and N. Asanjarani, *Journal of Environmental and Chemical Engineering* **2** (2014) 988-1000.
[18] T. Moriguchi, K. Yano, S. Nakagawa, F. Kaji, *Journal of Colloidal Interface Science* **260** (2003) 19-25.

[19] M. Ghaedi, A. Hassanzadeh, S.N. Kokhdan, *Journal of Chemical Engineering and Data* **56** (2011) 2511-2520.

[20] C. Saez, M. Panizza, M.A. Rodrigo, G. Cerisola, *Journal of Chemical Technology and Biotechnology* **82(6)** (2007) 575 - 581.

[21] M. Panizza, M.A. Oturan, *Electrochimica Acta* **56** (2011) 7084 – 7087.

[22] F. Yi, S. Chen, C. Yuan *Journal of Hazardous Materials* **B157** (2008) 78-87.

[23] M. Panizzan, G. Cerisola, *Water Research* **43** (2009) 339.

[24] N. Daneshvar, A. Oladegaragoze, N. Djafarzadeh, *Journal of Hazardous Materials* **B129** (2006) 116-122.

[25] I. Langmuir, *Journal of* American *Chemical Society* **40** (1918) 1361-1403.

[26] H.M.F. Freundlich, Journal of Physical Chemistry **57** (1906) 385-471.

[27] M.I. Tempkin, V. Pyzhev, *Acta Physica Chimca USSR* **12** (1940) 327–356.

[28] M.M. Dubinina, L.V. Radushkevich, *Izvestiya Akademii nauk, fizicheskoy khimii sektsii USSR* **55** (1947) 331-333.

[29] R. Sips, *Journal of Chemical Physics* **16** (1948) 490-495.

[30] O. Redlich, D.L. Peterson, *Journal of Physical Chemistry* **63** (1959) 1024-1026.

[31] T.S Aniruldhan, P.G. Radhakrishnan, *Journal of Chemical Thermodynamics* **40** (2008) 702-709.

[32] N. Daneshvar, A. Oladegaragoze, N. Djafarzadeh *Journal of Hazardous Materials* **B129** (2006) 116-122.

[33] N. Daneshvar, D. Salari, A. R. Khataee, *Journal of Photochemistry and Photobiology* **A 157** (2003) 111-116.

[34] M.Y.A. Mollah, P. Morkovsky, J.A.G. Gomes, M. Kesmez, J. Parga, D.L. Cocke, *Journal of Hazardous Materials* **B 114** (2004) 199-210.

[35] M. Bayramoglu, M. Kobya, O. Can, T.M. Sozbir, *Separation Purification Technology* **37** (2004) 117-125.

[36] M. Kobya, E. Demirbas, A. Dedeli, M.T. Sensoy, *Journal Hazardous Material* **173** (2010) 326-334.

[37] P. Saha, S.K. Sanyal, *Desalination,* **59** (2010) 131-139.

[38] A.I. Adeogun, B.R. Balakrishnan, *Applied Nanoscience* (2015) Doi: 10.1007/s13204-015-0484-9.

16

P(o-chlorophenol–co-o-hydroxyphenol): kinetic formation studies and pH-sensor application

Said M. Sayyah[1], Sayed S. Abd-Elrehim[2], Rehab E. Azooz[2,3,✉], Fatma Mohamed[1]

[1]*Polymer Research Laboratory, Chemistry Department, Faculty of Science, Beni-Suef University 62514, Beni-Suef City, Egypt*
[2]*Chemistry Department, Faculty of Science, Ain Shams University, 11566 Abbassia, Cairo, Egypt*
[3]*Present Address: Chemistry Department, Faculty of Science, Jazan University, 2097 Jazan, Saudi Arabia*

✉Corresponding author: e_azooz@yahoo.com

Abstract
Electrochemical copolymerization of o-chlorophenol (o-ClPh) with o-hydroxyphenol (o-HOPh) was conducted in aqueous H_2SO_4 using cyclic voltammetry technique at the Pt electrode. The reaction rate was found to be of the second order in the monomer concentration and first order in the acid concentration. The activation energy, enthalpy, and entropy for the copolymerization were found to be 20.20 kJ mol^{-1}, 19.24 kJ mol^{-1} and -281.47 J K^{-1} mol^{-1}, respectively. The obtained copolymer films show smooth feature with amorphous nature. Copolymer films adhere Pt electrode very well and show less reactivity in the H_2SO_4 medium. The pH sensitivity of the poly(oClPh-co-HOPh)-modified electrode has been investigated potentiometrically using different polymer thicknesses. The potentiometric responses to pH change of the poly(oClPh-co-HOPh)-modified electrode appeared reversible and linear in the range from pH 2-11 with a maximum sub-Nernstian potentiometric response slope of 40.7 mV/pH (30 °C). The slope became close to 56.2 mV/pH in the range from pH 4 to 9 at (30 °C). The poly(oClPh-co-HOPh)-modified electrode readily responded to pH change but was not stable with time.

Keywords
Pt-anode; poly(o-chlorophenol-co-o-hydroxyphenol); electrocopolymerization; kinetic study; activation parameters; potentiometric pH sensor

---

Introduction

Phenolic compounds (Phs) are a class of pollutants which, when absorbed through the skin and/or mucous membranes, can cause damage to numerous organs of living bodies, like the lungs, liver, kidney, and genitourinary tract [1]. Phs are widely used in wood preservatives, textiles, herbicides and pesticides, and released into the ground and surface water. The identification and quantification of these compounds represent an important issue in the environmental monitoring. o-hydroxyphenol (o-HOPh) and o-chlorophenol (o-ClPh) are two phenolic compounds, which are present as contaminants in medical, food and environmental matrices. Different methods were used for identification and clearing of them. Some of these methods are time-consuming, of low sensitivity with complicated pretreatments and expensive. Electrochemical methods, however, provide an easy and fast alternative for the analysis of such materials [2,3].

Electropolymerization is a good approach for preparing polymer-modified electrodes by adjusting the electrochemical parameters that control film thickness, permeation, and charge transport characteristics. In the field of electropolymerization, two topics are of great importance: the first is the study of the electropolymerization kinetics [4–9], which provides information about the nature of the reactions taking place at the electrode surface and the chemical structure of the polymer film, while the second is the potential use of polymer modified electrodes as sensors for qualitative and quantitative analyses of hazardous and biologically active compounds [4–6, 9-11].

There have already been a few reports on the kinetics and mechanism of the electrochemical polymerization [12-18]. Sayyah *et. al.* [19] studied the kinetics of homopolymerization of o-HOPh and o-ClPh by cyclic voltammetry (CV) in aqueous H_2SO_4 solution at the Pt anode. These authors used different tools for characterization of the obtained films. The obtained films with a smooth lamellar surface were characteristic for poly(o-chlorophenol), P(o-ClPh), and a smooth feature with uniform thickness for poly(o-hydroxyphenol), P(o-HOPh). In addition, they discussed the possible applications of P(o-ClPh) in the field of sensors and P(o-HOPh) in the field of dye removal from water. Previously, we showed that a block copolymer film was deposited by polymerizing o-ClPh with o-HOPh using CV methods [20]. The copolymer formation was directly affected by monomer and acid concentrations and temperature. The copolymer consisted of higher fraction of o-HOPh than o-ClPh units suggesting that o-HOPh is much more reactive than o-ClPh.

Determination of pH is one of the most common measuring procedure in a wide variety of industries and industrial applications [21]. A glass pH electrode, as one of the traditional pH electrodes, has excellent electrode performances with respect to response slope straightforward operation, selectivity, and long-term stability. However, the electrode with the internal liquid system exhibits some drawbacks, such as mechanical fragility, high cost, and limited miniaturization for use in clinical or biological applications [22, 23]. Much interest in the development of pH sensors based on polymers has been created [24]. The positive charges in the oxidized state of the polymer are compensated by counter-ions from the solution, giving ion-exchange properties of the polymer films. Therefore, polymers are applicable in potentiometric pH sensors [25]. The conducting polymer-based pH sensors, however, have suffered from long response time, non-reproducibility and salt effect [26].

The aim of the present work is to study the kinetics of electrooxidation of the binary mixture of o-HOPh and o-ClPh in the aqueous acid solution. The reaction orders with respect to the electrolyte and monomer will be determined from electrochemical data and the prepared films/electrode will be tested as a potentiometric pH sensor.

Experimental

Materials

o-ClPh (+97 %, Hopkin & Williams, *Dagenham, Essex, UK*), sulfuric acid, K_2HPO_4, potassium hydrogen phthalate (KHP) , Borax, NaOH, hydrochloric acid, $NaHCO_3$ (Merck, *Darmstadt, Germany*) and o-HOPh (99 %, Aldrich, *Germany*). All chemicals were of analytical pure grade and used as received. All solutions were prepared by using freshly double-distilled water.

Instrumentation

Electrochemical experiments were performed using a Potentiostat/Galvanostat Wenking PGS 95. The *i-E* curves were recorded by the computer software from the same company (Model ECT). All experiments were conducted at a given temperature (± 0.5 °C) with the help of circular water thermostat.

Electrochemical polymerization was carried out in a home-made three-electrode single-compartment cell with platinum plates as working and counter electrodes and a saturated calomel electrode (SCE) as the reference electrode. The control temperature (± 0.5 °C) was achieved by means of a jacketed cell through which water was circulated from a Polytemp thermostat. All electrochemical reactions were carried out at 25 °C unless otherwise specified in the text. The areas of the working and counter electrodes were 0.5 cm^2 each. The oxidation potentials were measured by linear sweeping voltammetry as the anodic peak values ($i_{p(II)}{}^{a}$).

The obtained copolymer was peeled off from Pt electrode and washed with H_2SO_4 (0.6 M), then washed with bi-distilled water and finally heated to 95 °C for 3 h and stored for further characterization.

Characterizion by UV–Vis spectrum was achieved using Shimadzu UV spectrophotometer (M160 PC) at room temperature in the range 200–700 nm. Dimethylformamide (DMF) was used as a solvent and reference (200 µL of the solution was diluted to 3 mL in a quartz cuvette of 1.0 cm optical path). IR-spectra was recorded using a Shimadzu FTIR-340 Jasco spectrophotometer (Japan) from 400 to 4,000 cm^{-1} with a spectral resolution of 4 cm^{-1} using KBr tablets (Merck, *Darmstadt, Germany*, 99 %) with 1:100 sample: KBr proportion. TGA was performed using a Shimadzu DT-30 thermal analyzer (Shimadzu, *Kyoto, Japan*) - the weight loss was measured from ambient temperature up to 600 25 °C, at the rate of 20 °C min^{-1} and nitrogen flow rate of 50 ml min^{-1} to determine the degradation rate of the copolymer. SEM is performed by a JXA-840A Electron Probe Microanalyzer (JEOL, *Tokyo, Japan*). XRD was done using (Philips 1976 Model 1390, *Netherlands*) - X-ray tube: Cu; current: 30 mA; preset time: 10 s; scan speed: 8°min^{-1}; voltage: 40 kV. The pH measurements were carried out using a pHS-3B pH meter and a commercial SCE. A series of buffer solutions were used for solutions of various pHs. The buffers were adjusted by adding diluted HCl or NaOH solution to the solution of KH phthalate, KH_2 phthalate, borax and/or $NaHCO_3$.

Results and discussion

Electrochemical copolymerization

The first cycle cyclic voltammogram (CV) recorded during the copolymerization a mixture of o-ClPh and o-HOPh with a molar feed ratio kept at 1:1 in the potential range from 0 to 1.7 V *vs.* SCE in the absence and presence of monomers is represented in Figure 1. From this figure, the adsorption peak of H_2 gas on Pt surface was observed at -0.3 V *vs.* SCE [19] in absence or presence of monomers, peak (I), whiile the oxidation peak of comonomers to yield copolymer, peak (II) was developed at

0.70 *vs.* SCE. By increasing number of cycles $i_{p(I)}$ was not changed while $i_{p(II)}$ decreased significantly and disappeared after 20 cycles.

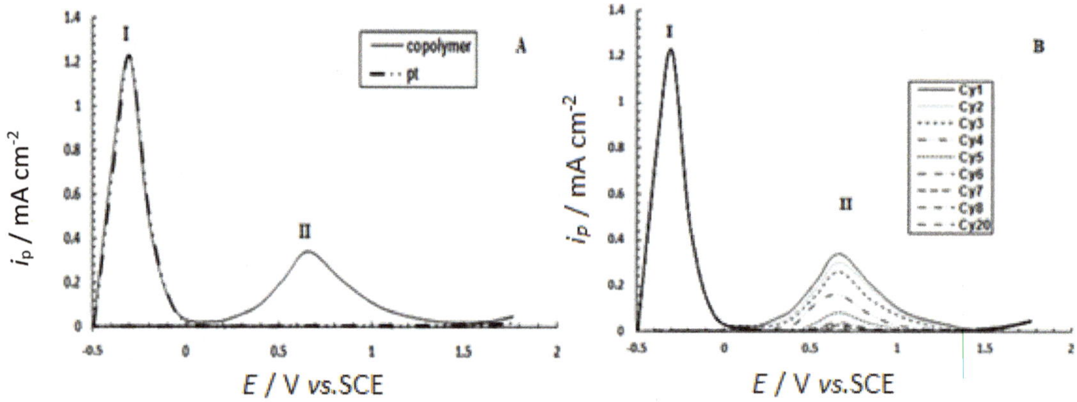

*Figure **1**: Cyclic voltamigrams of Pt electrode in aqueous solution containing 0.6 M H_2SO_4 at 303 K using 25 mV s^{-1} in absence and presence of 0.05 M comonomer (1:1 molar ratio) ,*
***A** - first cycle and **B** - repetative cycling up to 20 cycles.*

The mechanism of polyphenol formation was proposed elsewhere [20,27-30]. Similarly, we could propose that a removal of an electron from hydroxyl groups (O-atoms) to form free radicals adsorbed on the Pt electrode surface occurs at 0.70 *vs.* SCE. Adsorbed radicals react with each other or with a comonomer molecules via head-to-tail (C-O)-coupling to form dimeric radical what is followed by formation of the copolymer film.

On reversing the potential scan, the anodic current is very small indicating the presence of a copolymer film well adhered to the Pt surface [30,31]. The absence of any cathodic peaks indicates totally irreversible system as a result of the strong adhesion of the copolymer film to Pt surface. Insulating property of the copolymer film, no electroactive species, and/or the nature and kind of oxide species formed on the Pt surface during the electrooxidation is the reason of observed irreversibility [32,33].

CVs recorded for o-ClPh polymerization showed one oxidation peak at 0.86 V and one reduction peak at 0.25 V *vs.* SCE) while CVs for o-HOPh polymerization showed one oxidation peak at 0.62 V and one reduction peak at 0.20 V *vs.* SCE, respectively [19,20]. As seen in Figure 1, the obtained CVs for their mixtures differ, while no reduction peaks confirmed the copolymer formation.

By increasing the number of cycles repetitively (*i.e.* see Figure 1B), a fouling of Pt electrode is seen as in other poly phenols [30,33,34]. The reason of this phenomenon may be due to the formation of insulating and good adhered films on Pt surface which causeed a decrease of the peak current densities $i_{p(II)})$ with repetitive cycling [35]. E_p^a of the formation process does not shift with increasing number of cycles, indicating that the oxidation reactions were independent of the copolymer thickness [31,36,37].

Effect of the scan rate

The influence of the scan rate on the anodic current density $i_{p(II)}$ of the Pt electrode was studied in the range from 15 to 40 V *vs.* SCE for 0.05 M co-monomer (1:1 molar ratio) in an aqueous solution containing 0.6 M H_2SO_4 at 303 K. It is obvious from Figure 2 that almost linear relationship was observed for the dependence of $i_{p(II)}$ on both scan rate (v) and the square root of the scan rate ($v^{0.5}$). This behavior may be explained as the mixed diffusion-kinetically controlled process [20]. According to Randless [38] and Sevick [39] (relation between i_{pII} and ($v^{0.5}$ is linear see Fig 2B), the average diffusion coefficient (D^{diff}) is calculated to be 6.53 10^{-7} m^2s^{-1}. The values of D^{diff} are seen to be constant over the range of sweep rates, which again shows that the oxidation process is diffusion-controlled [33,40-43].

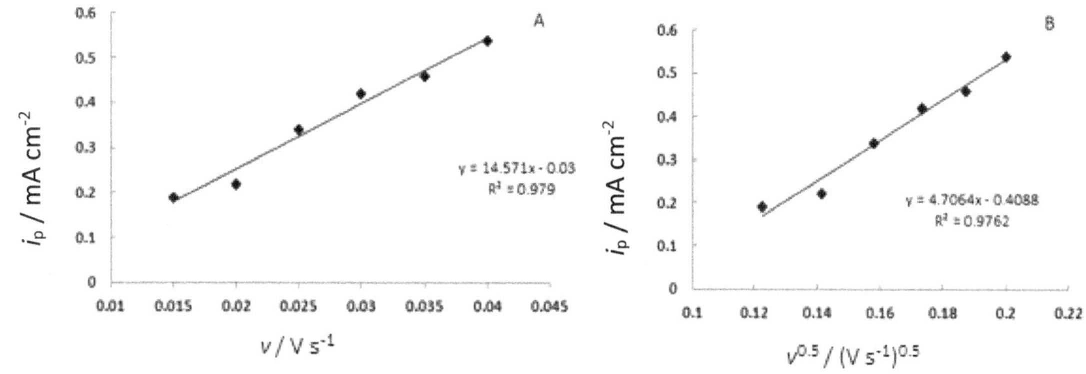

Figure 2. *Relation between* i_{pII} *and* ***A*** *- scan rate,* ***B*** *- square root of scan rate for Pt electrode and 0.05 M comonomer (1:1 molar ratio) in aqueous solution containing 0.6 M* H_2SO_4 *at 303 K.*

Increasing the temperature by 35 °C, the D^{diff} increased by a factor of ~6. The Walden product (the product of diffusion coefficient and the viscosity, $D \times \eta$) was constant over the temperature range studied suggesting that decrease of viscosity at higher temperature caused the observed increase of D^{diff}. Similarly to the Arrhenius law, the following equation allows the activation energy of diffusion, E^{diff}, to be calculated using a linear plot of Ln D^{diff} versus T^{-1}:

$$D + D^{\circ} e^{-\frac{E_a^{diff}}{RT}}$$

Depending on the above equation and data obtained from Figure 3 the linearity of relation >0.95 confirm good stimulation of Arrhenius relation and E_a^{diff} was found to be 38.25 kJ mol^{-1} at 25 mV s^{1}.

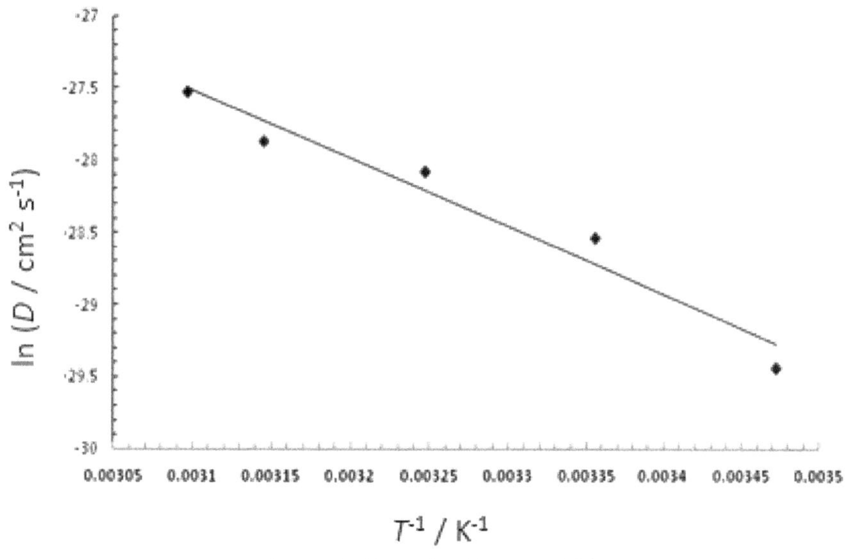

Figure 3. *Plot of ln* D versus T^{-1}.

Kinetics of the electropolymerization process

The electropolymerization kinetics was evaluated using deoxygenated aqueous solution containing monomer in the concentration range between 0.03 and 0.08 M and H_2SO_4 concentration in the range between 0.2 and 0.6 M at 303 K.

Assuming that the polymerization follows the following equation:

$$M + E \rightarrow P$$

where M is the monomer, E is the electrolyte and P is the polymer, then the kinetic equation can be formulated as follows:

$$R_p = \frac{d[W]}{dt} = k[E]^a[M]^b$$

where R_p is the polymerization rate, which represents the polymer weight, W, per unit time and cm^2 of the electrode surface area, a and b are the reaction orders with respect to the electrolyte and monomer, respectively, and k is the specific rate constant of the electropolymerization process.

The electrochemical study of the copolymer formation is used instead of gravimetric study since the polymer yield on Pt electrode is so small to be used in kinetic studies.

The value of the anodic current density $i_{p(II)}$ is proportional to the electropolymerization rate (R_p) at given concentrations of the monomer, acid, and electrolyte, and then we can replace the electropolymerization rate by the anodic current density [31,44-46]:

$$R_p = \frac{d[W]}{dt} = i_p = k[E]^a[M]^b$$

or expressed in logarithmic form as:

$$\log i_p = \log k + a\log[E] + b\log[M]$$

For the electropolymerization of the comonomer in aqueous solutions, the kinetic equation can be represented by:

$$R_p = k[H_2SO_4]^a[comonomer]^b$$

Comonomer mixture and sulfuric acid concentrations were varied keeping one of them constant to evaluate their respective reaction orders. The comonomer concentration varied from 0.04 to 0.08 mol L^{-1} (1:1 molar ratio) at a constant 0.6 mol L^{-1} H_2SO_4 concentration. Then, the H_2SO_4 concentration varied from 0.2 to 0.6 mol L^{-1} at a constant comonomer concentration of 0.08 mol L^{-1} (1:1 molar ratio).

Figure 4A presents the polymerization *i-E* plots corresponding to the copolymer formation from a constant 0.6 mol L^{-1} H_2SO_4 concentration and varying monomer concentrations (with 1:1 molar ratio). The log $i_{p(II)}$ *versus* log $c_{comonomer}$ relation is presented in Figure 4B. The slope of the linear relation was found to be 1.80 which means that the polymerization reaction is of the second order with respect to the comonomer concentration.

Figure 5A shows the effect of the change of H_2SO_4 concentration on the polymerization $i_{p(II)}$ at a constant concentration of comonomer 0.08 mol^{-1} using (1:1 molar ratio). The plot of log $i_{p(II)}$ against log $c_{H_2SO_4}$ is presented in Figure 5B. The linear relation has a slope of 0.80 which means that the polymerization reaction is of the first order with respect to H_2SO_4 concentration.

Based on the above data, the kinetic equation of the copolymerization process is defined as:

$$R_p = k[H_2SO_4]^{0.80}[comonomer]^{1.8}$$

From the above equation, the order of the copolymerization reaction is first order with respect to H_2SO_4 and second order with respect to comonomer concentration.

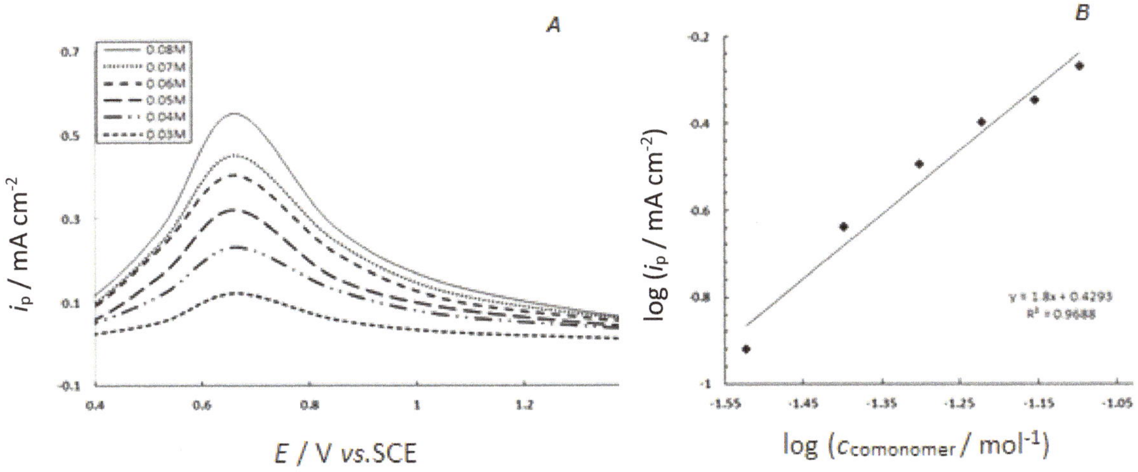

Figure 4. ***A***- i-E *curves for the effect of comonomer concentration on the electropolymerization of oHP and oCP from solution containing 0.6M H_2SO_4 at 303K with scan rate 25 mV s^{-1} on Pt electrode;* ***B*** *- Double logarithmic plot of* $j_{p(II)}$ versus *comonomer concentrations).*

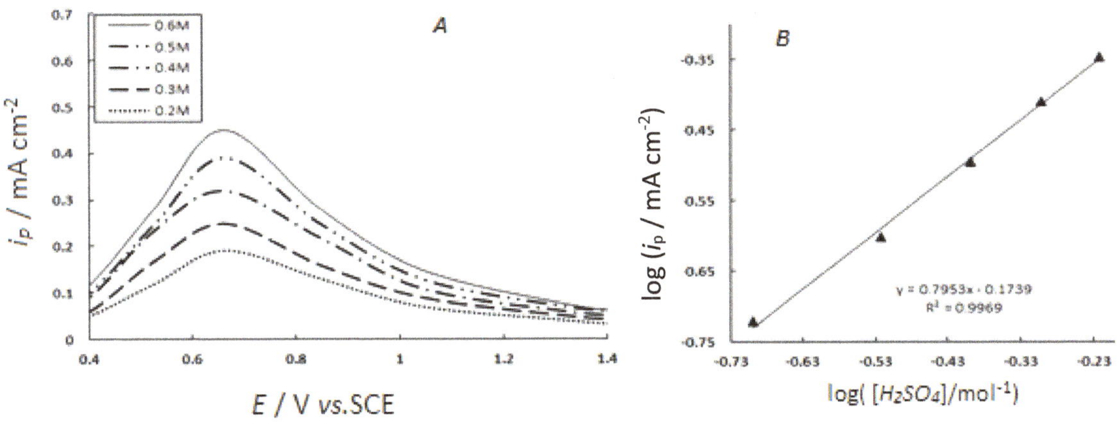

Figure 5. ***A*** - i-E *curves for the effect of acid concentration on the electropolymerization of oHP and oCP (1:1molar ratio) from solution containing 0.08M comonomer at 303 K with scan rate 25 mV s^{-1} on Pt electrode;* ***B*** *- Double logarithmic plot of* $i_{p(II)}$ versus *acid concentrations*

Effect of temperature

Figure 6A shows the potentiodynamic polarization curves as a function of the solution temperature in the range between 288 and 323 K. *i-E* curves were recorded in the solution of 0.6 M H_2SO_4 and 0.09 M monomer concentration using scan rate of 25 mV s^{-1}. From Figure 6A it appears that an increase of the reaction temperature resulted in a progressive increase of the charge included in the anodic peaks.

In general, the rate of polymerization increases with the increase of the reaction temperature. The Arrhenius-like plot exhibits a linear relationship between the logarithm of the rate constant of the electropolymerization reaction (log $i_{p(II)}$) and inverse absolute temperature (T^{-1}),

$$\ln i_p = \ln A - \frac{E_a}{RT}.$$

According to the Arrhenius equation, the apparent activation energy (E_a) and pre-exponential factor (A) can be determined from the slope and intercept of the plot respectively [47].

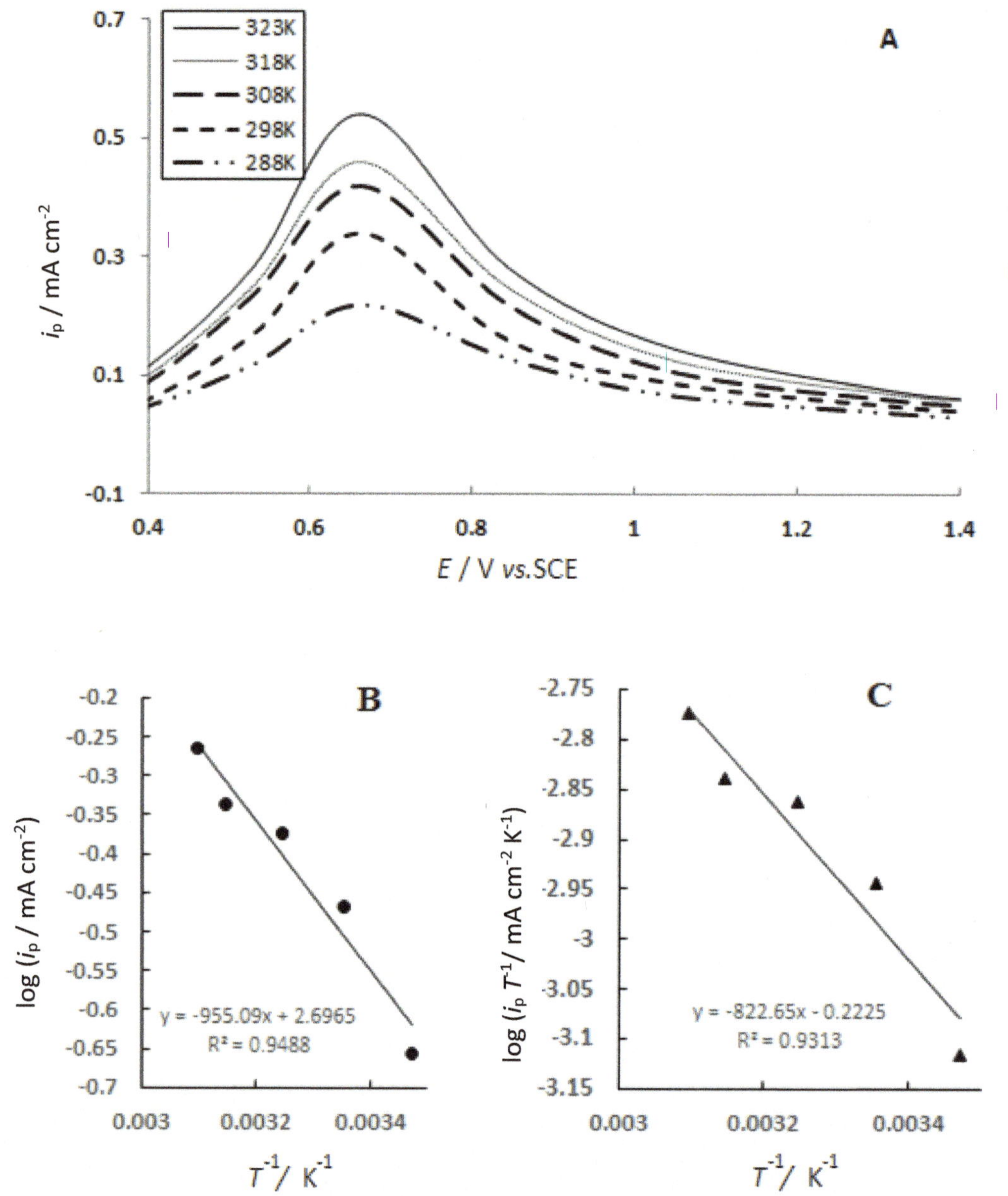

Figure 6. A - i-E *curves for the effect of temperature on the electropolymerization of oHP and oCP (1:1 molar ratio) from solution Containing 0.09 M comonomer and 0.6 M H_2SO_4 at 303 K with scan rate 25 mV s^{-1} on Pt electrode;* ***B*** *- Arrhenius plot of the electropolymerization of oHP and oCP (1:1 molar ratio);* ***C*** *- Eyring equation plot of the electropolymerization of a binary mixture of oHP and oCP (1:1 molar ratio).*

At temperatures higher than 323 K the log $i_{p(II)}$ deviates slightly from the straight line (*i.e.* the value is lower than expected) owing probably to the excessive diffusion of radicals away from the vicinity of the electrode. Such diffusion is evidenced by the color change in the electrolyte solution at higher temperatures.

The plot of log $i_{p(II)}$ *versus* T^{-1} yields a straight line (Figure 6B), indicating that the copolymerization also obeys the Arrhenius equation. The apparent activation energy was calculated using Arrhenius equation and found to be 20.2 kJ mol^{-1}.

Other activation parameters, *i.e.* activation enthalpy (ΔH^*) and activation entropy (ΔS^*) were determined from the slope and intercept of the plot of log ($i_{p(II)}$ T^{-1}) against T^{-1}, respectively, according to the following equation [48]:

$$\ln\frac{i_p}{T} = -\frac{\Delta H^*}{RT} + \frac{\Delta S^*}{R} + \ln\frac{k}{h}$$

where *R* is the gas constant, *k* is Boltzmann's constant, and h is Planck's constant. The change in ΔH^* and ΔS^* of activation for the electropolymerization reaction can be calculated from the Eyring equation plot at different temperatures (Figure 6C). We obtained a linear relationship with a slope of $-\Delta H^*/2.303\ R$ and intercept of {*log (k/h) +* $\Delta S^*/2.303$ R}. From the slope and intercept the values of ΔH^* and ΔS^* were found to be 19.24 kJ mol^{-1} and -281.47 kJ mol^{-1}, respectively.

The negative ΔS^* value suggests the formation of ordering activated complexes. There are several possible reasons for the negative ΔS^* which are consistent with the proposed mechanism [49]. First, the radicals generated during the polymerization are much more solvated in the media in comparison with the corresponding neutral reactants. Second, the polymer chain growth involves bimolecular reactions between the radicals and the incoming monomers. Both factors may have contributed to the observed entropy decrease in the activated complexes.

Electroactivity of the Pt-copolymer film

The obtained polymer film prepared on Pt electrode at the optimum conditions (0.08 M comonomer and 0.6 M H_2SO_4 at 303 K using 25 mV s^{-1}) was transferred to another cell with only 0.6M H_2SO_4 and cycled from - 600 to +2000 mV, *vs.* SCE (Figure 7). Figure7 shows that both oxidation peak (developed at +700 mV *vs.* SCE) and the H_2 peak (developed at -300 mV *vs.* SCE) disappeared completely.

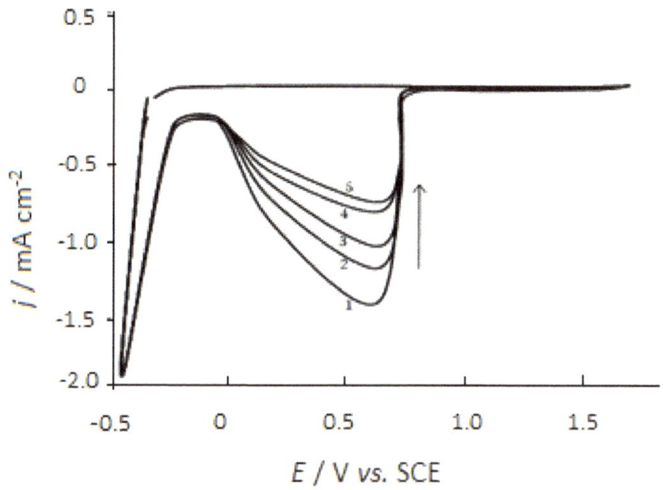

Figure 7. *Electroactivity behavior of copolymer coated Pt electrode in 0.6 M H_2SO_4 with different cycle numbers. Scan rate=25 mV s^{-1}.*

The lack of oxidation peaks confirms the insulating properties and the lack of active species inside the copolymer film. The appearance of a new broad reduction peak can be explained as follows: at the absence of other cations (comonomers cations) H^+ will accumulate on the copolymer/platinum surface forming local charge densities which are considered as a new copolymer form (Copolymer H). During the reduction process, the formed Copolymer H will reduce to Copolymer and H^+ *via* a chemical process. This phenomenon is the basis for the construction of polymeric pH sensors. By repetitive cycling, the current of cathodic peak decreases as a result of decreasing the copolymer amount on the Pt surface. For P(o-ClPh) and P(o-HOPh), adsorption peak of H_2 is observed and not affected by repeating cycles, no oxidation peaks were formed and one broad reduction peak is developed at -300 and -200 mV *vs.* SCE respectively [19].

UV-Vis and IR spectroscopy

The UV-Visible spectrum of P(o-HOPh -co-o-ClPh) shows that the π- π* transition ((E2 band) of the benzene ring and the β-band ($A_{1}g$ to $B_{2}u$)) appear at 257, 307 and 317 nm where the π-polaron transition (conjugation of the aromatic polymeric chain) appears at 485 nm [20,45,50]. This band doesn't appear at both homopolymers [19] which confirm the copolymer formation.

The assignments of the absorption bands of the obtained infrared spectra (KBr pellets) of copolymer sample is recorded and represented in Table 1.

Table 1. *IR - wavenumbers and thier assignments for the prepared copolymer sample.*

Assignments	Wavenumber, cm^{-1}
Electronic transition from the valence band to the conduction band	3700 – 1800
O-H group (phenolic) streching vibration	3320 and 3420
Aromatic ring C–H stretching vibration	3044 and 3105
out-of- plane bending of =C–H bonds of an aromatic ring	830 and 860
C=C stretching vibration bands	1500 and 1665
C–O–C stretching frequency of phenyl ether [51]	1230
dopant SO_4^{-2} incorporation [20]	1180
stretching vibration of the C-Cl bond	912

Thermogravimetric analysis of the prepared sample

The presence of water molecules in the repeated unit is confirmed by the TGA. The TGA steps of the prepared P(o-ClPh-co-o-HOPh) are shown in Figure 8. From the Figure 8 the loss of (water and doped anions) in the temperature ranging between 25 and 220 °C can be observed. At higher temperatures the polymer begins to decompose. It is suggested that P(o-ClPh -co-o-HOPh) is thermally stable until ~220 °C where it starts to decompose resulting in only 27 % polymer weight remaining at 600 °C. However, the remaining amount of P(o-HOPh) at 600 °C is ~65 % and for P(o-ClPh) ~70 % [19] which confirms the copolymer formation.

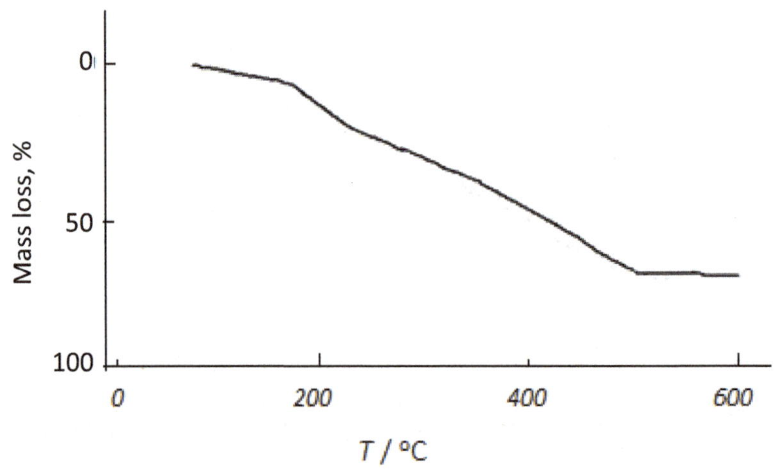

Figure 8. *TGA curve for P(o-ClPh-co-o-HOPh)*

Surface morphological studies

Based on data obtained from XRD and SEM techniques, we previously [19] showed that in a case of o-ClPh, a homogeneous, smooth, brown and well-adhering P(o-ClPh) films were electrodeposited on Pt electrode surface with a lamellar surface feature of the amorphous nature. In a case of o-HOPh, a smooth, black and well-adhering P(o-HOPh) films were electrodeposited on the Pt-surface with uniform thickness and amorphous structure.

In the case of the binary mixture, a homogeneous, smooth, black and well-adhering P(o-ClPh -co-o-HOPh) films were electrodeposited on the Pt- electrode surface. The XRD- pattern shows that the prepared copolymer was amorphous as shown in Figure 9A.

The surface morphology of the copolymer obtained at the optimum conditions was examined by SEM that shows smooth and amorphous in nature as seen in Figure 9B.

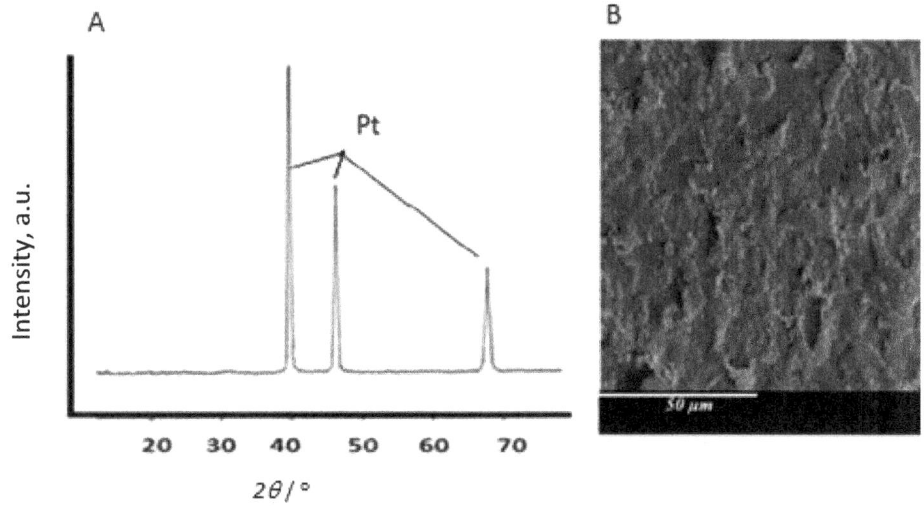

Figure 9. A *- XRD patterns of P(o-ClPh-co-o-HOPh);* ***B*** *- SEM image of P(o-ClPh-co-o-HOPh)*

The solubility of the obtained copolymer and the two homopolymers were examined in different solvents. Both P(o-HOPh) and P(o-ClPh) were found to be soluble well in DMSO, DMF and THF, while their copolymer is soluble in DMF, and HF but not soluble in DMSO. This confirmed the copolymer formation.

Potentiometric study of P(o-ClPh-co-o-HOPh)- Pt modified electrode

The pH measurements were carried out using a pHS-3B pH meter and a commercial SCE. A series of buffer solutions was used for various pH solutions which were adjusted by adding diluted HCl or NaOH solution to the solution of KH phthalate, KH_2 Phthalate, borax and/or $NaHCO_3$. Figure 10 shows that P(o-ClPh-co-o-HOPh)-Pt modified electrodes gave linear responses over the pH range from 2 to 12, with the slope values ranging from 32.5 to 56.7 mV/pH, and a linear correlation coefficients (R^2) with a value of > 0.93 as summarized in Table 2.

Table 2. *Slopes and R^2 values obtained from Figure 10*

No of cycles	pH range 2-12	
	-slope, mV pH^{-1}	R^2
3	56.29	0.99
5	48.80	0.99
10	44.90	0.97
15	29.71	0.99

Figure 10 and Table 2 show that the best slope value was obtained at a moderate thickness (no of cycles 3) while the smallest deviation of the straight line is found at pH < 4 and also at pH > 9. This suggests that the P(o-ClPh -co-o-HOPh)/Pt modified electrode may not be an effective pH sensor for more basic or more acidic solutions.

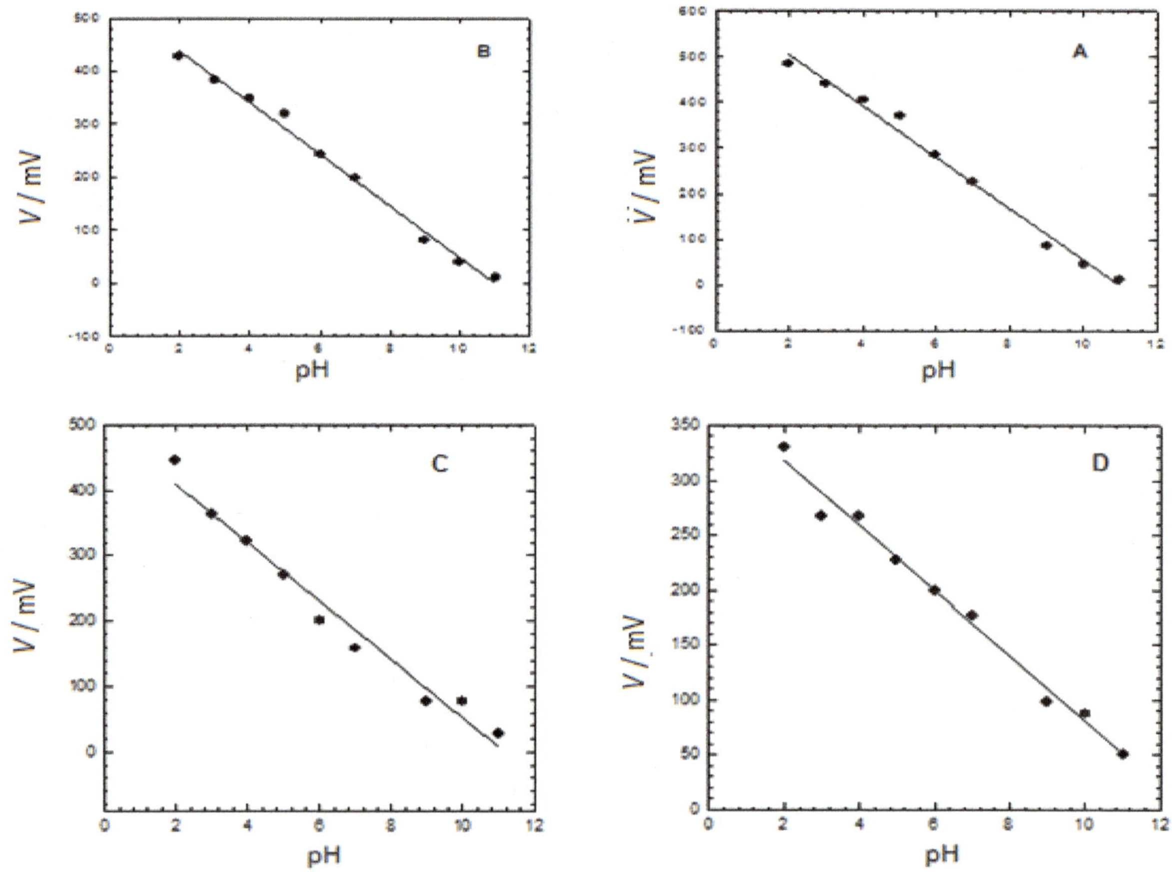

Figure 10. *P(o-ClPh-co-o-HOPh) response at different pH values, the prepared copolymer films after **A** - 3 cycles, **B** - 5 cycles, **C** - 10 cycles and **D** - 15 cycles*

Figure 11 shows the responses of P(o-ClPh-co-o-HOPh)-Pt modified electrode at different pH values (from 5 to 9). The best slopes were improved *i.e.* 56.19 mV pH^{-1} with R^2=0.98 for the film prepared after 3 cycles.

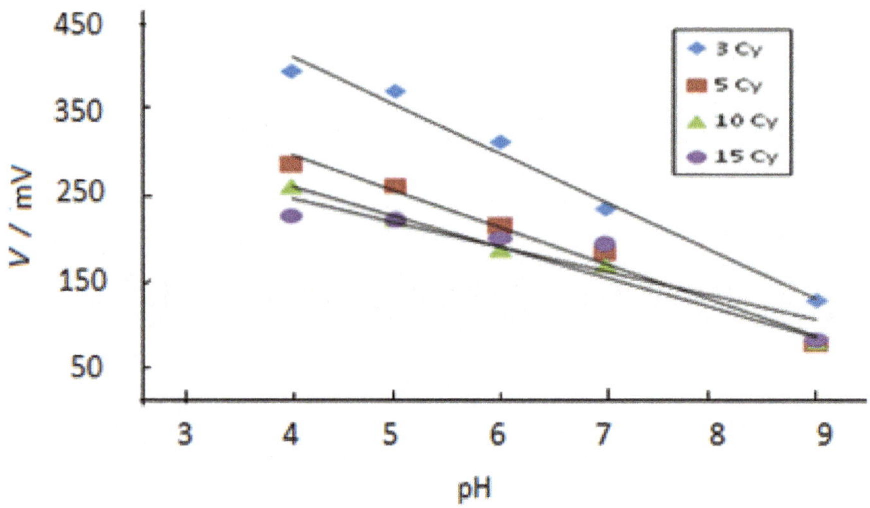

Figure 11. *The responses of P(o-ClPh-co-o-HOPh)-Pt modified electrode having different thicknesses at pH range (4-9).*

Electrode stability

Potentiometric response of the P(o-ClPh-co-o-HOPh)-Pt modified electrode was examined over a period of 9 days in order to study the stability of electrode. As seen in Figure 12 and Table 3, a significant decrease in the response slope is observed for the P(o-ClPh-co-o-HOPh)-Pt modified electrode. Also, the calibration curves of the P(o-ClPh-co-o-HOPh)-Pt modified electrode lost their linearity and were compared to the fresh electrode (1st day) even after repeatedly using it for 9 days.

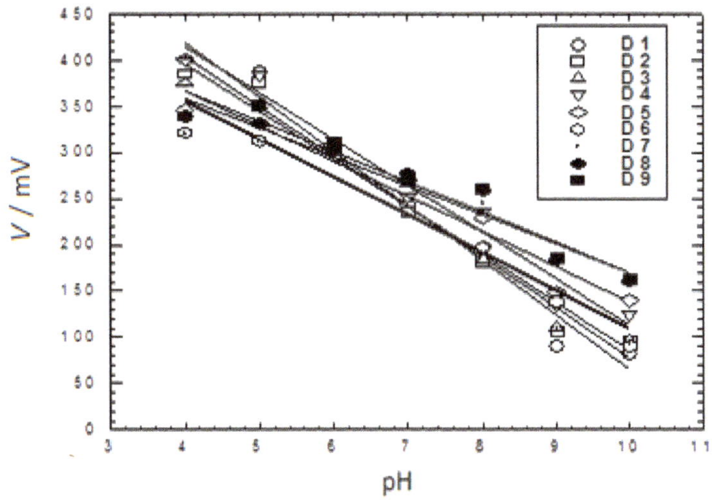

Figure 12. *P(o-ClPh-co-o-HOPh) response from the first to the ninth day*

Table 3. *Slopes and R^2 values obtained from Figs 12*

Time, day	-slope, mV pH^{-1}	R^2
1	58.7	0.97
2	54.5	0.97
3	51.5	0.97
4	49.9	0.97
5	37.9	0.95
6	40.9	0.91
7	40.5	0.91
8	31.3	0.94
9	32.5	0.93

Response mechanism

We have shown that the potentiometric responses to pH changes of the different modified electrodes are linear in the range 2-12. These responses must be mainly attributed to the copolymer films rather than the platinum substrate. Possible explanation is the affinity of the numerous hydroxyl groups and Cl atoms to the protons in solutions. The binding of H^+ onto copolymer creates local charge density excess at the electrode surface. Surface reactions seemed to take place on the copolymer film, essentially protonation and deprotonation of superficial OH groups of the polymers as symbolically described by:

$$\text{Copolymer} + H^+ \rightarrow \text{CopolymerH}^+.$$

When the equilibrium is attained at the copolymer/solution interface, we can write the equilibrium expression of the surface reaction and the equilibrium potential E as:

$$k = \frac{[\text{Copolymer}\,H^+]}{[\text{Copolymer}][H^+]}$$

$$E = E_\circ + \frac{RT}{F}\ln\frac{[\text{Copolymer}\,H^+]}{[\text{Copolymer}]} = E_\circ^{'} + \frac{RT}{F}\ln[H^+]$$

According to this mechanism of the reaction, we expect the slope of 59 mV pH^{-1} at 25 °C at all pH values. But our electrodes showed lower response slope. The presence of anionic and cationic species in the vicinity of the modified electrode surface such as (K^+, Na^+, Cl^-, *etc.*) in the buffer solutions which have an effect on the modified electrode response as mentioned in the literature [52–54].

Conclusion

P(o-ClPh-co-o-HOPh) films are prepared by the electropolymerization of a mixture of o-ClPh and o-HOPh (with 1:1 molar ratio) on Pt substrates from an aqueous medium. The process is fast and economic. The electropolymerization reaction was found to be first order with respect to the H_2SO_4 concentration, but of the second order for the comonomer concentration. The prepared P(o-ClPh-co-o-HOPh) films are stable up to 220 °C and exhibit amorphous and smooth structure with uniform distribution on the Pt substrate. IR and UV-Vis spectroscopy confirm the copolymer formation. The calculated activation energy, enthalpy, and entropy for the copolymerization were found to be 20.20 kJ mol^{-1}, 19.24 kJ mol^{-1} and -281.47 J K^{-1} mol^{-1}, respectively. The diffusion coefficient D is 6.53 10^{-7} m^2 s^{-1} in the scan rate of 15 up to 40 mV s^{-1} at 25 °C and increased with increasing temperature where E_a^{diff} was calculated to be 38.25 kJ mol^{-1} at 25 mV s^{-1}.

The potentiometric sensitivity to pH change of the P(o-ClPh-co-o-HOPh)-modified electrode exhibits a response slope of 52.16 mV/pH (30 °C), a linearity range from pH 4 to 9 and the response time less than 10 seconds. The electrode lost its response with time. The potentiometric responses to pH change were suggested to be related to the protonation and deprotonation of the polymer film atoms. Easy fabrication and low production cost of the P(o-ClPh-co-o-HOPh)-modified electrodes offer an alternative to other polymer-based pH sensors. Further development will focus on modification of the P(o-ClPh-co-o-HOPh) surfaces in order to improve the potentiometric responses to pH change and stability with time.

Acknowlegments: *Authors are highly indebted to Dr. M. Abd El-Gawad for his highly-appreciated efforts and valuable comments in the proofreading stage of the article*

References

[1] R. S. Freire, N. Duran, L. T. Kubota, *J. Braz. Chem. Soc.* **13(4)** (2002) 456-462.
[2] M. A. Ghanem, *Electrochem. Commun.* **9(10)** (2007) 2501-2506.
[3] J. Yu, W. Du, F. Zhao, B. Zeng, *Electrochim. Acta* **54(3)** (2009) 984-988.
[4] P. Kalimuthu, S. A. John, *Bioelectrochem.* **77(1)** (2009) 13-18.
[5] R. Zhang, G-D. Jin, D. Chen, X-Y. Hu, *Sensor. Actuat. B-Chem.* **138(1)** (2009) 174-181.
[6] O. Arrigoni, M. C. Tullio, *Biochim. Biophys. Acta* **1569(1/3)** (2002) 1-9.
[7] S. P. Arya, M. Mahajan, P. Jain, *Anal. Chim. Acta.* **417(1)** (2000)1-14.
[8] A. Malinauskas, R. Garjonyte, R. Mažeikiene, I. Jurevičiūte, *Talanta* **64(1)** (2004) 121-129.
[9] S. S. Medany, K. M. Ismail, W. A. Badawy, *JAR* **3(3)** (2012) 261-268.
[10] J. Li, J. Zhao, X. Wei, *Sensor. Actuator. B-Chem* **140(2)** (2009) 663.
[11] D-M. Zhao, X-H. Zhang, L-J. Feng, L. ia, S-F. Wang, *Colloid. Surface B* **74(1)** (2009) 317-321.
[12] E. M. Genies, G. Bidan, A.F. Diaz, *J. Electroanal. Chem.* **149 (1/2)** (1983)101-113.

[13] J. Bargon, S. Mohmand, R.J. Walkman, *IBM J. Res. Dev.* **27(4)** (1983) 330-331.
[14] A. F. Diaz, C J. rowley, J. Bargon, G. P. Gardini, J.B. Torrance, *J. Electroanal. Chem.* **121** (1981) 355-361.
[15] R. J. Waltman, J. Bargon, A. F. Diaz, *J. Phys. Chem.* **87(8)** (1983) 1459-1463.
[16] G. Zotti, G. Schiavon, A. Berlin, G. Pagani, *Chem. Mater.* **5(4)** (1993) 430-436.
[17] M. Satoh, K. Imanishi, K. Yoshino, *J. Electroanal. Chem.* **317(1/2)**(1991) 139-151.
[18] R. John, G. G. Wallace, *J. Elecrroanal. Chem.* **306** (1991) 157-167.
[19] S.M. Sayyah, A.B. Khaliel, R.E. Azooz and F. Mohamed. Electropolymerization of Some Ortho-Substituted Phenol Derivatives on Pt-Electrode from Aqueous Acidic Solution; Kinetics, Mechanism, Electrochemical Studies and Characterization of the Polymer Obtained, Electropolymerization, Dr. Ewa Schab-Balcerzak (Ed.), InTech, (2011).
[20] S. M. Sayyah, S. S. Abd-Elrehim, R. E. Azooz, F. Mohamed, *J. Korean Chem. Soc.* **58(3)** (2014) 289- 296.
[21] M. L. Pourciel-Gouzy, W. Sant, I. Humenyuk, L. Malaquin, X. Dollat, P. Temple-Boyer, *Sensor. Actuator. B-Chem* **103**(2004) 247-251.
[22] W. Natedungta, W. Prissanaroon-Ouajai, *Adv. Mat. Res.* **93-94** (2010) 591-594.
[23] A. Eftekhari, *Sensor. Actuator. B-Chem* **88(3)** (2003) 234-238.
[24] J. Xu, X. Sun, B. LIU, F. Xu, *Analytical Sciences*, **17 (ICAS2001)** (2001) i1363-i1365.
[25] N. Paradee, S. Noreechai, W. Prissanaroon-Ouajai, P. J. Pigram, *NU Science J.* **6(S1)** (2009) 86-95.
[26] O. Korostynska, K. Arshak, E. Gill, A. Arshak, *Sensors (Basel)* **7(12)** (2007) 3027-3042.
[27] M. Gattrell, D.W. Kirk, *J. Electrochem. Soc.* **140(4)** (1993) 903-911.
[28] P. I. Iotov, S.V. Kalcheva, *J. Electroanal. Chem.* **442** (1998) 19-26.
[29] M. Gattrell, D. W. Kirk, *J. Electrochem. Soc.* **140(6)** (1993) 1534-1540.
[30] M. Gattrell, D. W. Kirk, *Can. J. Chem. Eng.* **68(6)** (1990) 997-1003.
[31] S. M. Sayyah, S. S. Abd El-Rehim, M. M. El-Deeb, S. M. Kamal, R. E. Azooz, *J. Appl. Polym. Sci.* **117(2)** (2010) 943-952.
[32] S. Andreescu, D. Andreescu, O. A. Sadik, *Electrochem. Commun.* **5(8)** (2003) 681-688.
[33] G. Arslan, B. Yazici, M. Erbil, *J. Hazard. Mat.* B **124** (2005) 37-43.
[34] M. Gattrell, D.W. Kirk, *J. Electrochem. Soc.* **139(10)** (1992) 2736-3744.
[35] D. Mahanta, G. Madrus, S. Radhakrishnan, S. Pati, *Phys.Chem B.* **113(8)** (2009) 2299.
[36] F. Scholz, S. Schadel, A. Schultz, F. Schauer, *J. Electroan. Chem.* **480**(2000) 241-248.
[37] S. M. Sayyah, M. M. El-Deeb, S. M. Kamal, R. E. Azooz, *J. Appl. Polym. Sci.* **112(6)** (2009) 3695-3706.
[38] J. E. B. Randles, *Trans. Faraday Soc.* **44** (1948) 327-338.
[39] A. Sevick, *Collect. Czech. Chem. Commun.* **13** (1948) 349-377.
[40] S .M. Sayyah, M. M. El-Deeb, *J. Appl. Polym. Sci.* **103(6)** (2007)4047-4058.
[41] M. M. Ardakani, A. Sadeghiane, S. Moosavizadeh, M. A. Karimi, *Anal. Bioanal. Electrochem.* **1(3)** (2009) 224-238.
[42] R. S. Nicholson, *Anal. Chem.* **37(11)** (1965) 1351-1355.
[43] M. S. Ureta-Zanatru, P. Bustos, C. Berrios, M. C. Diez, M. L. Mora, C. Gutierrez, *Electrochim. Acta* **47(15)** (2002) 2399-2406.
[44] S. Mu, C. Chen, J. M. Wang, *Synth. Met.* **88 (3)** (1997) 249-254.
[45] S. M. Sayyah, S. S . Abd El-Rehim, S. M. Kamal, M. M. El-Deeb, R. E. Azooz, *J. Applied. Polym. Sci.* **119** (2011) 252-264.
[46] S. M. Sayyah, M. M. El-Deeb, S. M. Kamal, R. E. Azooz, *J Applied Polym. Sci.* **112** (2009) 3695-3705.
[47] K. A. Connor, *Chemical Kinetics*, VCH Publishers: New York, (1990).
[48] T. Sismanoglu, A. Ercag, S. Pura, E. Ercag, *J. Braz. Chem.* Soc. **15(5)** (2004) 669-675.

[49] Y. Wei, J. Tian, D. Glahn, B. Wang, D. Chu, *J. Phys. Chem*. **97(49)** (1993) 12842-12847.
[50] S. S. Abd El Rehim, S. M. Sayyah, R. E. Azooz, *Port. Electrochim. Acta* **30(1)** (2012) 67-80.
[51] S. Dubey, D. Singh, R.A. Misra, *Enzyme. Microbiol. Technol.* **23(7/8)** (1998) 432-437.
[52] A. Cadogan, A. Lewenstam, A. Ivaska, *Talanta* **39(6)** (1992) 617-620.
[53] V. M. Jovanović, L. Markičević, S. Stanković, R. Stanković, M. S. Jovanović, *Electroanalysis* **7(6)** (1995) 574-578.
[54] A. Michalska, K. Maksymiuk, A. Hulanicki, *J Electroanal Chem* **392(1/2)** (1995) 63-68.

17

EDTA as a corrosion inhibitor for Al in 0.5 M HCl: adsorption, thermodynamic and theoretical study

Rehab E. Azooz✉

Chemistry Department, Faculty of Science, Jazan University, 2097 Jazan, Saudi Arabia

✉Corresponding Author: re_azooz@yahoo.com

Abstract

In this study; EDTA is used in very small amount (10^{-10} M) as an inhibitor for the Al corrosion in 0.5 M HCl. Thermodynamic and adsorption parameters are calculated. The result shows that, in this range of concentrations, EDTA is chemisorbed at the Al surface, forming a stable complex with Al and give inhibition efficiency up to 89 %. For more concentration, unstable complex is formed and acceleration of corrosion occurs. The adsorption fit well to Langmuir, Temkin isotherms and El-awady model. Density functional theory (DFT) is used to study the geometrical optimizations of EDTA. From the obtained optimized structure, The highest occupied molecular orbital (E_{HOMO}), the lowest unoccupied molecular orbital (E_{LUMO} and their energy difference (ΔE), the total energy (TE), electronegativity (χ), dipole moment (μ), global hardness (η), global softness (σ), electron affinity (A), ionization potential (I), the fraction of electrons transferred (ΔN) and were determined using B3LYP/6-31G(d,p) basis set.

Keywords

EDTA; Inhibition efficiency; Adsorption isotherms; Thermodynamic parameters; Theoretical parameters

Introduction

The study of Al corrosion is of great importance; various industrial operations depend mainly on Al. Most investigations on the corrosion of Al have been carried out on. The development of corrosion inhibitor is a good branch based on a functional organic compound. The structure and function groups of used organic compounds are useful for obtaining a good inhibitors [1-4]. Depending upon excellent conductivity (electrically and thermally) of Al; application of Al is varied and widespread. [5]. Adsorption of inhibitor on the charged metal surface is the main process to inhibit corrosion, on this basis; multiple bond(s), an electron rich atom as, S, N or P or a ring is a main centers for the adsorption processes. In aqueous media, inhibitors are used to prevent or reduce the corrosion of metals [6–11]. It was shown that, compounds containing N or/and O atoms exhibit a good inhibiting effect. A polyprotic acid, *i.e.* Ethylenediaminetetraacetic acid (EDTA), with a lone pair of electrons in its amino groups and two carboxylic acid groups is used for complexation with the charged metal ions [6,12]. Complexation occur between (free or π) -electrons from inhibitor and the vacant d-orbital of a metal through the formation of donor–acceptor surface [6,14-16]. In the last years, EDTA has been studied to protect metals from corrosion in different environments [16–19], it was found that, different parameter affects the inhibition effect of EDTA including, the pH value, temperature, concentration, and type of the metal. Nahle [20] has found that the Sn(II)–EDTA complex increased the dissolution rate of Sn in a basic medium. Milošev *et al.* [16] have investigated the corrosion of stainless steel in physiological solutions, while, EDTA prevents the formation of a passive layer and increases the solubility of the metal. Gadiyar *et al.* [22] have discovered that EDTA inhibits the corrosion of carbon steel. However, its inhibiting effect is imperfect. Alhaji and Reda [23] have stated that EDTA is effective in decreasing the corrosion rate of copper-nickel alloy in seawater contaminated with sulfur. S. Zor *et. al.* [24] observe that, the corrosion of Al is higher in 0.1 M NaCl solution in higher concentration of EDTA, and become slower at 10^{-4}M EDTA

The molecular structure, electronic structure and reactivity of Inhibitors are determined well by quantum chemical methods [25]. A powerful framework is provided by DFT [25,26] that help in understanding a lot of chemical processes [27-31]. Concepts as, electronegativity hardness or softness *etc.* are used to describe chemical reactivity [28], are appear naturally within DFT. The local electron density/population displacements represented the inflow of a single electron is measured using Fukui function [30] and is representing the relative local softness of the electron gas. In the present study the inhibition effect of EDTA for the corrosion of Al in 0.5 M HCl has been done using both weight loss and electrochemical methods. The temperature effect and adsorption isotherms will be studied in details. Also analyzing the inhibitive properties of EDTA using DFT calculations will be done.

Experimental

Chemical and reagents

Al strips have a rectangular form (4.5×3.5×0.2 cm), with the composition 99.11 % of Al, 0.019 % of Zn, 0.036 % of Cu, 0.001 % of Mg, 0.834 % of Si and, were mechanically polished using different grades of emery sheets, washed with acetone and distilled water and dried. EDTA disodium salt (Analar grade) and HCl were obtained from Fluka AG, Switzerland. All solutions were prepared using freshly prepared bidistilled water. Stock solution of EDTA was prepared, from which all used concentrations are prepared via dilution.

Methods

Weight loss measurements

The Al samples (coupons) were weighed before immersion in 250 ml beaker containing 50 ml of the respective prepared test solutions at room temperature and desired temperatures. The setups were exposed for a period of 100 min. Corrosion reaction is quenched in concentrated HNO_3

digressed in CH_3COCH_3 washed under water, dried and weighed. A mean value triplicate experiments is reported in each case. The values of weight loss in the presence and the absence of EDTA is used to calculate efficiency at the end of definite intervals of time.

Temperatures effects

The same procedure adopted where the temperature of the study was varied, in the range (303-333 K), from at the end of each experiment. The specimens were taken out, washed both in running tap water and into distilled water. They were dried and their weights were recorded.

The loss in weight was calculated. Each experiment was duplicated to get good reproducibility. Weight loss measurement was performed in 0.5 M HCl with and without the addition of EDTA in the range $(6.4 - 10.07)\times10^{-10}$ M.

Electrochemical methods

All electrochemical experiments were recorded using a potentiostate/galvanostate (EG&G 326A, U.S.A). The potential was scanned at the scan rate 10 mV s^{-1}. All experiments were repeated to ensure reproducibility. Fresh solution was used for each experiment. The cell used is a three compartment home-made one, with a reference saturated calomel electrode (SCE), an auxiliary (Pt-foil) electrode and a working (Al) electrode with 0.4 cm^2 area exposed to corroded solution was used.

Adsorption isotherms

The adsorption of inhibitor at a metal /solution interface is the main source of inhibition effect, accordingly, the isotherms of adsorption can be determined. In order to obtain the isotherm the fractional surface coverage values (θ) as a function of inhibitor concentration must be obtained. The values of θ can be easily determined from the weight loss measurements by the ratio;

$$\theta = \frac{I.E.}{100}$$

where *IE* is inhibition efficiency obtained by a weight loss method. So, it is necessary to determine empirically which isotherm fits best to the adsorption of inhibitor on the Al surface.

Scanning electron microscopy (SEM)

After a period of 100 min, Al coupons was removed from solution, rinsed with a double distilled water, dried and observed in a Scanning Electron Microscope (JSM-T20 Electron Probe Micro-analyzer (JEOL, Tokyo, Japan)) to examine the surface morphology. The following cases were examined, to understand the morphology of the Al surface in the absence and presence of inhibitors, (i) aluminum coupon after polishing, (ii) aluminum coupon dipped in 0.5 M HCl for 100 min. at 303 K and (iii) aluminum coupon dipped in 0.5 M HCl containing 2.7×10^{-10} M of EDTA inhibitor 100 min

Quantum chemical calculations

DFT is used to obtain the complete geometrical optimizations of EDTA, with Beck's exchange functional along with nonlocal correlation functional (B3LYP) of Lee–Yang–Parr [32–34] with 6-31G* basis set in Gaussian 03 program package [35]. From the obtained optimized structure, serval quantum chemical parameters were calculated; E_{HOMO}, E_{LUMO}, ΔE_{gap}, the dipole μ and *TE*.

Results and discussion

The molecular structure of an organic compound used in the present study is given in Scheme 1.

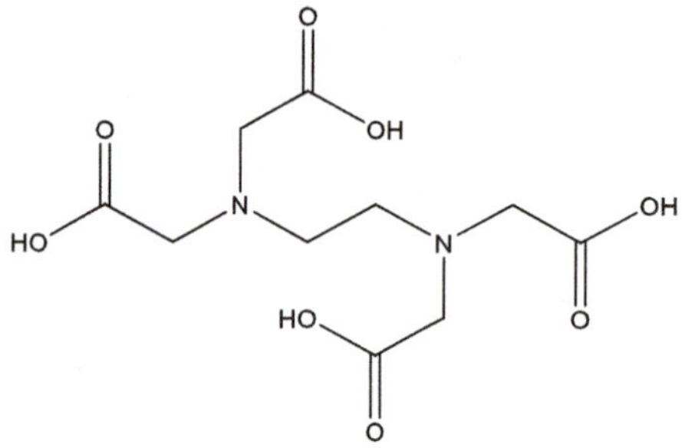

Scheme 1. *Structure of EDTA*

Open circuit potential

Potential-time curves was recorded for 60 minutes of immersion of the Al specimens in aqueous 0.5 M HCl solution without and with EDTA at required concentrations. As seen in Figure 1. From Figure 1, when Al is immersed in the HCl solution E_{OCP} drop sharply, then began to increase to more positive value and reached a stationary value after 25 minutes of immersion. The aggressiveness of the corroded solution may cause the differences in E_{OCP} values at the beginning of Al exposure, It was suggested that, adsorption of EDTA molecules on the Al surface is the reason for the initial negative shift.

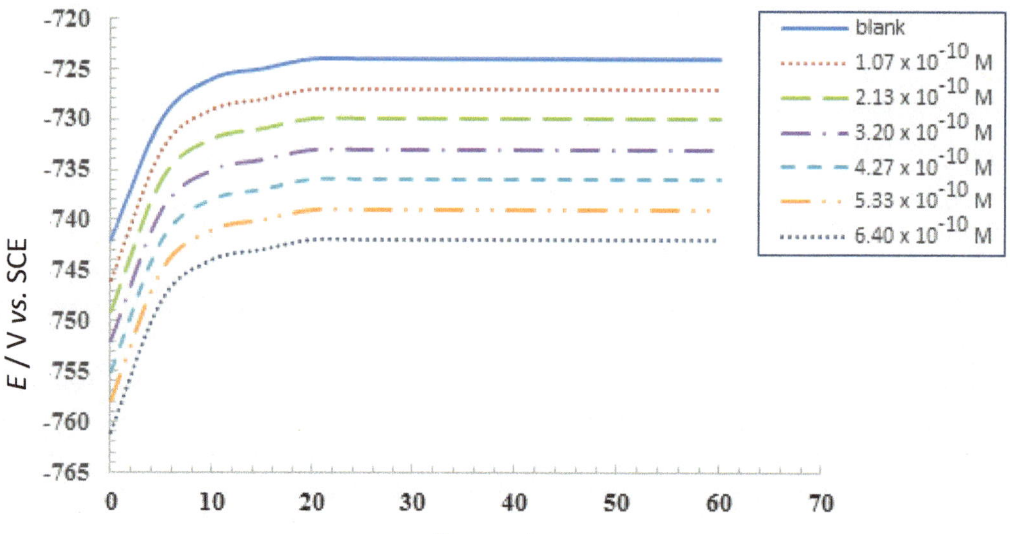

Figure 1. *Potential - time curves for Al in 0.5 M HCl in absence and presence of different concentrations of (EDTA) at 303 K.*

The results have shown that the addition of EDTA molecules at the beginning shifts E_{OCP} to more negative values. And then become more positive with time, due to oxide film growth [36]. In particular, initial values are more negative than steady state values, also the dependence of the E_{OCP} on concentration suggests that, the inhibitor molecules are strong and rapidly adsorbed at the steady state potentials [36].

Potentiodynamic polarization studies

The cathodic and anodic polarization curves of Al in 0.5 M HCl in the absence and presence of different EDTA concentrations at 303 K are shown in Fig. 2. The electrochemical kinetic parameters (in the potential range ±50 mV from E_{corr}), namely, corrosion current (i_{corr}), corrosion potential (E_{corr}), and Tafel slopes, (β_c and β_a), have been determined simultaneously and are listed in Table 1. Data infer that, the addition of EDTA to the acid solutions increases both the anodic and cathodic overpotentials, decreases the corrosion current density, i_{corr}, and shifts the E_{corr} to more positive values. This means that the presence of EDTA inhibits the partial anodic dissolution of Al and also retards the partial cathodic reduction of hydrogen ion.

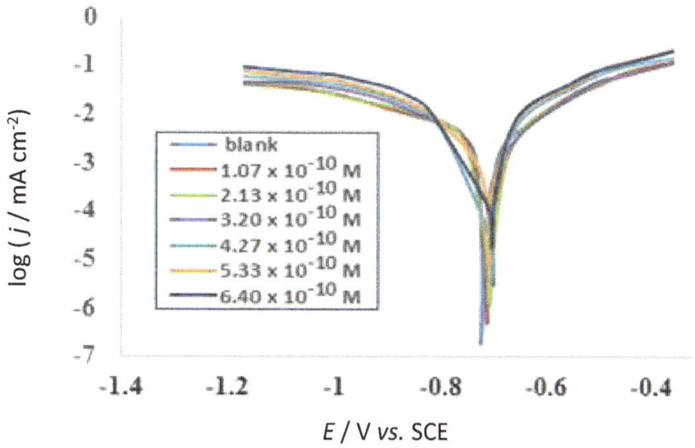

***Figure 2**. Potentiodynamic polarization curves for Al in 0.5M HCl at 303 K with scan rate of 10 mV s^{-1} with and without different concentrations of EDTA.*

These results reveal that EDTA acts as a mixed type inhibitor. The inhibition efficiency *IE*, at different inhibitor concentrations at 303 K for Al electrode in 0.5 M HCl solution was calculated from the Equation 1 [37-39]:

$$IE\,/\,\% = \left(1 - \frac{i_{corr}}{i^0_{corr}}\right) \times 100 \qquad (1)$$

where, i°_{corr} and i_{corr} are corrosion current density for uninhibited and inhibited solutions respect-tively.

***Table 1**. The electrochemical kinetic parameters (i_{corr}, E_{corr}, β_c and β_a) and inhibition efficiency (IE) obtained from polarization curves of Al electrode in 0.5M HCl at 303 K in the absence and the presence of EDTA.*

c_{EDTA} / M	i_{corr} / mA cm^{-2}	$-E_{corr}$ / mV	$-\beta_c$ / mV dec^{-1}	$-\beta_a$/ mV dec^{-1}	IE / %
Blank	0.89	670	122	0.69	--
1.07 10^{-10}	0.47	660	118	0.60	47.2
2.13 10^{-10}	0.46	650	116	0.53	48.3
3.20 10^{-10}	0.35	630	118	0.52	60.7
4.27 10^{-10}	0.21	580	112	0.51	76.4
5.33 10^{-10}	0.12	530	115	0.50	86.5
6.40 10^{-10}	0.10	500	114	0.50	88.8

Mass loss

The mass losses of Al in 0.5 M HCl solution, with and without different concentrations of the EDTA were recorded after 100 min. of immersion at different temperatures. The corrosion rates of Al alloy were calculated using Equation 2 [37].

$$CR = \frac{87.6\Delta m}{Atd} \qquad (2)$$

where Δm is the mass lost (g), 87.6 is a constant, A is the surface area of the coupon (cm^2), d is the density (g cm^{-3}), t is the time of exposure (h). The calculated CR fits into the range (less than 0.50 mm year^{-1}) at which the application is acceptable [39]. Figure 3 (A and B) shows the variation in mass loss for Al coupons in the absence and the presence of EDTA. The mass loss in the presence of inhibitor is much smaller than the blank solution. The significant difference shows reduce impact on the CR of Al in 0.5 M HCl.

Both of the surface coverage (θ) and the inhibition efficiency (IE) were calculated using mass loss data according to Equations 3 and 4, respectively [38].

$$\theta = \left(1 - \frac{w_{inh}}{w_{blamk}}\right) \quad (3)$$

$$IE\,/\,\% = \theta \times 100 \quad (4)$$

where, w_{blank} is the corrosion rate in the uninhibited environments. w_{inh} is the corrosion in the inhibited environment. The high inhibition efficiency as the inhibitor concentration increases could be understood to be due to the reduction in corrosion rate. Thus, EDTA could be considered as an inhibitor of Al in 0. 5 M HCl solution given the high level of inhibition efficiency. The inhibitor efficiency, increased with the inhibitor concentration.

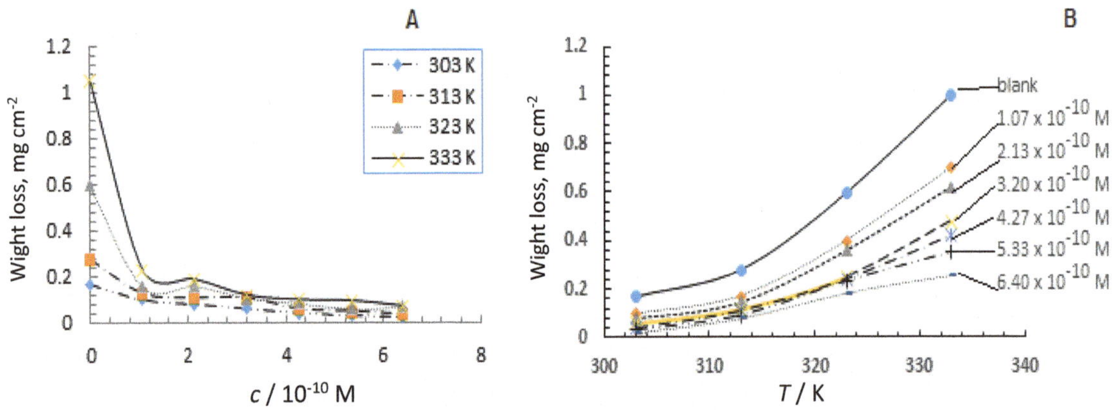

***Figure 3**. Mass loss of Al immersed for 100 min. in 50ml HCl in the presence or absence of EDTA at different temperatures and EDTA concentrations.*

Figure 4 shows the inhibition efficiency in different concentration of the EDTA and it is seen that the IE increases linearly with the inhibitor concentration.

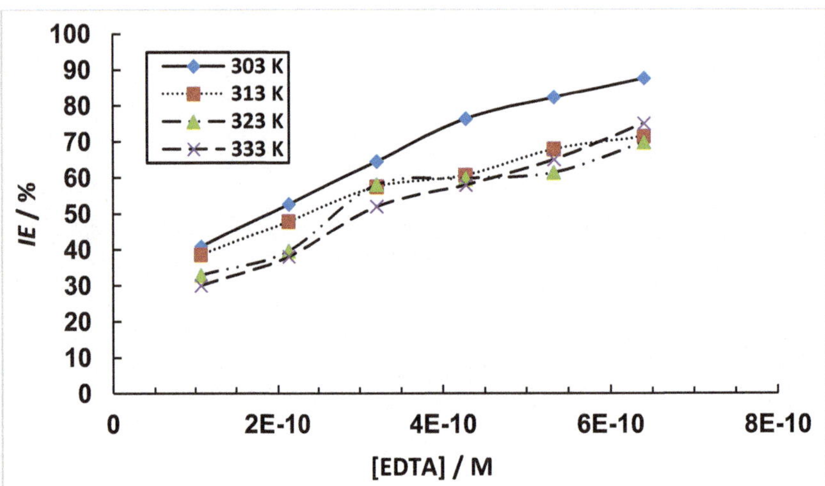

***Figure 4.** IE after 100 min. in 50 ml HCl at different [EDTA] at different temperatures*

Adsorption studies

Inhibition efficiencies of Al in 0.5 M HCl was increase with increasing additive concentrations of EDTA, this phenomenon can be explained on the basis of adsorption. Adsorption of the inhibitor can explain the nature of Al/EDTA interaction. In the acid solution, firstly inhibitor is adsorbed on the metal surface and cover certain area from corroded solution and decrease or prevent this area from dissolution, whereas corrosion reactions normally occurred on inhibitor-free areas. Accordingly, the area covered with inhibitor species (θ), can follow as a function of inhibitor concentration and/or solution temperature. When θ is tested as a function of the concentration (at constant temperature), the adsorption isotherm can be evaluated at the equilibrium condition. Four adsorption isotherms were tested using data from both weight loss and electrochemical techniques;

A. Langmuir's isotherm

The dependence of θ at the concentration of the inhibitor, was fitted to Langmuir's isotherm, assuming that, a fixed number of adsorption sites is present on Al surface, each one of these sites holds only one adsorbed species. Figure 5 shows linear plots of c/θ versus c with $R^2 \geq 0.90$, the average correlation coefficient, which suggests that adsorption was fitted to Langmuir's isotherm as in Equation 5 [37].

$$\frac{c}{\theta}=\frac{1}{K_{\mathrm{ads}}}+c \tag{5}$$

where c is inhibitor concentration, K_{ads} adsorptive equilibrium constant representing the degree of adsorption (*i.e.* if K_{ads} having higher value, the inhibitor is strongly adsorbed on metal surfaces).

As shown in Table 2, the value of K_{ads} which was obtained from the reciprocal of the intercept of a Langmuir plot lines, and R^2 of all lines were near unity. This means that obtained results is fit well with Langmuir isotherm. The higher values of K_{ads} indicating a strong interaction between EDTA and the Al surface. It seemed, therefore, that electrostatic interaction (physisorption) between inhibitor molecules existing as cations should prevail over molecular interaction, and this often results in strong interactions (chemisorption).

***Table 2**. Data obtained from Figure 3*

Temperature, K	303		313		323		333	
Technique used*	Wt.-loss	Elec.	Wt.-loss	Elec.	Wt.-loss	Elec.	Wt.-loss	Elec.
R^2	0.99	0.93	0.99	0.96	0.98	0.97	0.98	0.97
ΔG^{o}_{ads}, kJ mol^{-1}	-66.38	-66.38	-67.25	-70.95	-70.78	-70.76	-71.83	-75.16

* Wt.-loss - Weight loss measurements; Elec. - Electrochemical methods

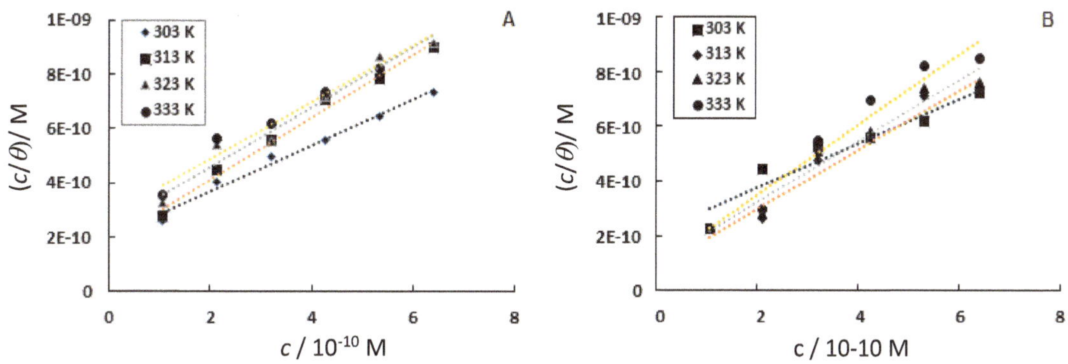

***Figure 5**. Plots of c/θ versus c of Langmuir's adsorption isotherm for the corrosion of Al in 0.5 M HCl at different temperatures. A: From weight loss technique and B: from electrochemical technique*

The Equilibrium constant of adsorption K_{ads} is related to the standard adsorption free energy (ΔG^{o}_{ads}) by Equation 6:

$$K_{ads} = \frac{1}{55.5}\exp\left(-\frac{\Delta G^{0}_{ads}}{RT}\right) \quad (6)$$

where 55.5 is the concentration of water in the solution expressed in, R is the gas constant and T is the absolute temperature. From Table 1, the average value of standard adsorption free energy (ΔG^{o}_{ads}) > -40 kJ mol^{-1}. The negative value of ΔG^{o}_{ads} ensures spontaneity of the adsorption process and the stability of the adsorbed layer on metal surfaces. In general, the values of (ΔG^{o}_{ads} up to -20 kJ mol^{-1} are consistent with the electrostatic interaction between the charged molecules and the charged metal (physisorption), while those around -40 kJ mol^{-1} or higher are associated with chemisorption as a result of sharing or transferring of electrons from organic molecules to metal surface to form a coordinate type of bond. In the present work, the calculated value of $\Delta G°$ in all studied temperatures in both techniques are > -40 kJ mol^{-1} indicating that the adsorption mechanism of EDTA on Al surfaces in 0.5 M HCl solution was typical of chemisorptions.

B. Temkin isotherm

The nature of the interaction at metal/solution interface is studied by Temkin isotherm. By assuming a uniform distribution of the adsorption energy that increases with the increase of the θ. Temkin isotherm model are given by the Equation (7a and 7b).

$$\exp(f,\theta) = K_{ads}c \quad (7a)$$

and it is rearranged

$$\theta = (1/f)\log c + (1/f)\log K_{ads} \quad (7b)$$

where K_{ads} is the equilibrium constant, c is the inhibitor concentration, θ is the surface coverage, f is the interaction term parameter, a lateral attraction between the adsorbing molecules is assumed if $f > 0$, but if $f < 0$, there is a lateral repulsion. The plot of θ versus log c, yields curve with linear correlation coefficient $R^2 \geq 0.90$, close to unity, in all cases. The obtained value of K_{ads}(average) ≈ 4.1×10^4 and ≈4.4×10^4 in case of weight loss and polarization techniques repetitively, $f > 0$ indicating a strong lateral attraction between the adsorbing molecules of EDTA and the surface of the Al.

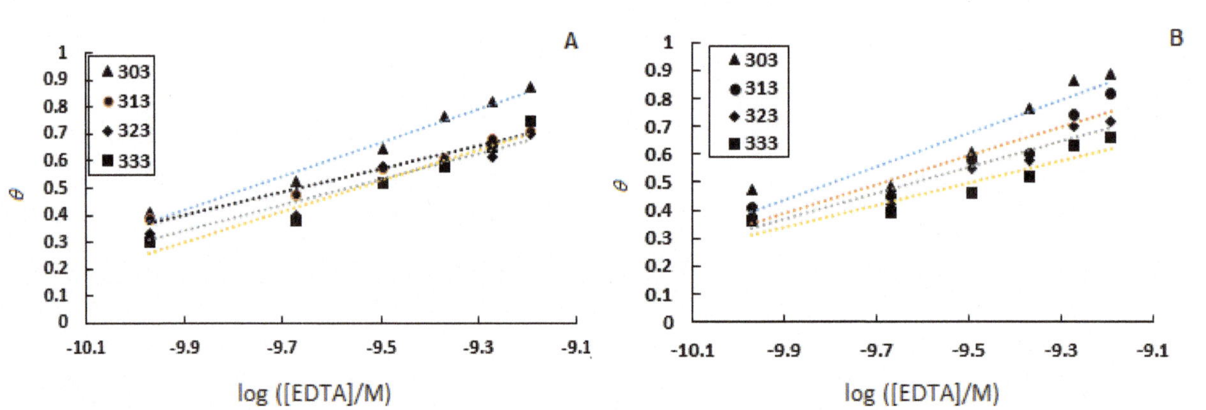

Figure 6*. Plots of c versus θ of Temkin's adsorption isotherm for the corrosion of Al in 0.5 M HCl at different temperatures, A: from weight loss technique and B: from electrochemical technique*

***Table 3.** Data obtained from Figure 4*

T / K	303		313		323		333	
Technique used*	Wt.-loss	Elec.	Wt.-loss	Elec.	Wt.-loss	Elec.	Wt.-loss	Elec.
f	1.6	1.7	2.3	1.93	2.1	2.15	1.8	2.49
K_{ads}	3.9 x10^4	4.1 x10^4	5.1 x10^4	4.2 x10^4	4.1 x10^4	4.4 x10^4	3.3 x10^4	4.7 x10^4
R^2	0.95	0.90	0.96	0.90	0.95	0.91	0.95	0.91

* Wt.-loss - Weight loss measurements; Elec. - Electrochemical methods

C. Flory-Huggins isotherm

The amount of the inhibitor molecules that could displace the water molecules from the metal surface is studied using Flory-Huggins isotherm, which is showed by equation. 8

$$\log(\theta/c) = \log k + x \log(1-\theta) \tag{8}$$

where x is the size parameter that measure the number of adsorbed water molecules replaced by a given inhibitor molecule. Figure 7 shows the plot of log (θ/c) *vs.* log (1-θ), linear relationships with R^2 > 0.8 is obtained, and indicating Flory-Huggins isotherm was obeyed. The obtained $(K_{ads})_{avg}$ = 1.5×10^4 and the calculated ΔG_{ads} > -34 kJ/mol. The size parameter is approximately 1.

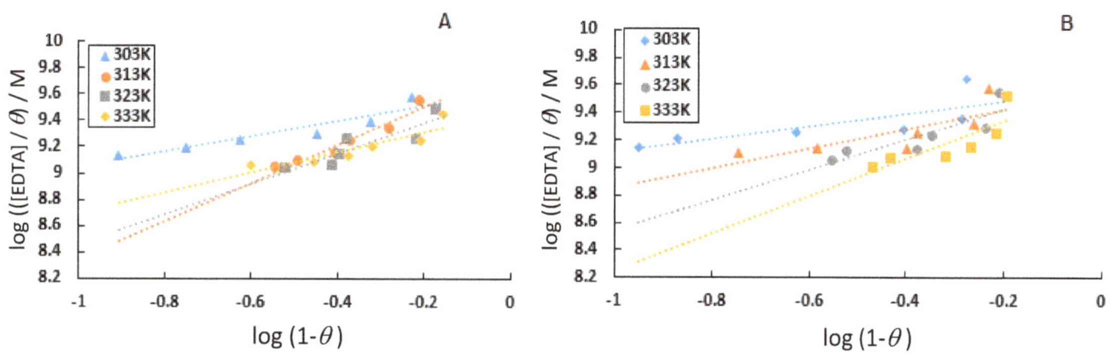

***Figure 7**. Plots of log (1- θ) versus log (θ/c) of Flory Huggin's adsorption isotherm for the corrosion of Al in 0.5 M HCl at different temperatures, A: From weight loss technique and B: from electrochemical technique*

***Table 4**. Data obtained from Figure 5*

T / K	303		313		323		333	
Technique used*	Wt.-loss	Elec.	Wt.-loss	Elec.	Wt.-loss	Elec.	Wt.-loss	Elec.
x	0.58	0.47	1.45	0.71	1.15	1.1	0.76	1.4
K_{ads}	1.5 x10^4	1.5x10^4	1.8 x10^4	1.4x10^4	1.5 x10^4	1.5x10^4	1.3 x10^4	1.5 x10^4
R^2	0.96	0.90	0.98	0.90	0.95	0.91	0.95	0.91

* Wt.-loss - Weight loss measurements; Elec. - Electrochemical methods

D. Thermodynamic-kinetic model

The surface coverage values obtained from the gravimetric and polarization measurements were also fitted into the adsorption isotherm of the thermodynamic-kinetic model of El-Awady et al. are represented in Equation. 9

$$\log\left(\frac{\theta}{1-\theta}\right) = \log K'c + y \log c \tag{9}$$

where c is the concentration of the exudates, θ is the degree of surface coverage, K_{ads} is the Equilibrium constant of adsorption process, and K_{ads}= $K^{1/y}$. 1/y is the number of inhibitory molecules occupying one active site (or the number of water molecules replaced by one molecule of EDTA. Curves fitting of the data in the thermodynamic-kinetic model is shown in Fig. 8. This data gave straight lines, the values of 1/y and K_{ads} calculated from the El-Awady *et al.* curve model is given in

Table 4. The values of $1/y$ (average) obtained are more than unity in all cases, indicating that each molecule EDTA involved in the adsorption process is attached to more than one active site on the metal surface.

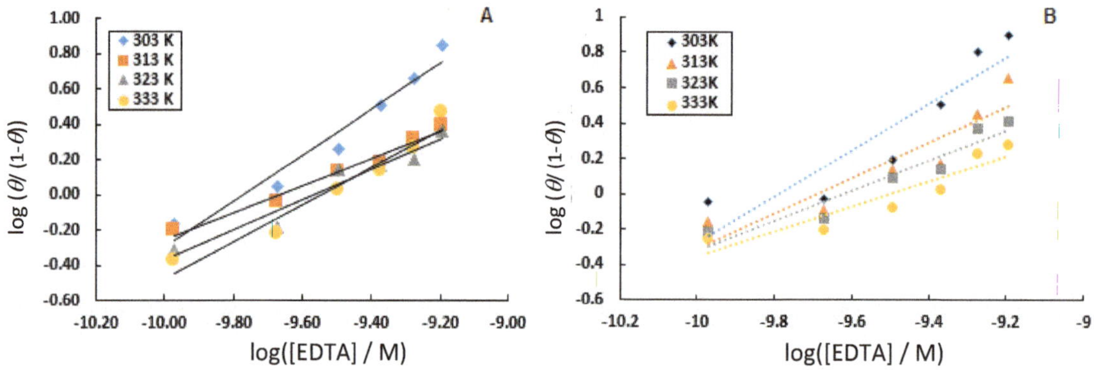

Figure 8. *Plots of log c versus log (θ / 1-θ) of thermodynamic-kinetic model for the corrosion of Al in 0.5 M HCl at different temperatures, A: From weight loss technique and B: from electrochemical technique*

Table 5. *Data obtained from Figure 5*

T / K	303		313		323		333	
Technique used*	Wt.-loss	Elec.	Wt.-loss	Elec.	Wt.-loss	Elec.	Wt.-loss	Elec.
y	1.31	1.32	0.77	1	0.86	0.84	1.05	*0.72*
1/y	0.77	0.76	1.31	1	1.22	1.18	0.92	*1.39*
K'	7.7 x10^{14}	9.1x10^{14}	4.1 x10^{12}	3.7 x10^{13}	8.9 x10^{12}	7.9x 10^{12}	5.3x 10^{13}	*2.1x1012*
R^2	0.95	0.90	0.94	0.90	0.94	0.92	0.93	*0.91*

* Wt.-loss - Weight loss measurements; Elec. - Electrochemical methods

By rearrangement of Gibbs-Helmholtz equation we obtain Equation 10, which is used to calculate the enthalpy of adsorption (ΔH_{ads})

$$\Delta G_{ads}/T = (\Delta H_{ads}/T)\ K \tag{10}$$

A plot between the variations of ($\Delta G_{ads}/T$) and ($1/T$) gave a straight line whose slope is ΔH_{ads} as shown in Figure 9. The entropy of adsorption ΔS_{ads} was calculated using the following thermosdynamic Equation (Equation 11):

$$\Delta S_{ads} = (\Delta H_{ads}\ \Delta G_{ads})\ /\ \mathrm{T} \tag{11}$$

where, data of ΔG_{ads} were taken from Langmuir isotherm results (from its R^2 value, it is the best fit model)

The obtained date of the calculated ΔH_{ads} and ΔS_{ads} was tabulated in Table 6.

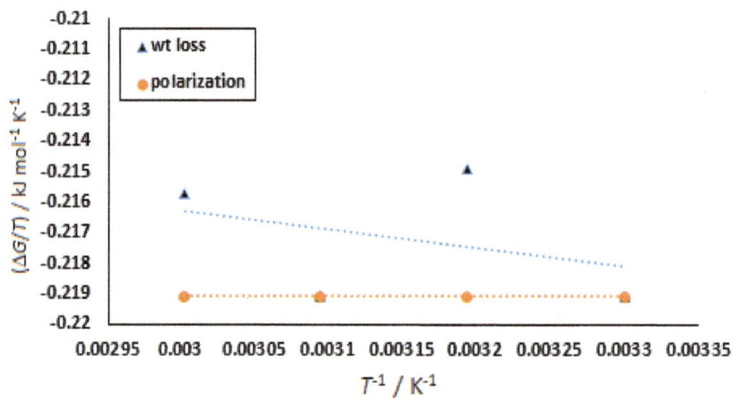

Figure 9. *Gibbs-Helmholtz rearranged relation between (ΔG_{ads}/T) and (1/T)*

Table 6. *Adsorption thermodynamic parameters obtained using Langmuir isotherm*

T / K	303		313		323		333	
Technique used*	Wt.-loss	Elec.	Wt.-loss	Elec.	Wt.-loss	Elec.	Wt.-loss	Elec.
$-\Delta G_{ads}$ / kJ mol^{-1}	66.38	66.38	67.25	70.95	70.78	70.76	71.83	75.16
$-\Delta H_{ads}$ / kJ mol^{-1}	6.02	0.02	6.02	0.02	6.02	0.02	6.02	0.02
ΔS_{ads} / J mol^{-1} K^{-1}	199.2	196.1	195.6	203.7	200.4	196.1	197.6	202.7

* Wt.-loss - Weight loss measurements; Elec. - Electrochemical methods

The negative sign of ΔH_{ads} indicated the exothermic process of adsorption of the inhibitor on aluminum surface in HCl. The positive value of ΔS_{ads} in the presence of inhibitor can be attributed to the increase in the solvent entropy and more positive desorption entropy. It is also interpreted that the increase of disorderness is due to more water molecules which can be desorbed from the metal surface by one inhibitor molecule. Therefore, it is revealed that decrease in the enthalpy is the driving force for the adsorption of the inhibitor on the surface of aluminum [28,29].The calculated values of heat of adsorption and entropy of adsorption are listed in Table (6).

Effect of temperature

Based on temperature effect, inhibitors may be classified into three groups:

1. Inhibitors whose inhibition efficiency (*IE*) decreases with temperature increase. The value of the apparent activation energy E_a, found is greater than that in the uninhibited solution;
2. Inhibitors in whose *IE* does not change with temperature variation. The apparent activation energy E_a, does not change with the presence or absence of inhibitors;
3. Inhibitors in whose presence the *IE* increases with temperature increase while the value of E_a for the process is smaller than that obtained in the uninhibited solution.

Thus, in examining the effect of temperature on the corrosion process in the presence of EDTA, the Arrhenius Equation (Eq. 12) is helpful

$$\log CR = \frac{-E_a}{2.303\,RT} + \log A \tag{12}$$

where CR is the corrosion rate, E_a is the apparent activation energy, R is the molar gas constant, T is the absolute temperature, and A is the frequency factor. Figure 10 represents the Arrhenius plot as log CR *vs.* 1/T for Al corrosion in 0.5 M HCl in free so inhibited solution, linear plots were obtained. The values of E_a were obtained from the slope of the Arrhenius plot and are presented in Table 8. From the table, it is seen that E_a increases in the presence of the inhibitors compared to the blank. The higher value of the activation energy of the process in an inhibitor's presence when compared to that in its absence is attributed to its physisorption, while the opposite is the case with chemisorption.

According to Eyring relationships (Eq. 13), both of S^* and H^* could be obtained,

$$-\ln\frac{Rh}{NT}R_c = \frac{\Delta H^*}{RT} - \frac{\Delta S^*}{R} \tag{13}$$

where h is the Planck's constant (6.626176×10^{-34} J s), N is the Avogadro's number (6.02252×10^{-23} mol^{-1}), R is the universal gas constant, ΔH is the enthalpy of activation and ΔS is the entropy of activation. The kinetic results were found to fit the Arrhenius and Eyring equation, where plots of $1/T$ *vs.* ln Rc/T or $1/T$ *vs.* $-\ln(hRc/k_BT)$ (k_B is Boltzman constant and equation the term R/N) resulted in good straight lines. The activation parameters ΔH^* and ΔS^* can be evaluated from the slopes and intercepts of the straight line.

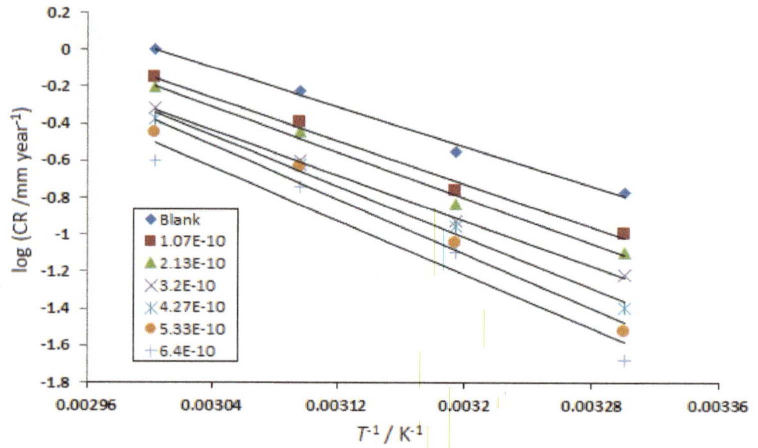

Figure 10*. Arrhenius plot as log CR* vs. *1/T for Al corrosion in 0.5 M HCl in the absence and presence of various concentrations of EDTA. Data obtained from weight loss technique*

Table 7*. Activation energy, E_a for aluminum corrosion in the presence of EDTA in 0.5 M HCl.*

c / M	$-E_a$ / kJ mol^{-1}
0	51.05
1.07 x10^{-10}	56.01
2.13 x10^{-10}	59.11
3.20 x10^{-10}	58.55
4.27 x10^{-10}	65.93
5.33 x10^{-10}	70.02
6.40 x10^{-10}	69.64

Figure 11 shows Eyring plot and all lines are straight from which ΔH and ΔS were evaluated and their values are put in Table 8.

The positive values of ΔH reflect the endothermic dissolution of Al in the presence and absence of the inhibitor. The increase in ΔH_a with the increase in the concentration of the inhibitor for Al corrosion reveals that, the decrease in Al corrosion rate is mainly controlled by kinetic parameters of activation. The negative values of ΔS may reflect the association mechanism of corrosion, *i.e.*, the decrease in disorder takes place on going from reactants to the activated state.

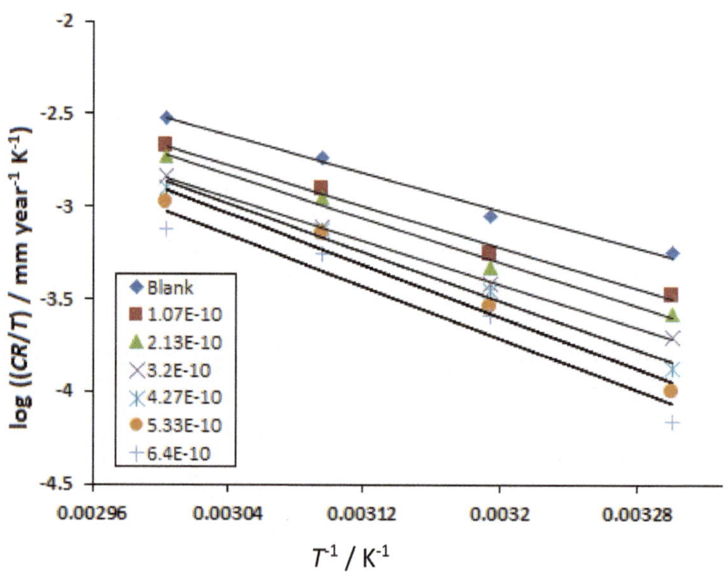

Figure 11*. Eyring plot as log CR/T vs. 1/T for Al corrosion in 0.5 M HCl in the absence and presence of various concentrations of EDTA. Data obtained from weight loss technique*

Table 8*. Thermodynamic parameters, ΔH and ΔS (for aluminum corrosion in the presence of EDTA in 0.5 M HCl)*

c / M	-ΔS^* / J mol^{-1} K^{-1}	ΔH^* / kJ mol^{-1}
0	100.49	21.02
1.07 10^{-10}	88.52	23.18
2.13 10^{-10}	80.07	24.52
3.20 10^{-10}	84.13	24.28
4.27 10^{-10}	62.26	27.48
5.33 10^{-10}	50.85	29.26
6.40 10^{-10}	54.26	29.10

SEM

SEM analysis of Al metal surface, The SEM image of the aluminum specimen before and after immersing in 0.5M HCl for100 min in the absence and presence of inhibitor system are shown in Figures 12 (A, B and C) repetitively. The SEM micrographs of aluminum surface after polishing (Fig. 8A) shows a smooth surface of the Al with no corrosion products on its surface. The SEM micrographs of the Al surface immersed in 0.5 M HCl (Fig. 12B) Shows its roughness which indicate the corrosion of Al in HCl. Fig. 12C indicates that in the presence of 10^{-10} M of EDTA, the surface coverage increases, which in turn results in the formation of insoluble complex on the surface of the metal (EDTA/inhibitor complex) and the surface is covered by a thin layer of inhibitor which effectively control the dissolution of aluminum.

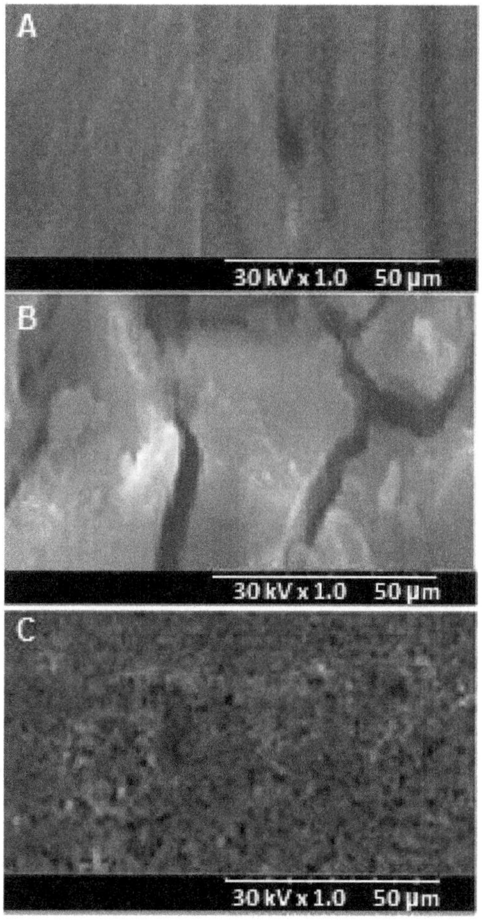

Figure 12*. SEM of Al surface at 30 °C; A) After polishing, B) after immersion in 0.5 M HCl for 100 min., and C) the same as B but in the presence of 2.7x10^{-10} M EDTA*

Quantum chemical calculations

The activity properties of an inhibitor is related to its geometry as well as the nature of its Frontier Molecular Orbitals, FMO, namely, the HOMO and LUMO. Therefore, in this study, quantum chemical calculations were performed to investigate the relationship between molecular structure of this compound and their inhibition effect. The optimized molecular structure and the FMO density distribution of the studied molecule are shown in Figs. 13 and 14, and the calculated quantum chemical parameters are given in Table 9 .

Adsorption centers of the inhibitor molecules are predicted by FMO. These centers are responsible for the interaction with surface metal atoms [42,43]. It was reported that, inhibitors with high HOMO energy offering electrons to unoccupied d orbital of the metal. Where, inhibitors with lower LUMO energy accept electrons from metal surface, as the ΔE_g decreased, the efficiency of inhibitor improved [44]. The dipole moment (μ) of EDTA is 5.0542 Debye (1.69×10^{-29} C m), which is higher than that of H_2O ($\mu = 6.20 \times 10^{-30}$ C m = 1.856 Debye). The high value of μ probably increases the adsorption between EDTA and Al surface [45]. Accordingly, the adsorption of EDTA from the aqueous solution can be regarded as a quasi-substitution process between the EDTA in the aqueous phase [$EDTA_{sol}$] and water molecules at the electrode surface [H_2O_{ads}]. Analysis of Fig. 13 shows that the distribution of two energies HOMO and LUMO localized in the nitrogen and oxygen atoms, consequently this is the favorite sites for interaction with the metal surface. The total energy of the EDTA is equal to -691282.91 kcal mol^{-1}. This result indicated that EDTA is favorably adsorbed through the active centers of adsorption.

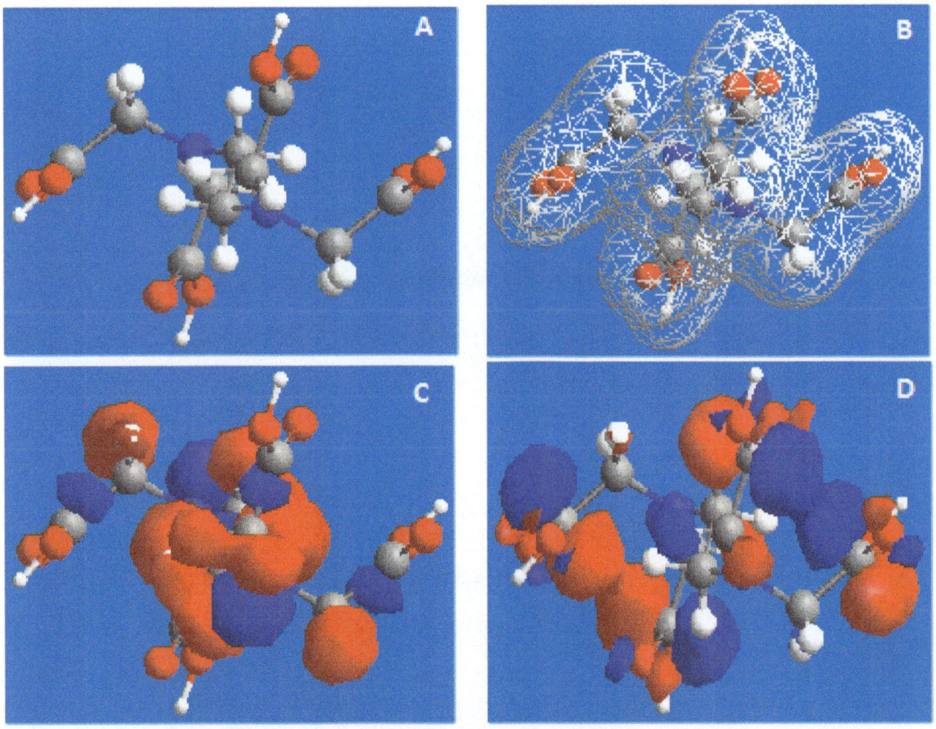

Figure 13.*Optimized structure (A), total energy (B) Frontier molecular orbital diagrams; HUMO (C) and LUMO (D) of the EDTA by B3LYP/6-31G (d,p)*

The number of transferred electrons (ΔN) was also calculated according to Eq. (14) [46,47]

$$\Delta N = \frac{\chi_{Al} - \chi_{EDTA}}{2(\eta_{Al} + \eta_{EDTA})} \tag{14}$$

Where χ_{Al} and χ_{EDTA} denote the absolute electronegativity of Al and EDTA molecule, respecttively; η_{Al} and η_{EDTA} denote the absolute hardness of Al and EDTA molecule, respectively. These quantities are related to electron affinity (A) and ionization potential (I)

$$\chi = \frac{I+A}{2} \text{ and } \eta = \frac{I-A}{2}$$

where, I and A are related in turn to E_{HOMO} and E_{LUMO}

$$I = -E_{HOMO} \text{ and } A = -E_{LUMO}$$

Values of η_{EDTA} and χ_{EDTA} were calculated by using the values of I and A obtained from quantum chemical calculation. The theoretical values of χ_{Al} and η_{Al} are 3.230 and 2.77 eV mol^{-1}, respectively [46]. The fraction of electrons transferred from inhibitor to the iron molecule (ΔN) was calculated. According to other reports [46,47], value of ΔN showed inhibition effect resulted from electrons donation. Also the softness is calculated depending upon the following relation:

$$\sigma = 1/\eta$$

In this study, the EDTA was the donators of electrons while the Al surface was the acceptor. The EDTA was bound to the Al surface, and thus formed inhibition adsorption layer against corrosion.

The adsorption centers of EDTA are estimated by Mulliken population analysis [48]. Authors believe that, the heteroatom with more negatively charged, is adsorbed on the metal surface through the donor-acceptor type reaction [43].

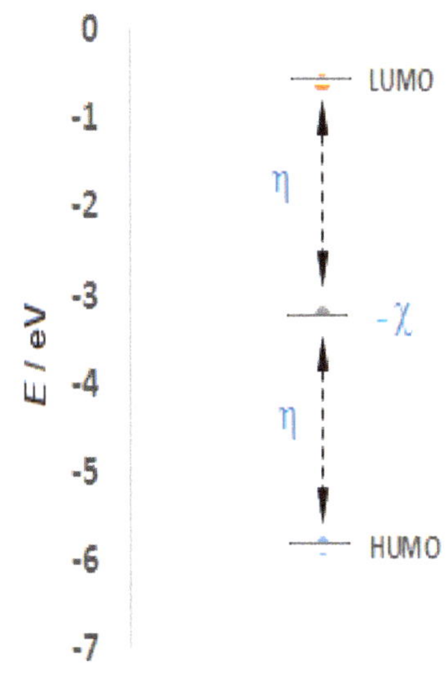

Figure 14*. Energy distribution of EDTA using B3LYP/6-31G (d,p)*

Table 9*. Calculated quantum chemical data for EDTA by B3LYP/6-31G (d,p)*

T.E. / kcal mol^{-1}	E_{HUMO} /eV	E_{LUMO} / eV	ΔE_{gap}/ eV	μ / Debye	I / eV	A / eV	χ / eV	η / eV	σ / eV	ΔN / eV
-691282.91	-5.849	-0.613	5.236	5.0542	5.849	0.613	3.231	2.618	0.382	-9.28x10^{-5}

The Mulliken charge of EDTA was shown in Table 10. It can be seen that the most favorable sites for the interaction with the Al surface were the following atoms: N32, N24, N30, N13, O23, O31, O15 and O14. Because these atoms have larger negative charge, that donate electron. This being the preferred zone for nucleophilic attack. For EDTA, the HOMO is localized over the nitrogen N and oxygen O atoms, consequently this is the favorite sites for interaction with the metal.

Table 10. *Mulliken charge of EDTA by B3LYP/6-31G (d,p)*

Atom	Charge	Atom	Charge	Atom	Charge
1 C	0.214984	**22 C**	0.325515	**43 H**	0.121164
2 C	-0.086817	**23 O**	-0.567231	**44 H**	0.133160
3 C	-0.127361	**24 N**	-0.721935	**45 H**	0.133125
4 C	-0.144686	**25 C**	0.210742	**46 H**	0.140598
5 C	-0.066095	**26 C**	-0.010327	**47 H**	0.139131
6 C	0.015268	**27 C**	-0.083902	**48 H**	0.140004
7 C	0.216462	**28 C**	-0.148904	**49 H**	0.151805
8 C	0.040626	**29 C**	-0.123161	**50 H**	0.116621
9 C	-0.006822	**30 C**	0.326666	**51 H**	0.379970
10 C	0.000002	**31 O**	-0.567208	**52 H**	0.337115
11 C	-0.073095	**32 N**	-0.732883	**53 H**	0.147454
12 C	0.028970	**33 C**	0.185485	**54 H**	0.162757
13 N	-0.306615	**34 C**	0.001330	**55 H**	0.130880
14 O	-0.413216	**35 C**	-0.084377	**56 H**	0.379729
15 O	-0.419654	**36 C**	-0.146845	**57 H**	0.336862
16 N	-0.302210	**37 C**	-0.105158	**58 H**	0.154020
17 C	0.220038	**38 C**	0.388987	**59 H**	0.167164
18 C	-0.031616	**39 O**	-0.588086	**60 H**	0.137476
19 C	0.044560	**40 N**	-0.863854	**61 H**	0.351188
20 C	-0.021716	**41 H**	0.134813	**62 H**	0.302186
21 C	-0.140200	**42 H**	0.120365	**63 H**	0.346753

Conclusions

1. EDTA acts as inhibitors for aluminum corrosion in acidic medium at very low concentrations 10^{-10} M.
2. Inhibition efficiency of EDTA increases with increase in concentration of the inhibitors, but decreases with increase in temperature.
3. The values of ΔG^{o}_{ads} are negative, which suggests that the inhibitors were strongly adsorbed on the Al surface. The values obtained support the chemisorption adsorption mechanism.
4. EDTA is found to obey Langmuir, Temkin adsorption isotherm and kinetic-thermodynamic model of El-Awady et al for both weight loss and polarization techniques, from the fit of experimental data.
5. Thermodynamic parameters revealed that the adsorption process is spontaneous.
6. Quantum chemical parameters such as E_{HOMO}, E_{LUMO}, ΔE (E_{LUMO} - E_{HOMO}), dipole moment (μ), number of transferred electrons (ΔN), and total energy (TE) were found to give good correlation with experimentally determined inhibition efficiency

References

[1] J. W. Diggle, T. C. Downie, C. W. Goulding, *Chem. Rev.* **65** (1969) 365-405

[2] C. M. A. Brett, *J. Appl. Electrochem.* **20** (1990) 1000-1003

[3] J. B. Bessone, D. R. Salinas, C. E. Mayer, W. J. Lorenz, *Electrochim. Acta.* **37** (1992) 2283-2290

[4] G. R. T Schueller, S. R Taylor, E. E. Hajcsar, *J. Electrochem. Soc*, **139** (1992) 2799-2805

[5] T. D. Burleigh, A. T. Smith, *J. Electrochem. Soc.* **138(8)** (1991) L34-L35

[6] Q. Qu, S. Jiang, W. Bai, L. Li, *Electrochim. Acta* **52(24)** (2007) 6811-6820

[7] E. M. Sherif, S. M. Park, *Electrochim. Acta* **51(22)** (2006) 4665-4673

[8] M. A. Quraishi, R. Sardar, *Corrosion* **58(9)** (2002) 748-755

[9] S. Zor, P. Doğan, B. Yazici, *Corros. Sci.* **47(11)** (2005) 2700-2710

[10] S. Zor, P. Doğan, *Corros. Rev.* **22(3)** (2004) 209-217

[11] S. Zor, *Turkish J. Chem.* **26(3)** (2002) 403-408

[12] S. T. Selvi, V. Raman, N. Rajendran, *J. Appl. Electrochem.* **33(12)** (2003)1175-1182

[13] Y. K. Agrawal, J. D. Talati, M. D. Shah, *J. Appl. Electrochem.* **46(3)** (2004) 633-651

[14] M. A. Quraishi, H. K. Sharma, *Mater. Chem. Phys.* **78(1)** (2003) 18-21

[15] N. Ochoa, F., Moran N. Pebere, B. Tribollet, *Corros. Sci.* **47(3)** (2005) 593-604

[16] G. M. Treacy, A. L. Rudd, C. B. Breslin, *J. Appl. Electrochem.* **30(6)** (2000) 675-683

[17] N. S. Gaikwad, C. H. Bhosale, *Mater. Chem. Phys.* **76(2)** (2002) 198-203

[18] L. R. B. Holzle, D. S. Azambuja, C. M. S. Piatnicki, G. E. Englert, *Mater. Chem. Phys.* **91(2-3)** (2005) 375-380

[19] L. Di Palma, R. Mecozzi, *J. Hazard. Mater.* **147(3)** (2007) 768-775

[20] A. Nahle, *Bull. Electrochem.* **14(2)** (1998) 52-56

[21] I. Milošev, *J. Appl. Electrochem.* **32** (2002) 311-320

[22] H. S. Gardiyar, N. S. D. Elayathu, *Corrosion* **36(6)** (1980) 306-312

[23] J. N. Alhajji, M. R. Reda, *J. Electrochem. Soc.* **141(6)** (1994)1432-1439

[24] S. Zor, H. Ozkanzanc, M. Bingul, *Mater. Sci.* **44(6)** (2008) 850-856

[25] E. Kraka, D. Cremer, *J. Am. Chem. Soc.* **122 (2000)** 8245-8264

[26] R. G. Parr, W. Yang, *Density Functional Theory of Atoms and Molecules*; Oxford University Press, New York, NV, USA, 1989

[27] M. H. Cohen. *In Topics in Current Chemistry*; R. F. Nalewajski, Ed.; Springer-Verlag: Heidelberg, Germany, **183** (1996) 143-173

[28] R.T Sanderson, *J. Am. Chem. Soc.* **74 (1952)** 272-274

[29] M. K. Awad, *J. Electroanal. Chem.* **567** (2004) 219-225

[30] R. G. Parr, W. Yang, *J. Am. Chem. Soc.* **106** (1984) 4049-4050

[31] R. G. Pearson, *J. Am. Chem. Soc.* **85** (1963) 3533-3543

[32] A. D. Becke, *J. Chem. Phys.*, **96** (1992) 9489-9502

[33] A. D. Becke, *J. Chem. Phys.* **98** (1993) 1372-1377

[34] C. Lee, W. Yang, R. G. Parr, *Phys. Rev. B* **37** (1988) 785-789

[35] M. J. Frisch, *et al.*, *Gaussian 03, Revision B.01*, Gaussian, Inc., Pittsburgh, PA, USA, 2003.

[36] F. El-Hajjaji, R. A. Belkhmima, B. Zerga, M. Sfaira, M. Taleb, M. Ebn Touhami, B. Hammouti, S. S. Al-Deyab, E. Ebenso, *Int. J. Electrochem. Sci.*, **9** (2014) 4721-4731

[37] S. S. Abd El Rehim, S. M. Sayyah, M. M. El-Deeb, S. M. Kamal, R. E. Azooz, *Mater. Chem. Phys.* **123** (2010) 20-27

[38] S. S. Abd El Rehim, S. M. Sayyah, R. E. Azooz, *Port. Electrochim. Acta* **30(1)** (2012) 67-80

[39] D. William, Jr. Callister, *Materials Science and Engineering: an Introduction*, 7th edition, John Wiley & Sons, Inc., New York, NY, USA, 2007

[40] B. Bouklah, M. Hammouti, F. L. Bentiss, *Corr. Sci.* **48(9)** (2006) 2831-2842

[41] J. I. Bhat, D. P. Vijaya, *J. Korean Chem. Soc.* **55(5)** (2011) 835-841

[42] J. Fang, J. Li, *J. Mol. Struct. THEOCHEM* **593** (2002) 179-185

[43] G. Bereket, E. Hur, C. Ogˇretir, *J. Mol. Struct. THEOCHEM* **578** (2002) 79-88

[44] K.F. Khaled, *Appl. Surf. Sci.* **255** (2008) 1811-1818

[45] K. Ramji, D.R. Cairns, S. Rajeswari, *Appl. Surf. Sci.* **254** (2008) 4483-4493

[46] I. Lukovits, E. Kalman, F. Zucchi, *Corrosion* **57** (2001) 3-8

[47] H. Ju, Z. Kai, Y. Li, *Corros. Sci.* **50** (2008) 865-871

[48] M. Sahin, G. Gece, E.K. Arci, S. Bilgic, *J. Appl. Electrochem*, **382** (2008) 809-815

18

Electrochemical behaviour and voltammetric determination of p-phenylenediamine at carbon paste electrode

Abdelaziz Ait Sidi Mou✉, Aicha Ouarzane, Mama El Rhazi

Laboratory of Physical Chemistry and Bioorganic Chemistry Mohammedia (FSTM), Morocco

✉Corresponding author: E-mail: aitsidimou_aziz@yahoo.fr

Abstract

An electrochemical method was developed for voltammetric determination of p-phenylenediamine using a carbon paste electrode (CPE). The electrode exhibited the highest electrocatalytic activity toward oxidation and reduction of p-phenylenediamine in the phosphate buffer, pH 7. After optimizing the experimental conditions and square wave voltammetry parameters, a linear current response toward concentration of p-phenylenediamine was obtained in the range of 0.12–3.00 μM with detection limit of 0.071 μM. The proposed procedure was successfully applied for determination of the total p-phenylenediamine content in the takeout extract sample.

Keywords

p-phenylenediamine; electroanalysis; voltammetry

Introduction

p-phenylenediamine (p-PD) is an aromatic diamine and derivative of the aniline with the chemical structure presented in Figure 1. It is commonly used in the manufacture of dyestuffs, as hair dye and as adjuvant of henna [1,2], in several industries, dyeing furs, photochemical processes, cosmetic and fabrication of household goods [3,4]. Toxic effects of p-PD, however, have also been reported by several authors [5-8].

Many analytical methods, including gas chromatography coupled mass spectrometry (GC-MS) [9,10], gas chromatography (GC) [11], and high-performance liquid chromatography (HPLC) [2] have already been reported for the determination of p-PD. Although chromatographic methods are sensitive and reliable, they have some disadvantages such as being time and labour consuming, expensive and demanding for sample pre-treatments and qualified personnel. Promising alternatives in this regard have been found in the electroanalytical methods which can offer high sensitivity, rapid response, easy operation and low cost [12].

Though the electrochemical properties of p-PD have been investigated at glassy carbon [13,14] and carbon paste electrodes [15-18], most of these studies were mainly concerned on the mechanism of p-PD polymerization. Although several studies on the detection of some similar compounds in medical plants [19,20] and drugs [21,22] were already published, the direct electroanalytical determination of p-PD on carbon paste electrode has not been reported previously.

In this paper, we report on the electrochemical properties of p-PD on the surface of carbon paste electrode and this electrode will also be applied for the determination of p-PD content in the plant of takeout.

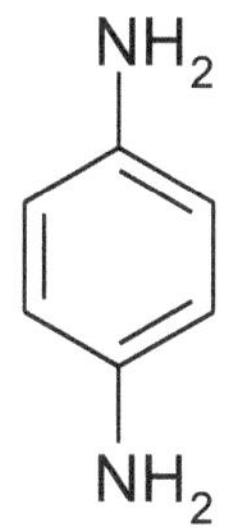

Figure 1. *Chemical structure of p-phenylenediamine*

Experimental

Apparatus

Electrochemical experiments were carried out with an Autolab (Metrohm-Autolab, Utrecht Netherlands) PGSTAT302N potentiostat/galvanostat controlled by GPES 4.9 software. A three-electrode electrochemical cell was employed for all electrochemical measurements. Carbon paste electrode (4 mm diameter) served as the working electrode, a platinum plate as the counter electrode, and Ag/AgCl (3 mol/L KCl) as the reference electrode, respectively. The pH of the buffer solutions was measured using a Fisher Scientific Accumet AB15 BASIC pH-meter.

Reagents

All chemicals used were of analytical grade. P p-PD and graphite were purchased from Sigma and Fluka, respectively. Sodium hydroxide, NaOH, $Na_2HPO_4\times 2H_2O$ and $NaH_2PO_4\times 2H_2O$ powders were purchased from Riedel-de Haën. All aqueous solutions were prepared in distilled water. The pH was adjusted with NaOH. A p-PD stock solution of 1 mM was prepared by dissolving p-PD in the phosphate buffer solution (PBS) before use in measurements.

Preparation of the carbon paste electrode

The carbon paste electrode (CPE) was prepared by mixing 1 g of graphite powder and 0.3 mL of paraffin oil using a mortar and pestle until a homogenous paste was obtained. The paste was then incorporated into the electrode cavity (4 mm diameter) and polished by a smooth paper. Prior each measurement, the electrode surface was renewed.

Sample preparation

Takeout plant (*Tamarix aphylla*) samples were obtained from Zagora region, Morocco. The samples were firstly grounded by a crusher, then 1 g of the powder was weighed, dissolved in 50 mL of ethanol, and placed in a mounting reflux for 2 h. After filtration, the solvent was evaporated by a rotavapor and an amount of the filtrate (108 mg) was dissolved directly in 100 mL of 0.1 M phosphate buffer solution (PBS), pH 7.0 for the final determination of the p-PD content.

Analytical procedure

The electrochemical cell containing 1 mM of p-PD dissolved in 0.1 M PBS, pH 7.0 was prepared. After that, the CPE electrode was placed in the test solution and the cyclic voltammetry was applied between – 0.2 and 0.8 V at a scan rate of 50 mV s^{-1}. The parameters of SWV of all measurements were, step potential 5 mV, amplitude 50 mV and frequency 50 Hz. The calibration curve of p-PD concentrations was illustrated by SWV in the range of 0.12–3.0 μM under the optimized conditions, and then used for directly determination to the amount of p-PD in the sample of takeout extracts.

Results and discussion

Electrochemical behaviour of p-phenylenediamine

The electrochemical behaviour of 1 mM p-PD on a CPE in the 0.1 M, pH 7.0 PBS was investigated using cyclic voltammetry at a scan rate of 50 mV s^{-1}. Whilst not any response is observed for the CPE in the blank solution (Fig. 2a), a pair of well-defined redox waves are observed in presence of 1 mM p-PD in 0.1 M, pH 7 PBS (Fig. 2b). The anodic, E_{pa}, and cathodic, E_{pc}, potentials are situated at E_{pa} = 0.10 V and E_{pc} = 0.03 V, respectively. In general, presence of a peak in the reverse potential scan indicates reversibility of the oxidation process on the surface of the electrode.

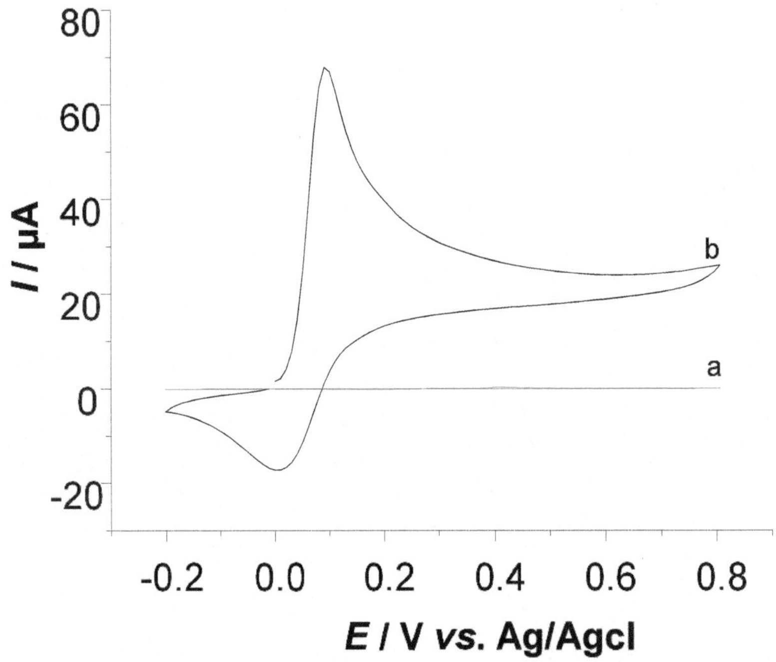

Figure 2. *Cyclic voltammograms of CPE recorded at 50 mV s^{-1} in blank (a) and for 1 mM p-PD in 0.1 M, pH 7.0 PBS (b).*

The oxidation mechanism of p-phenylenediamine on the electrode surface

The redox reaction of p-PD belongs to a class of two-electron and two-proton processes [23]. Possible oxidation mechanism of p-PD is shown in the Scheme 1. After formation of radical cation from the amino group, electrons are delocalized over the cycle forming another cation from the second amino group, while by elimination of two protons, the oxidized p-PD is formed. The mesomeric effect of nitrogen atom plays an important role, and electron density increases at π-bond and hence oxidation becomes easy.

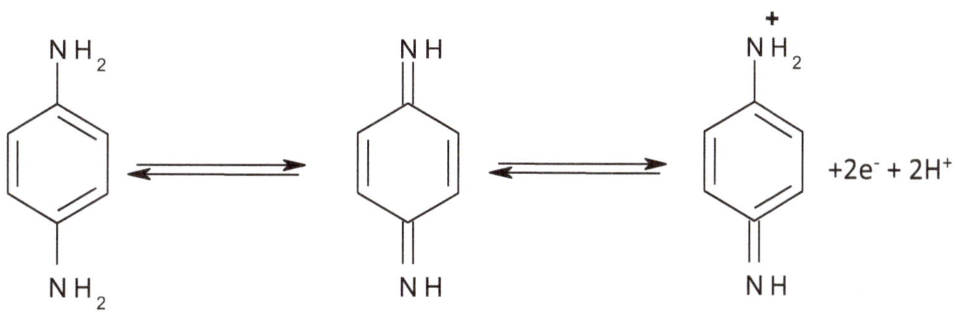

Scheme 1. *The oxidation mechanism of p-phenylenediamine*

Effect of buffer pH

The effect of pH of the PBS on the electrochemical behaviour of p-PD was studied in the pH range of 3.0–10.0. Fig. 3 shows that as pH value of the solution was increased, the redox peak is shifted negatively what indicates involvement of protons in the redox reaction [24]. Depending on a pH ranging from 3 to 7, the formal potential (E^o) for the anodic and cathodic peaks changed linearly. The redox current values increased from pH 3.0 and reached a maximum at pH 7.0 (curve *e* in Fig. 3). Therefore, the PBS, pH 7.0 was selected as the optimum working solution which will offer a higher peak resolution of p-PD and relatively higher current response.

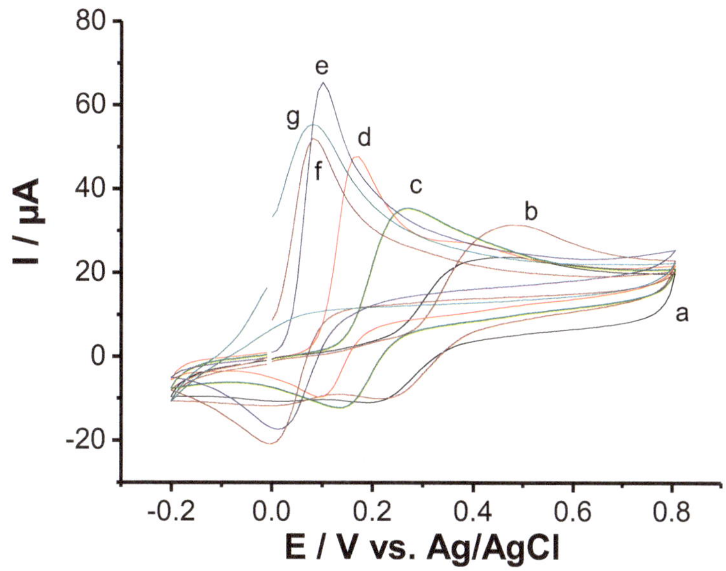

Figure 3. *Cyclic voltammograms of CPE recorded at 50 mV s^{-1} for 1 mM p-PD in PBS having pH 3–10: a - 3, b - 4, c - 5, d -d 6, e - 7, f - 8, g - 10.*

Effect of potential scan rate

Fig. 4 shows the cyclic voltammograms of p-PD at CPE in the PBS, pH 7.0, recorded at the scan rates varied from 20 to 400 mV s^{-1}. As is seen in the inset of Fig. 4, the redox peak currents at the CPE in the solution containing p-PD increased linearly with the scan rate in the range from 50 to 250 mV s^{-1}, what indicates the adsorption-controlled redox process. The linear regression equations can be defined as:

I_{pa} / µA = 71.76 + 0.36 υ (mV s^{-1}) (R^2 = 0.995)

I_{pc} / µA = 18.30 – 0.37 υ (mV s^{-1}) (R^2 = 0.996)

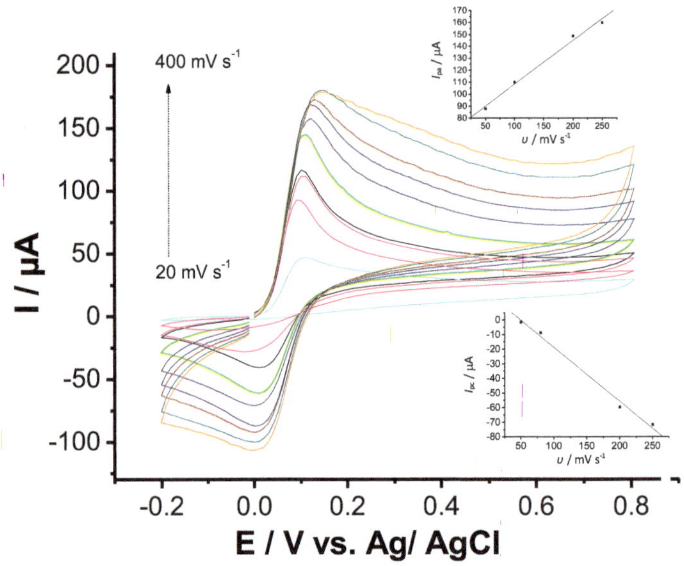

Figure 4. *Cyclic Voltammograms of CPE for 1 mM p-PD in PBS, pH 7.0, recorded at scan rates (υ) of 20, 50, 80, 100, 150, 200, 250, 300, 350 and 400 mV s*$^{-1}$*. Inset: plots of peak currents (*I_{pa} *and* I_{pc}*) vs. scan rate (υ) in the range from 50 to 250 mV s*$^{-1}$*.*

Effect of the square wave voltammetry (SWV) parameters and accumulation conditions

The effect of the accumulation potential in SWV measurements was investigated between –0.2 V and 0.1 V at the accumulation time of 30 s. As shown in Fig. 5A, the peak current values decreased significantly when the accumulation potential was shifted to potentials more positive than –0.1 V, because the p-PD adsorbed on the electrode becomes oxidized. Therefore, –0.1 V was used as the accumulation potential. Similarly, the effect of accumulation time was studied in the 30–240 s range at the accumulation potential of –0.1 V. Fig. 5B shows firstly a rapid increase in peak current values with increase of the accumulation time, then a maximum is reached at 120 s and after that, the constant values obtained indicate the saturation of the electrode surface by adsorbed layer. Therefore, 120 s was used as the proper accumulation time.

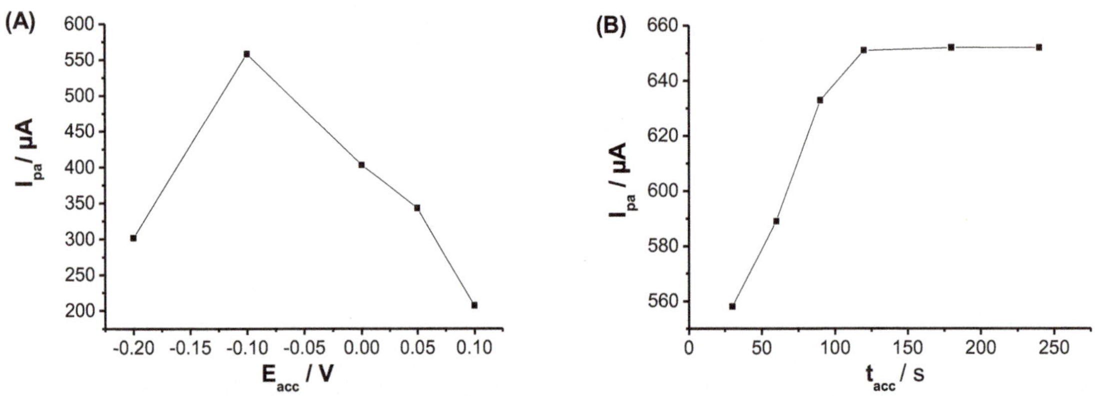

Figure 5. A *- Effect of accumulation potential (–0.2 V to 0.1 V) to* I_{pa} *values of SWV responses measured for the accumulation time of 30 s;* ***B*** *- Effect of accumulation time (30 s to 240 s) to* I_{pa} *values of SWV responses measured at the accumulation potential of – 0.1 V.*

Calibration curve

The relationship between the peak current and concentration of p-PD was studied at the CPE by using SWV under the optimized conditions. Fig. 6 shows that the peak current values were linearly related to the concentration of p-PD in the range of 0.12–3.0 μM. The linear regression equation was defined as: I_{pa} / μA = 0.181 + 0.241 C / μM, with the correlation coefficient of 0.999.

The detection limit calculated as ($3\sigma/p$) was 0.071 μM, where σ and p are the standard deviation of the blank and the slope of the calibration graph, respectively. The relative standard deviation (RSD) for repetitive measurements of 1.5 μM p-PD was found to be 3.49 % (five replicates), suggesting that the electrode demonstrated excellent repeatability.

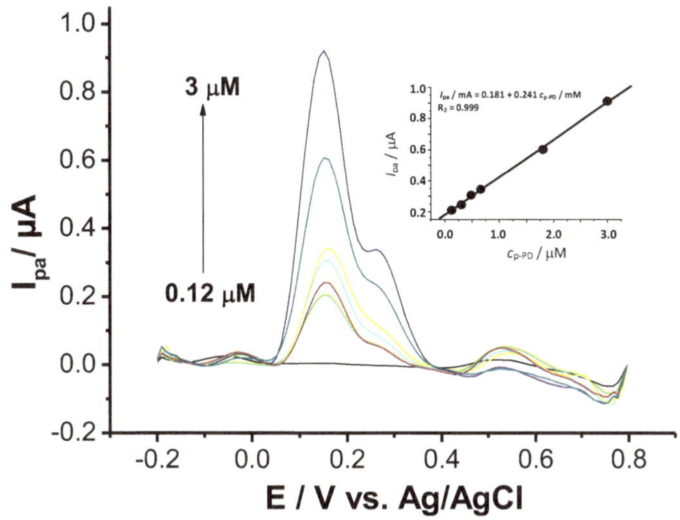

Figure 6. *SWV responses of CPE in 0.1 M PBS, pH 7.0 and various concentrations of p-PD in the range 0.12–3.0 μM. Inset: linear calibration curve of peak current (I_{pa}) versus p-PD concentration*

Interference study

To evaluate possible interference of some foreign species on the determination of p-PD, a systematic study was carried out. Data listed in Table 1 show the p-PD peak current ratios in the presence and absence of respective interferents. It seems that presence of 1.5 μM Mg^{2+}, K^+, Na^+ and Zn^{2+} did not affect the determination of p-PD. On the other hand, presence of 1.5 μM Cu^{2+} showed a slight interference. In general, despite the high redox potential used for p-PD detection, the interference study showed promising results for its selective determination.

Table 1. *Effect of some foreign species on the peak current response of 0.3 μM p-PD in PBS, pH 7 at a fivefold concentration of interferent cations.*

Interfering compound	Peak current ratio
Mg^{2+}	1.013
Zn^{2+}	0.760
K^+	1.050
Na^+	1.050
Cu^{2+}	1.330

Analytical application

The CPE was used to determine the amount of p-PD in the takeout extracts. The results are shown in Table 2. Recovery measurements were made by adding known amounts of the standard p-PD solution to the previously analyzed samples. The recovery for determination of p-PD was 110 %. The recovery indicated that the accuracy of the proposed method is acceptable.

Following the sample preparation, measurements were performed similarly to those of standard p-PD solutions. The total p-PD contents were determined by using the former calibration curve equation. The total measured p-PD content of the takeout sample (w/w) was determined as 1 %.

Table 2. *Determination of p-PD in plant takeout extract sample.*

	Amount of p-DP, µM				Recovery, %
Sample	Detected	Added	Expected	Found	
Takaout extract	0.54	1.5	2.04	2.25	110

Conclusions

In this paper, a simple, rapid, and sensitive determination of p-phenylenediamine in the phosphate buffer solution, pH 7.0 using the carbon paste electrode was proposed. A pair of well-defined redox peaks was obtained at the electrode. Under optimized conditions, the electrode showed excellent performance in terms of detection limit, linearity range, selectivity, repeatability, and recovery values. The proposed method was successfully applied for the determination of the total content of p-phenylenediamine in the plant of takeout sample.

References

[1] A. Ababou, K. Ababou, A. Mosadik, C. Lazreq, A. Sbihi, *Annales Françaises d'Anesthésie et de Réanimation* **19** (2000) 105-107.

[2] I. J. Kang, M. H. Lee, *Contact dermatitis* **55** (2006) 26–29.

[3] C. Mascres, J. Gaeton, *L'union Medicale Du Canada* **103** (1974) 672-677.

[4] K. S. Chugh, G. H. Malik, P. C. Singhal, *Journal of Medicine* **13** (1982) 131-137.

[5] A. Filali, I. Semlali, V. Ottaviano, C. Furnari, D. Corradini, R. Soulaymani, *African Journal of Traditional, Complementary and Alternative Medicine* **3** (2006) 142–149.

[6] S. Motaouakkil, B. Charra, A. Hachimi, H. Ezzouine, H. Guedari, H. Nejmi, A. Benslama, *Annales Françaises d'Anesthésie et de Réanimation* **25** (2006) 708–713.

[7] M. Abdelraheem, M. Hamdouk, E. E. Zijlstra, *Arab Journal of Nephrology and Transplantation* **3(1)** (2010) 39-43.

[8] N. Habti, G. Benzakour, Y. Zaid, A. Naya, N. Nourichafi, H. Mifdal, S. Motaouakkil, M. Oudghiri, *Journal of Toxicology and Environmental Health Sciences* **2(7)** (2010) 101-107.

[9] A.Stambouli, M. A. Bellimam, N. El Karni, T. Bouayoun, A. El Bouri, *Forensic Science International* **146S** (2004) S87–S92.

[10] M. L. Di Gioia, A. Leggio, A. Le Pera, A. Liguori, A. Napoli, F. Perri, C. Siciliano, *Journal of Chromatography A* **1066** (2005) 143–148.

[11] S. Madi-Jebara, G. Sleilaty, A. Yazigi, *Annales Françaises d'Anesthésie et de Réanimation* **25** (2006) 902–910.

[12] A. Geto, M. Pita, A. L. De Lacey, M. Tessema, S. Admassie, *Sensors and Actuators B* **183** (2013) 96– 101.

[13] V. K. Gupta, M. L. Yola, N. Atar, Z. Ustundag, A. O. Solak, *Electrochimica Acta* **112** (2013) 541– 548.

[14] L. Zhang, Z. Shi, Q. Lang, J. Pan, *Electrochimica Acta* **55(3)** (2010) 641–647.

[15] I. Adraoui, M. El Rhazi, A. Amine, L. Idrissi, A. Curulli, G. Palleschi, *Electroanalysis* **17(8)** (2005) 685-693.

[16] A. Madhankumar, N. Rajendran, *Synthetic Metals* **162** (2012) 176– 185.

[17] B. Lakard, G. Herlem, S. Lakard, B. Fahys, *Journal of Molecular Structure: THEOCHEM* **638** (2003) 177–187.

[18] M. Rani, R. Ramachandran, S. Kabilan, *Synthetic Metals* **160** (2010) 678–684.
[19] J. Li, J. Chen, X. L. Zhang, C. H. Lu, H. H. Yang, *Talanta* **83** (2010) 553–558.
[20] M. H. Parvin, *Electrochemistry Communications* **13** (2011) 366–369.
[21] X. Kang, J. Wang, H. Wu, J. Liu, I. A. Aksay, Y. Lin, *Talanta* **81** (2010) 754–759.
[22] M. Chao, X. Ma, X. Li, *International Journal of Electrochemcal Science* **7** (2012) 2201– 2213.
[23] S. A. da França, M. F. Dario, V. B. Esteves, A. R. Baby, M. V. R. Velasco, *Journal of Cosmetics* **2** (2015) 110-126.
[24] Q. Zheng, P. Yang, H. Xu, J. Liu, L. Jin, *Journal of Environmental Sciences* **24(9)** (2012) 1717–1722.

19

Electrocoagulation of textile wastewater containing a mixture of organic dyes by iron electrode

Borislav N. Malinovic[1,✉], Miomir G. Pavlovic[2] ,Tijana Djuricic[1]

[1]*University of Banja Luka, Faculty of Technology, Stepe Stepanovića 73, 78000 Banja Luka, Bosnia and Herzegovina*

[2]*University of Belgrade, ICTM-CEH, Njegoševa 12, Belgrade, Serbia*

✉Corresponding author: borislav.malinovic@tf.unibl.org

Abstract

This study focused on testing the efficacy of iron (Fe) electrode in an electrochemical treatment (electrocoagulation) of wastewater containing a mixture of organic dyes. The mixture consists of the following azo dyes: Acid Black 194, Acid Black 107 and Acid Yellow 116. The present organic dyes are toxic, cause skin and eye irritation and are extremely dangerous to aquatic organisms. The study was conducted on a synthetic wastewater prepared in a laboratory electrochemical reactor. During the research, the impact of the current density, various concentrations of dye and supporting electrolyte, electrolysis duration and pulsed current regime were tracked. *The results are shown through color removal efficiency, chemical oxygen demand (COD) removal efficiency, current efficiency, and specific energy consumption. At the initial concentration of dye (γ=200 mg/L) and concentration of supporting electrolyte (γ_{NaCl}=1 g/L) the color removal efficiency of 80.64% was achieved for 420 seconds of treatment (j=10 mA/cm^2). At the initial concentration of dye (γ=50 mg/L)* and *γ_{NaCl}= 8 g/L, the color removal efficiency of 96.01% was attained for 300 seconds of treatment (j=10 mA/cm^2).*

Keywords

organic dyes, electrochemical treatment; removal efficiency

Introduction

The textile industry is one of the world's biggest polluters according to the amount of wastewater where the most problems are generated by the textile dyeing process. General characteristics of textile dyeing wastewaters are related to, a high content of organic matter, presence of heavy metals and high coloration. As the annual global production of synthetic dyes exceeded 900,000 tonnes in 2009, it is expected to be well over a million tonnes per annum at present [1]. Since nearly two-thirds (60 – 70 %) of all synthetic dyes are azo dyes [2], azo dyes are considered as the most commonly used dyestuff in the dye industry.

During the dyeing processes in textile industry, up-to 50% of the used dyes may not be fixed to their fibre substrates and hence may be washed out to form highly coloured effluent streams [3]. According to Yu and Wen [4], many synthetic dyes belong to xenobiotic chemicals that are degraded with difficulty in nature. Difficult degradation of synthetic dyes is the main reason why their removal from aqueous effluents from the textile industry has received a considerable environment research attention [4]. Due to their complex structure and synthetic origin, azo dyes are difficult to be decolourized, what imposed an obligation for their removal from industrial effluents before being disposed into enviroment. Many studies have shown that azo dyes contribute to mutagenic activity of ground and surface waters polluted by textile effluents [5,6]. Also, discharge of azo dyes into surface water leads to aesthetic problems and obstructs light penetration and oxygen transfer into water bodies, affecting thus the aquatic life. Thus, the removal of color from textile effluents has grown as a topic of major concern [7].

In this paper we used a mixture that is used for dyeing polyamide fibers. The mixture contained synthetic dyes that belong to azo dyes and heavy metals cobalt and chromium (Co and Cr III-complex). An extensive literature, reporting the characteristics and applications of most important conventional technologies developed for this purpose, including physico-chemical and chemical methods, advanced oxidation processes (AOPs), adsorption, microbiological treatments, enzymatic decomposition and electrochemical technologies, has already been published [8-13]. Here, an increasingly present electrocoagulation (EC) treatment for removing color from wastewater is described. EC treatment implies formation of coagulants *in situ* by electrolytic dissolution of the anode made from either aluminum or iron. During the EC process, the anode leads to the formation of metal ions, while evolution of hydrogen gas at the cathode carries the flocculated particles to the surface of the water. The chemical reactions taking place at the anode are given as follows [14]:

For iron anode:

$$Fe - 2e^- \rightarrow Fe^{2+} \quad (4)$$

at alkaline conditions:

$$Fe^{2+} + 2OH^- \rightarrow Fe(OH)_2 \quad (5)$$

at acidic conditions:

$$4Fe^{2+} + O_2 + 2H_2O \rightarrow 4Fe^{3+} + 4OH^- \quad (6)$$

In addition, there is oxygen evolution reaction:

$$2H_2O - 4e^- \rightarrow O_2 + 4H^+ \quad (7)$$

The reaction at the cathode is:

$$2H_2O + 2e^- \rightarrow H_2 + 2OH^- \quad (8)$$

The EC treatment has received great attention in recent years due to its unique features, such as versatility, energy efficiency, automation and cost effectiveness [12,15]. Alaton *et al.* [16] have proven the effectiveness of using EC as a method of removing color from wastewater of textile industry. According to their study, EC proved to be a very promising alternative treatment which is increasingly used in wastewater treatment. Simulated wastewater polluted by acidic dyes has already been treated by EC using aluminum and steel electrodes. The results showed that the COD removal efficiency, color removal efficiency and the production of sludge were affected by the current intensity, concentration of supporting electrolyte and the initial pH-value. EC with aluminum electrodes is found able to remove almost 90 % of color and reduce COD-value to 40 %, while with steel electrodes, COD value was reduced to about 50 % [16].

The phenomenon of electrode passivation has limited further application of EC treatment. Application of *pulsed current* regime, also known as *pulsed electrocoagulation* - PE (it uses the interactions of electrochemical technology and polarity reversal in an electrical field), brought a significant impact on the electrochemical processes though higher activity and efficiency during the process [17-20]. Hence, the development of PE was explored in recent years [20].

Experimental

Electrochemical batch reactor (Fig. 1) of capacity 500 cm^3 is made of polypropylene and has a possibility of constant mixing (500 rpm/min). The reactor contains two electrodes of the same dimensions (area), P=32.8 cm^2, put at a distance, d=30 mm. Electrodes were connected to a digital power source (Atten, APS3005SI; 30V, 5A). *Pulsed current regime* is provided by a generator of variable periodic current (Selekt-automatika, Serbia).

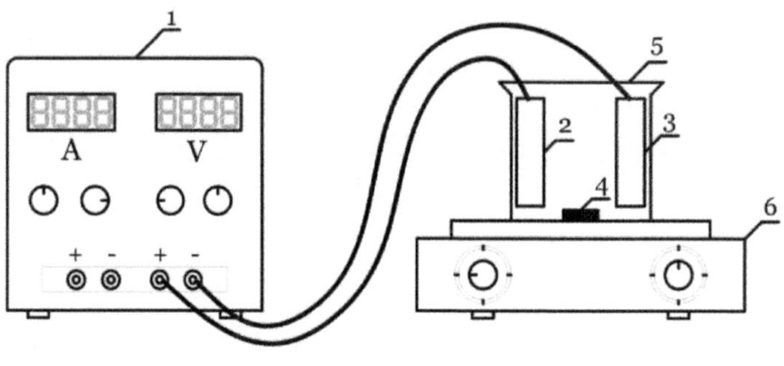

Figure 1. *Schematic view of electrochemical reactor: 1 – source of electric power; 2 – anode; 3 – cathode; 4 – magnetic stir bar; 5 – electrochemical cell; 6 – magnetic stirrer*

Electrode materials were metals of known compositions: iron (EN10130-91; max. 0,08 % C, max. 0.12 % Cr, max. 0.45 % Mn, max. 0.60 % Si) and stainless steel (EN 1.4301/AISI 304; max. 0.07 % C, 18.1 % Cr, 8.2 % Ni). Attention was put to the appropriate relations between electrode surface area (A_s) and volume (V) of the reactor. For used electrochemical reactor that relation, (A_s/V) / (m^2/m^3), is 16.4 m^2/m^3, which is in compliance with recommendations given by Mameri et al. [21] for current density values of 10 to 200 A/cm^2. All experiments were performed at an initial temperature of sample t=20°C. Before each treatment, current density was set at the desirable value and electrodes were mechanically cleaned and washed with detergent and acetone in order to remove surface grease. Electrode surfaces were additionally cleaned by emerging (5 min) in diluted (1:2) solution of HCl before each treatment.

In present experiments, commercially available 99.5 % sodium chloride, NaCl, 35 % hydrochloric acid, HCl, acetone, $(CH_3)_2CO$ (Lachner, Czech) and "Ostalan Black SR" dye (Synthesia a.s., Czech) were used. "Ostalan Black SR" is a mixture of synthetic dyes that belong to the group of azo dyes. According to Safety Data Sheet (SDS) [22], "Ostalan Black SR" consists of the following azo dyes: Acid Black 194 (the mixture containing 40-50 %), Acid Black 107 (mixture contains 40-50 %) and Acid Yellow 116 (mixture contains >1 %). All mentioned dyes are potentially harmful to humans and the environment and labeled as very toxic to living organisms and harmful to aquatic organisms [22].

COD was measured by the closed spectrophotometric method on COD Reactor (Hach, USA), colorimeter (COD CheckItDirect, Lovibond, Germany) by standard cuvette (Test Tube MR, Lovibond, Germany), and the dye concentration was determined spectrophotometrically (λ_{max} = 573nm) on UV-VIS spectrophotometer (Perkin Elmer, Lambda 25) according to standard methods [23]. Measurement in the **pulsed current** regime is defined by the cathode current density (j_k), time cathode deposition (t_k), anodic current density (j_a), and time of anodic dissolution (t_a). The period of pulsed current waves, (T), is the sum of the time of cathode deposition (t_k) and time of anodic dissolution (t_a) [24].

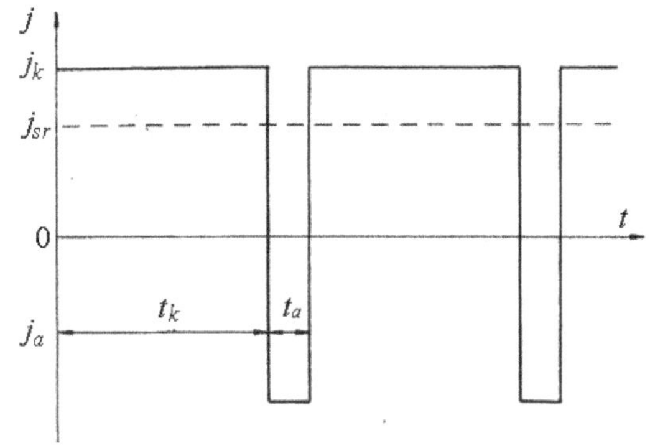

Figure 2. *Schematic view of PE* [24]

Results and discussion

Results of the electrochemical color removal are showed through decrease of mass concentration, γ/ (mg/L), and color removal efficiency, E_u /% defined as

$$E_U \ / \ \% = \frac{\gamma_i - \gamma_f}{\gamma_i} 100 \tag{9}$$

where γ_i and γ_f are initial and final dye concentrations, mg/L.

Current efficiency, η_I, has been calculated according to the following equation (10):

$$\eta_I \ / \ \% = \frac{(COD_i - COD_f)FV}{8It} 100 \tag{10}$$

In Eq. (10), COD_i and COD_f are initial and final chemical oxygen demand (COD) values, F is the Faraday constant, V is the volume of solution (wastewater), 8 is constant, I is the current intensity and t is electrolysis duration.

Selection of electrodes (soluble or insoluble anode) depends primarily on the mechanism of electrolytic reaction and according to the literature data and previous research it plays the most important role [25]. In the case of EC, selection of electrode material is narrowed to the iron and aluminum electrodes, respectively.

Figure 3 shows the color removal efficiency at different current densities (1; 2.5; 5; 10 mA/cm^2) for the initial dye concentration, γ_0 = 200 mg/L, supporting electrolyte concentration, γ_{NaCl} = 1 g/L, and electrolysis duration, t = 5 min. At the current density of j = 10 mA/cm^2, E_u = 67.75 %.

Change of current density to efficiency of color removal could be presented by the following equation.

$$E_u = 18.832 \ln j + 25.085 \tag{11}$$

During the study, NaCl has been used as the supporting electrolyte. Figure 4 shows the effect of NaCl concentration on color removal efficiency (γ_0 = 200 mg/L). Duration of electrolysis for all concentrations of the supporting electrolyte was t = 5 min at j =1 0 mA/cm^2. Figure 4 shows that removal efficiency is the highest at γ_{NaCl} = 8 g/L (E_u = 96.01 %), and the lowest at γ_{NaCl}=1 g/L (E_u = 85.3 %). Change of the supporting electrolyte concentration to efficiency of color removal could be presented by the equation (12).

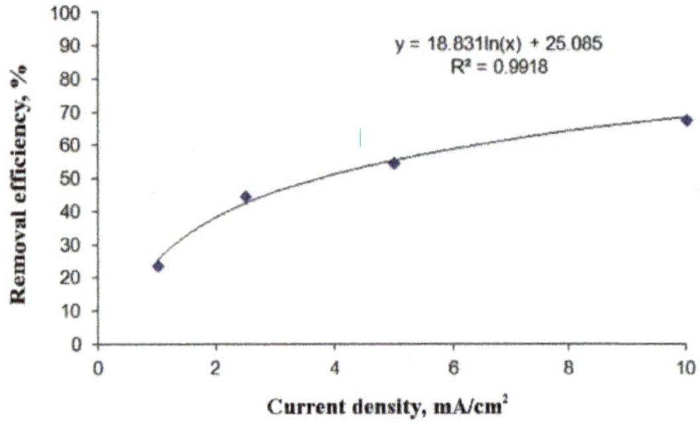

Figure 3. *Effect of current density on color removal efficiency*

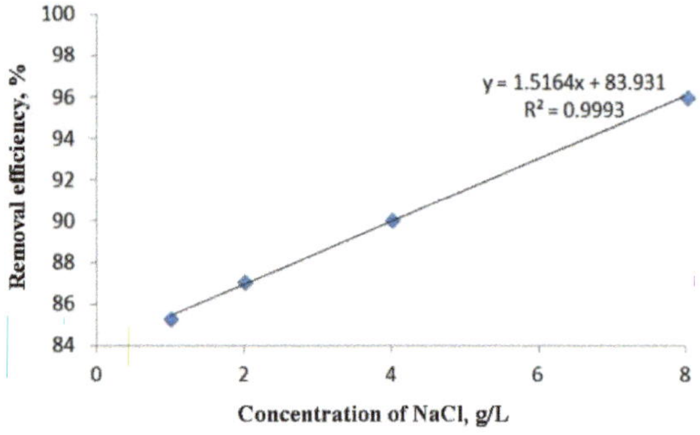

Figure 4. *Effect of NaCl concentration on color removal efficiency*

$$E_u = 1.5164\ \gamma_{NaCl} + 83.931 \qquad (12)$$

Since there is a little difference in E_u between γ_{NaCl} = 1 g/L and γ_{NaCl} = 8 g/L observed, γ_{NaCL}= 1 g/L will be used in the further part of research.

At the current density, j = 1 mA/cm^2, electrolysis time duration, t = 5 min, and the concentration of supporting electrolyte γ_{NaCl} = 1 g/L, the removal efficiency is increased by reducing the initial concentration of dye, which is shown in Figure 5. Removal efficiency is the highest at dye concentration, γ = 50 mg/L (E_u = 85.3 %) and the lowest at γ = 400 mg/L (E_u = 45.81 %.). Effect of the initial concentration of dye to the removal efficiency could be presented by the equation (13).

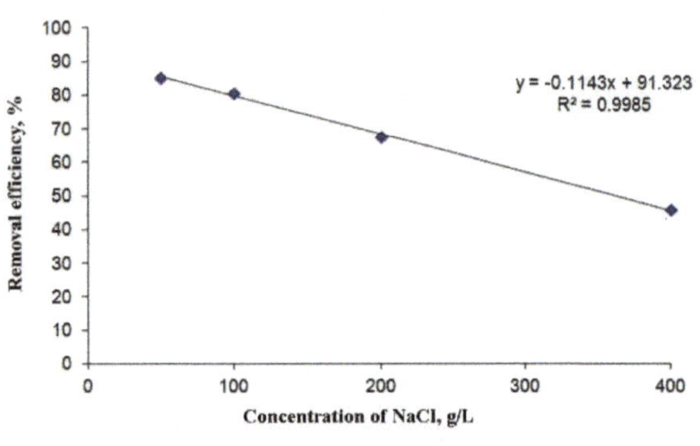

Figure 5. *Efficiency of color removal for different initial concentrations of dye*

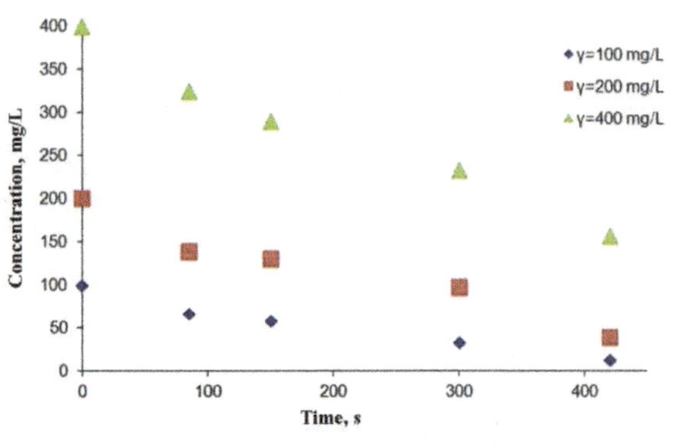

Figure 6. *Effect of electrolysis duration on concentration decrease for different initial dye concentrations.*

$$E_u = -0.1143\ \gamma_{dye} + 91.323 \qquad (13)$$

Figure 6 shows the effect of electrolysis time on decrease of dye concentration for different initial dye concentrations (j = 10 mA/cm^2, γ_{NaCl} = 1 g/L). In Figure 7, these results measured using the iron : iron (Fe : Fe) electrode pair are compared with the results obtained at same conditions (γ_0 = 400 mg/L, j = 10 mA/cm^2, γ_{NaCl} = 1 g/L) using the stainless steel (SS) as the cathode material. Data in Figure 7 suggest that only slightly lower efficiency is achieved for the Fe : SS *vs.* Fe : Fe electrode pair. It seems, however, that the whole process is easier to maintain using the SS cathode, what can be explained by much better anti-corrosion properties of SS than Fe.

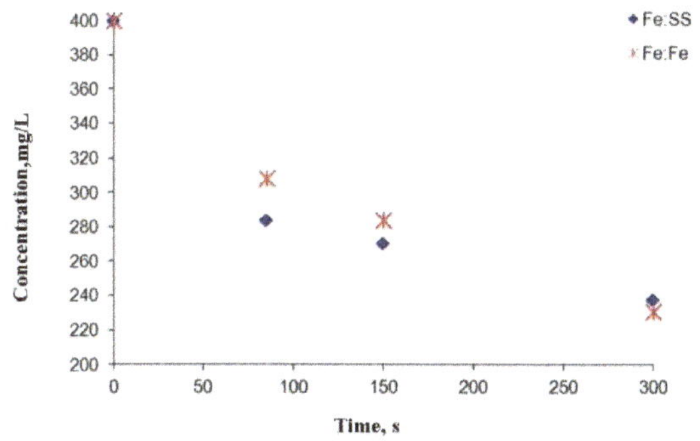

Figure 7. *Effect of application different electrode pairs*

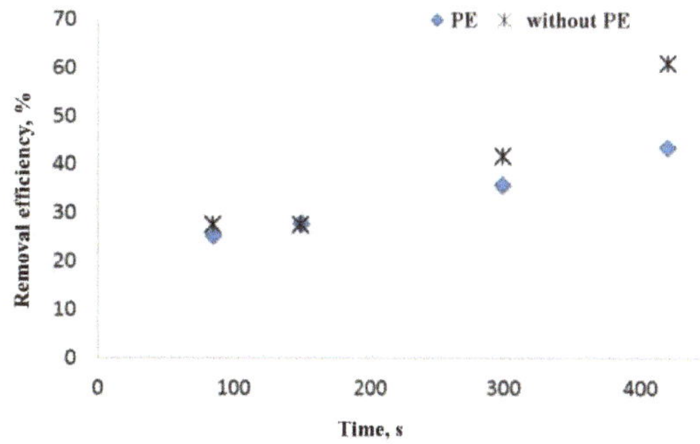

Figure 8. *Color removal efficiency with and without application of PE*

In the pulsed current regime (PE) the polarity of electrodes is reversed at a given time during the electrolysis (Fig. 2). Figure 8 shows differences between the efficiencies of color removal obtained by application PE and without PE in the same conditions (j = 5 mA/cm^2, γ_{NaCl} = 1 g/L, γ_0 = 400 mg/L, t_k = 42 s, t_a = 42 s). It is evident in Figure 8, that for longer treatment period, the removal efficiency is higher without applying PE. Contrary to already given predictions [23], such behavior could be a sign of no significant passivation and "dirtying" of the electrodes.

Effect of electrolysis duration (85, 150 and 300 s) at a current density j = 5 mA/cm^2 (γ_0 = 400 mg/L, γ_{NaCl} = 1 g/L) and j = 10 mA/cm^2 (γ_0 = 200 mg/L, γ_{NaCl} = 1 g/L) on COD values is shown in Figure 9. It is shown in Figure 9 that in both cases COD-values are linearly depended on time. For t = 300 s, COD-value is reduced from the initial 385 mgO$_2$/L to 224 mgO$_2$/L (γ_0 = 400 mg/L), and from the initial 125 mgO$_2$/L to 56 mgO$_2$/L (γ_0 = 200 mg/L), what is in full agreement with the study Alaton *et al.* [16].

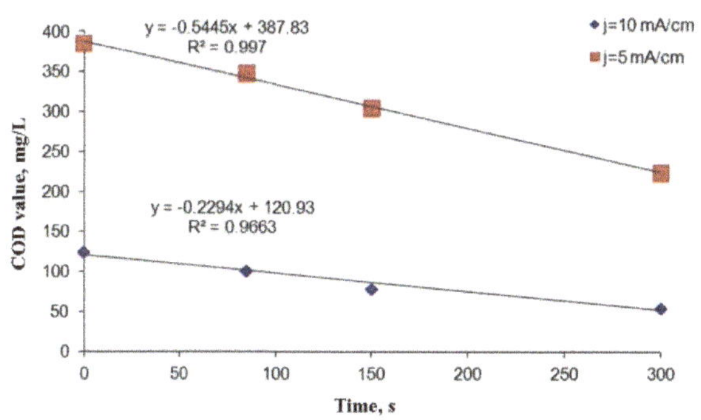

Figure 9. *Effect of electrolysis duration on COD value*

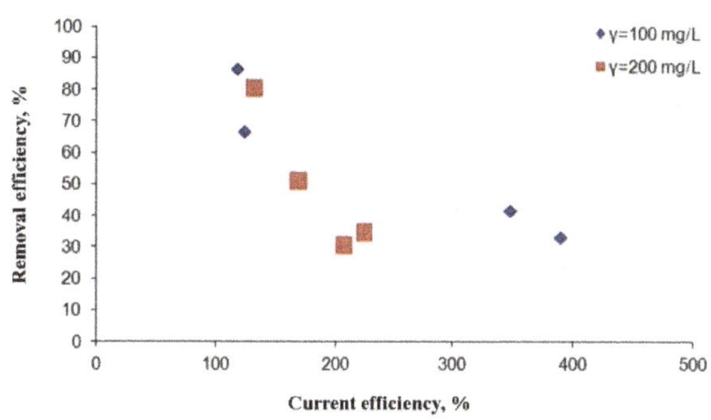

Figure 10. *Removal efficiency vs. current efficiency*

Changes of current efficiency on the color removal efficiency for different initial dye concentrations (100 and 200 mg /L) are shown in Figure 10. Measurements were made at the current density of j = 10 mA/cm^2 and the concentration of supporting electrolyte γ_{NaCl} = 1 g/L. It is obvious in Figure 10 that current efficiency exceeds 100 % (119-389 %), which means that in addition to electrochemical reactions some chemical reactions may take place. As expected from previous researches [20,26], for higher color removal efficiency from wastewater, the current efficiency decreases.

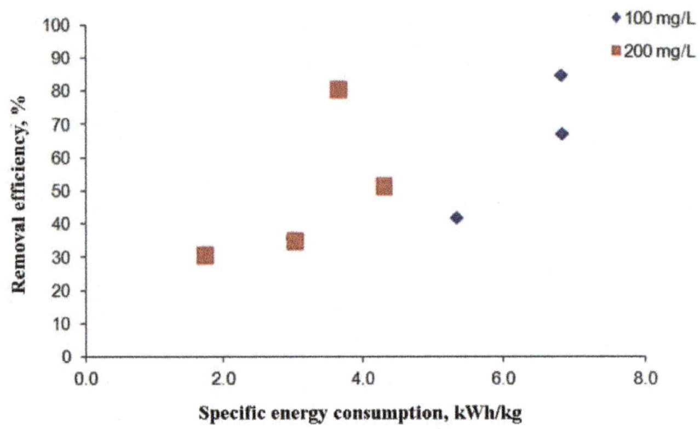

Figure 11. *Energy consumption as a function of color removal efficiency*

Dependence of energy consumption on the color removal efficiency at different initial concentrations is shown in Figure 11. At initial concentration, γ_0=400 mg/L, energy consumption (per kilogram of dye removed from the waste water) at removal efficiency, E_u = 54.76 %, is W_{sp}=3.8021 kWh/kg dye (j = 10 mA/cm^2, t = 600 s, γ_{NaCl} = 1 g/L).

Conclusions

This study shows that EC with Fe anode is the efficient process for removing the mixture of azo dyes from waste water. The process is influenced mostly by concentration of supporting electrolyte (NaCl), dye concentration, current density and reaction time. For 80.64 % color removal efficiency at initial dye concentration of solution γ_0 = 200 mg/L (j = 10 mA/cm^2, γ_{NaCl} = 1 g/L), up to 3.64 kWh/kg dye of energy has been consumed.

Acknowledgements: *This work was supported in part by the Ministry of Science and Technology of the Republic of Srpska under Project 19/6-020/961-171/14.*

References

[1] F. M. D. Chequer, J. P. F. Angeli, E. R. A.Ferraz, M. S. Tsuboy, J. C. Marcarini, M. S. Mantovani, D. P. De Oliviera, *Mutation Research/Genetic Toxicology and Environmental Mutagenesis,* **676(1-2)** (2009) 83-86.

[2] F. P. Van Der Zee, G. Lettinga, J. A. Field, *Chemosphere* **44(5)** (2001) 1169-1176.

[3] E. Fernando, *PhD thesis*, University of Westminster, Faculty of Science and Technology, 2014.

[4] Z. Yu, X. Wen, *Int. Biodeterior. Biodegrad.* **56(2)** (2005) 109-114.

[5] P. Rajaguru, L. J. Fairbairn, J. Ashby, M. A. Willington, S. Turner, L. A. Woolford, N. Chinnasamy, J. A. Rafferty, *Mutation Research* **444(1)** (1999) 175-180.

[6] G. A. Umbuzeiro, H. Freeman, S. H. Warren, F. Kummrow, L. D. Claxton, *Food and Chemical Toxicology* **43(1)** (2005) 49-56.

[7] A. Pandey, P. Singh, L. Iyengar, *International Biodeterioration & Biodegradation* **59(2)** (2007) 73-84.

[8] C. A. Martinez-Huitle, E. Brillas, *Applied Catalysis B: Environmental* **87(3-4)** (2009) 105-145.

[9] E. Bagda, E. Bagda, *Journal of Environmental Protection and Ecology* **13(2)** (2012) 517–531.

[10] C. Zaharia, D. Suteu, *Journal of Environmental Protection and Ecology* **15(4)** (2014) 1680–1689.

[11] G. Demir, *Journal of Environmental Protection and Ecology* **11(1)** (2010) 7–19.

[12] B. N. Malinovic, M. G. Pavlovic, *Journal of Electrochemical Science and Engineering* **6(1)** (2016) 67-75.

[13] C. E. Barrera-Díaz, G. Roa-Morales, P. B. Hernández, C. M. Fernandez-Marchante, M. A. Rodrigo, *Journal of Electrochemical Science and Engineering* **4(4)** (2014) 285-296.
[14] G. Chen, *Separation and Purification Technology* **38(1)** (2004) 11–41.
[15] P. Kariyajjanavarl, J. Narayanal, Y. A. Nayaka, *Hydrology Current Research* **2(1)** (2011) 1-7.
[16] A. Alaton, I. Kabdasli, Y. Sahin, *The Open Environmental & Biological Monitoring Journal* **1(1)** (2008) 1-7.
[17] J. Zhou, C. Deng, S. Si, Y. Shi, X. Zhao, *Electrochim. Acta* **56(5)** (2011) 2062–2067.
[18] K. Xie, L. Sun, C. Wang, Y. Lai, M. Wang, H. Chen, C. Lin, *Electrochimica Acta* **55** (24) (2010) 7211–7218.
[19] A. Molina, F. Martinez-Ortiz, E. Laborda, R.G. Compton, *Electrochimica Acta* **55** (18) (2010) 5163–5172.
[20] B. Malinovic, M. G. Pavlovic, *Yucorr XVII*, *Proceeding*, Tara, Serbia, 2015, 85–93 (in Serbian).
[21] N. Mameri, A. R. Yeddou, H. Lounici, D. Belhocine, H. Grib, B. Bariou, *Water Research* **32(5)** (1998) 1604-1612.
[22] Synthesia a.s. *Safety Data Sheet.* From http://dyes.synthesia.eu/organics-dyes/textile-dyes/ostalan-black-sr (2015, June 20).
[23] AMERICAN PUBLIC HEALTH ASSOCIATION (APHA). Standard Methods of Water and Wastewater. 18th ed. American Public Health Association, American Water Works Association, Water Environment Federation publication. APHA, Washington D.C., 1992.
[24] K. I. Popov, *Periodično promenljivi režimi u elektrohemijskom taloženju metala*, in: M. G. Pavlović, A.Dekanski [eds.] *Primena pulsnih režima u galvanskoj tehnici*: IHTM - Centar za elektrohemiju, Belgrade, Serbia, 1992, 49-73 (in Serbian).
[25] B. Malinovic, M. G. Pavlovic, J. Mandic, *Glasnik hemičara, tehnologa i ekologa Republike Srpske* **(9)** (2013) 21-27 (in Serbian).
[26] B. N. Malinovic, M. G. Pavlovic, N. Halilovic, *Journal of Environmental Protection and Ecology* **16(4)** (2015) 1273-1281.

20

Application of the quadratic logistic differential equation for the rationalization of methanol electrooxidation dynamics

Hossein Heli[✉], Fereydoon Gobal*

Nanomedicine and Nanobiology Research Centre, Shiraz University of Medical Sciences, Shiraz, Iran
**Department of Chemistry, Sharif University of Technology, Tehran, Iran*

[✉]Corresponding author: hheli7@yahoo.com, heli@sums.ac.ir

Abstract

The electrooxidation of methanol in both acidic and alkaline media on poly-crystalline platinum under the regime of cyclic voltammetry is analyzed by application of quadratic logistic equation. The current-charge curves in the anodic cycles fit the logistic differential equation reasonably well and are accounted on the basis of the non-linearity of the kinetics and the effect of positive feedback. In the reverse cycle however, no fit is observed, presumably due to the lack of correlation between the net faradaic current and the surface charge of adsorbates.

Keywords

Logistic differential equation; Feedback; Methanol electrooxidation; Electrocatalysis; Fuel cell

Introduction

Methanol is regarded as an attractive fuel in the fast-emerging fuel cell industry, but despite extensive electrocatalytic studies [1-4], there are major problems and kinetic limitations to its direct employment. Ample information concerning the kinetics and mechanism and reactive surfaces, as well as stable bulk intermediates involved in the methanol electrooxidation reaction (MEO) exist in the literature [5,6]. For further amplification of findings, studies on the non-linearity, multi-stability, and chaotic behavior of MEO are essential. MEO process is described very well by mathematical methods and modeling studies due to: (i) high quality data that either exist or can easily be obtained in a using relatively straight forward electrochemical measurements; (ii) the important factors, electrochemical or else, can easily be controlled and the responses of the system measured; (iii) multitude of correlations can be visualized in electrochemical measurements and in the present context, the current-charge (dQ/dt *vs.* Q) dependency seems to be most suitable. The quadratic logistic differential equation (QLE) [7-9] in the form of a difference for mapping [10,11] in the normalized form

$$dx / dt = r\,x\,(1\text{-}x) \qquad (1)$$

describes the effects of feedback, as well as the effect of parameter r on the value of normalized variable x ($0 < x < 1$). On the one hand, Eq. (1) is associated with the intrinsic rate of the process under study. On the other hand, it describes non-linearity and chaos in the dynamics of the process in a simple way, by illustrating these phenomena through working with only one variable. For $r > 3$ the value of x progressively turns to a periodic motion and for $3.5699... < r < 4$ it behaves chaotically.

Eq.-(1) can be alternatively written as:

$$x_{i+1} = r\,x_i\,(1\text{-}x_i) \qquad (2)$$

where consecutive values of x can be worked out on the basis of a set of initial values [12], while the term (1-*x*) surely guarantees non-linear effects.

In the present study, a new approach is presented in an order to understand the methanol electrooxidation dynamics, both in acidic and alkaline solutions. This approach is based on the quadratic logistic equation that has been used extensively, as an example of simple equation with only one degree of freedom that illustrates occurrence of bifurcation and chaos in dynamical systems [9].

Experimental

Materials and methods

Sodium hydroxide, sulfuric acid and methanol used in this work were of reagent grade of Merck origin and used without further purifications. All solutions were prepared with distilled water. Electrochemical experiments were carried out in a conventional three-electrode cell with platinum wire, exposing the surface area of 0.25 cm^2, as the working electrode and its potential monitored against the saturated Ag/AgCl reference electrode. A large platinum sheet was used as the counter electrode. All electrochemical measurements were performed at room temperature. The cell was powered by a µ-Autolab potentiostat/galvanostat run by computer through the GPES software. Charges were calculated by integrating the area under voltammograms left after correction for the background. All potential values were further converted and reported relative to the normal hydrogen electrode.

Results and discussion

Figure 1 shows typical cyclic voltammograms (CV) representing MEO on polycrystalline Pt in 0.5 M sulfuric acid (a) and 0.5 M sodium hydroxide (b) solutions. In both systems, methanol concentration and potential sweep rates were 0.1 M and 20 mV s^{-1}, respectively. The large peak observed in the anodic half cycle is believed [13,14] to be due to the oxidative stripping of protons from the adsorbed methanol molecules to form strongly adsorbed intermediates of COH_{ads} and CO_{ads}. Their subsequent reaction with OH_{ads} (or alternatively with OH^-_{aq}) at higher over-potentials yields the final oxidation products [3,15]. Carbon dioxide or carbonates are the final products and the surface remains largely covered by OH_{ads} species at the end of an anodic half-cycle. Desorption of this species in the reverse potential sweep creates vacant active sites and promotes continuation of methanol oxidation with a spurt of anodic current until the potential became so cathodic that neither oxidative stripping, nor the follow-up reactions can occur to any significant extent. Higher anodic currents met in the alkaline solution is due to abundance of available hydroxide ions. It has been suggested that formation of OH_{ads} from alkaline solutions is more kinetically favored over that from acidic media by energy roughly corresponding to that of water ionization [16,17]. Surely, the higher currents attainable in alkaline media are off-set by formation of carbonates, which is undesirable, as far as the fuel cell industry is concerned [18].

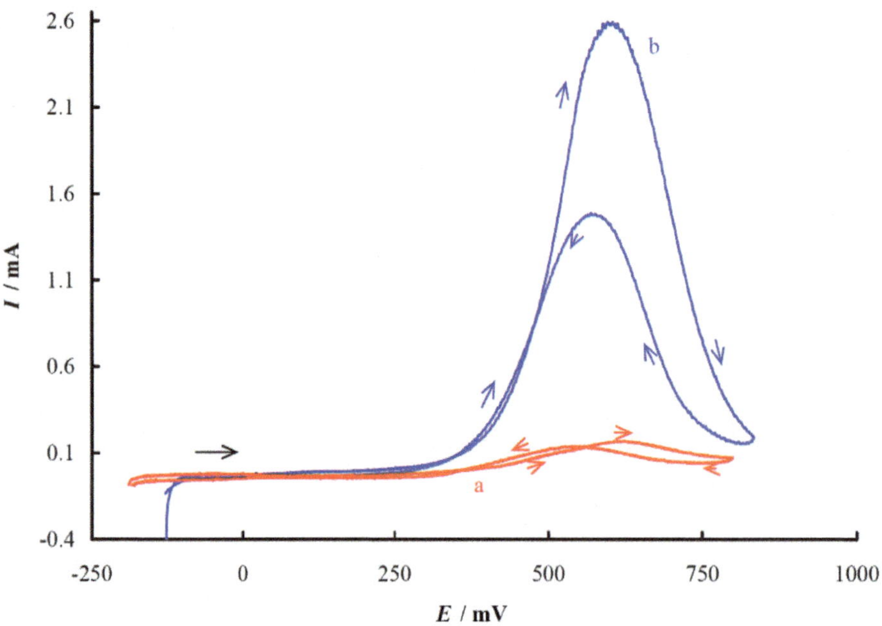

Figure 1. *Typical cyclic voltammograms representing MEO on polycrystalline Pt in 0.5 M sulfuric acid **(a)** and 0.5 M sodium hydroxide **(b)** solutions comprised 0.1 M methanol. Potential sweep rates are 20 mV s^{-1}.*

Figure 2a illustrates the galvanostatic scan of MEO on Pt electrode. Current sweep rate was 10 μA s^{-1}. A hard transition from oscillatory to stationary behavior is seen at high current, while a soft supercritical Hopf bifurcation is discerned at low current. The potential-time series of methanol electrooxidation on Pt electrode with the anodic current step of 300 μA is represented in Figure 2b. After approximately 3 s, a periodic pattern was observed, which was followed by an aperiodic oscillation as an indication of chaotic state.

In the mechanistic front, it is believed that the current at low potentials is controlled by the oxidative decomposition of adsorbed methanol and formation of COH_{ads} and CO_{ads} species [19-21]. It is conceivable [21] that three adjacent surface sites are initially required for the formation of COH_{ads} according to the following mechanism:

$$CH_3OH + Pt \rightarrow Pt\text{-}CH_3OH_{ads} \quad (3)$$

$$Pt\text{-}CH_3OH_{ads} \rightarrow Pt\text{-}CH_2OH_{ads} + H^+ + e \quad (4a)$$

$$Pt\text{-}CH_3OH_{ads} + OH^- \rightarrow Pt\text{-}CH_2OH_{ads} + H_2O + e \quad (4b)$$

$$Pt\text{-}CH_2OH_{ads} + Pt \rightarrow Pt_2{=}CHOH_{ads} + H^+ + e \quad (5a)$$

$$Pt\text{-}CH_2OH_{ads} + Pt + OH^- \rightarrow Pt_2{=}CHOH_{ads} + H_2O + e \quad (5b)$$

$$Pt_2{=}CHOH_{ads} + Pt \rightarrow Pt_3{\equiv}COH_{ads} + H^+ + e \quad (6a)$$

$$Pt_2{=}CHOH_{ads} + Pt + OH^- \rightarrow Pt_3{\equiv}COH_{ads} + H_2O + e \quad (6b)$$

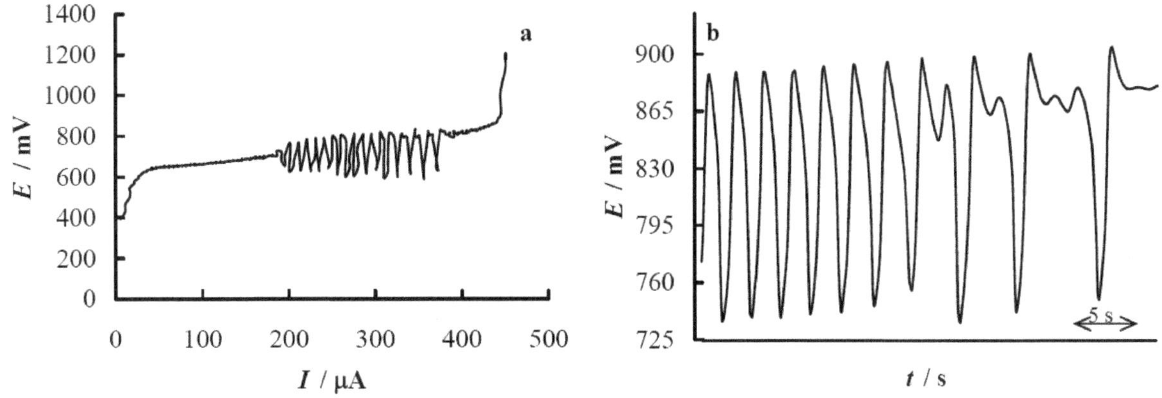

Figure 2. (a) *Onset of sustained potential oscillation in the galvanostatic scan on polycrystalline Pt in 0.5 M sulfuric acid comprised in 1 M methanol. Current sweep rate is 10 µA s^{-1},*
(b) *Potential-time response of methanol electrooxidation on polycrystalline Pt in 0.5 M sulfuric acid comprised 1 M methanol. Current step is 300 µA.*

Up to two of these sites, however, are subsequently liberated in the course of reactions yielding bridged or linearly bonded CO species [14]:

$$Pt_3{\equiv}COH_{ads} \rightarrow Pt_2{=}CO_{ads} + Pt + H^+ + e \tag{7a}$$

$$Pt_3{\equiv}COH_{ads} + OH^- \rightarrow Pt_2{=}CO_{ads} + Pt + H_2O + e \tag{7b}$$

$$Pt_3{\equiv}COH_{ads} + Pt\text{-}OH_{ads} \rightarrow Pt_2{=}CO_{ads} + 2Pt + H_2O \tag{8a}$$

$$Pt_3{\equiv}COH_{ads} + Pt\text{-}OH_{ads} \rightarrow Pt\text{-}CO_{ads} + 3Pt + H_2O \tag{8b}$$

Regeneration of active sites through the processes involving OH species can just partly promote continuation of methanol adsorption and more importantly, it seems that this regeneration needs three adjacent sites with a C3v symmetry as a pre-requisite [22]. Therefore, dissociative adsorption of methanol should follow a self-feeding mechanism, where surface diffusion of adsorbates accompanies flow of electrons into external circuit and subsequent regeneration of vacant active sites. Surface diffusion of reaction intermediates has also been proposed in other research article [22], especially for interpretation of higher electrocatalytic activities of alloys compared to pure platinum. In this regard, although CO is known to strongly adsorb, its surface mobility on platinum and some of its alloys has been reported [23-25]. Indeed, this feedback is more clearly visualized through the "cleaning" of the surface according to the following reactions [13]:

$$Pt_2{=}CO_{ads} + Pt\text{-}OH_{ads} \rightarrow PtCOOH_{ads} + 2Pt \tag{9}$$

$$Pt\text{-}COOH_{ads} + Pt\text{-}OH_{ads} \rightarrow 2Pt + CO_2 + H_2O \tag{10}$$

Overall, the anodic oxidation of methanol on unmodified polycrystalline platinum seems to be coupling of the faradaic processes and surface transport of adsorbates. This is characteristically a non-linear and dissipative process, which must be influenced by feedback. Non-linearity, dissipation and feedback in a dynamic system are clearly represented in the logistic map [26]. Figure 3a and 3b present the current-charge (time derivative of charge vs. charge) dependencies obtained from the anodic half cycle voltammograms of Figure 1. In order to make comparative plots like those in Figure 3, it was necessary to normalize the logistic equation by setting r=4, and to normalize other quantities by dividing them by their maximum values. It has been observed that data was fitted the logistic differential equation quite well (with the correlation coefficient of 0.98), what indicates chaotic nature of the MEO dynamics. Chaotic switching in the oxidation of methanol carried out under the regime of cyclic voltammetry has already been observed by Schell and Cai [27]. Figures 4a and 4b represent the current-charge dependencies of the cathodic half cycle measurements and the related attempted logistic equations. No fit has been observed. Apparently, in the cathodic half cycle the initial partial desorption of hydroxide ions surely promotes further dissociation and oxidation of methanol. However, in the further cathodic domains these two effects are no longer cooperative They are virtually independent and merely controlled by the potential itself with no feedback imposed by presence of one upon another. Consequently, no fit to the logistic equation with one degree of freedom is expected. MEO on polycrystalline platinum electrodes in both alkaline and acidic media seems to follow the non-linear electrocatalytic dynamics, which is characterized by regeneration of active sites in the course of an anodic potential sweep. This process complies with the logistic differential equation. It seems that the processes ensued in the cathodic half cycle of the potential sweep are not controlled by the feedback mechanisms and do not fit the logistic equation.

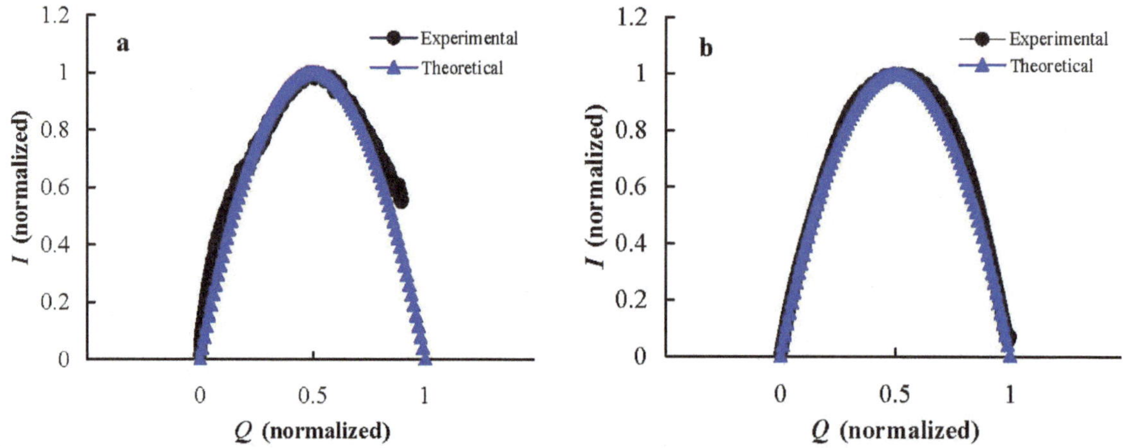

Figure 3. *Time derivative of charge (current) vs. charge dependencies of the anodic half cycle voltammograms of MEO on polycrystalline Pt, normalized with respect to corresponding maximum values and theoretical curves:* ***(a)*** *acidic solution,* ***(b)*** *alkaline solution*

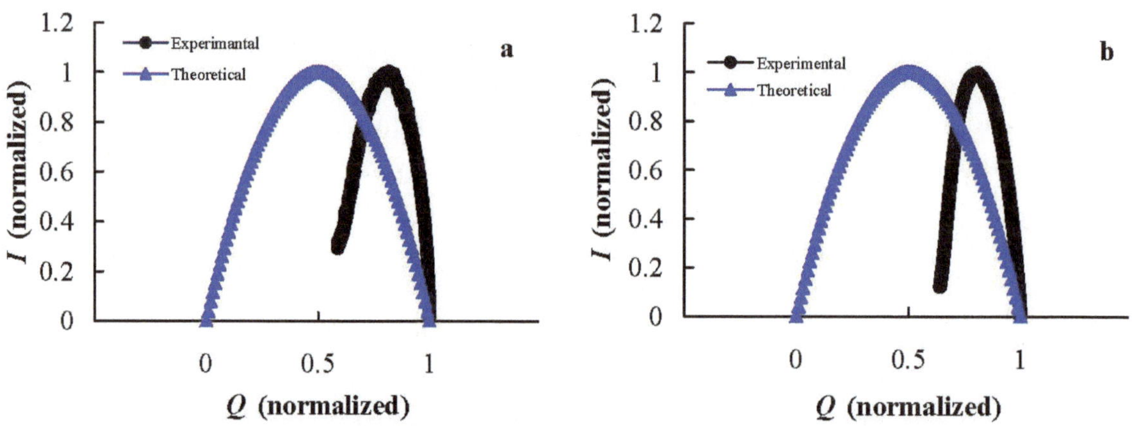

Figure 4. *Current-charge dependencies of the cathodic half cycle voltammograms of MEO and the related logistic maps:* ***(a)*** *acid solution,* ***(b)*** *alkaline solution.*

Conclusion

The electro-oxidation of methanol on platinum was analyzed by quadratic logistic equation indicating domination of two different behaviors in the anodic and cathodic (reverse) half cycles. In the anodic sweep, positive feedback controlled the reaction kinetics, while, the reverse sweep did not obey the logistic map. The results can be of importance in the analysis of the output of the hydrocarbon based fuel cells and the approach is expandable to other direct electrode processes.

References

[1] L. Lin, Q. Zhu, A. Xu, *Prog. Chem.* **27** (2015) 1147-1157. http://dx.doi.org/10.1016/j.nanoen.2013.06.009

[2] H. Bahrami, A. Faghri, *J. Power Sources* **230** (2013) 303-320. http://dx.doi.org/10.1016/j.jpowsour.2012.12.009

[3] H. Hoster, T. Iwasita, H. Baumgartner, W. Vielstich, *Phys.* Chem. *Chem. Phys.* **3** (2001) 337-346. http://dx.doi.org/10.1039/B004895J

[4] W. F. Lin, P. A. Christensen, A. Hamnet, Phys. *Chem. Chem. Phys.* **3** (2001) 3312-3319. http://dx.doi.org/10.1039/B102699M

[5] Z. Tang, G. Lu, *Prog. Chem.* **19** (2007) 1301-1312. http://www.progchem.ac.cn/EN/Y2007/V19/I9/1301

[6] J. M. Leger, *J. Appl.* Electrochem. **31** (2001) 767-771. http://dx.doi.org/10.1023/A:1017531225171

[7] H. Varela, R. M. Torresi, E. R. Gonzalez, *Quim. Nova* **25** (2002) 99-106. http://dx.doi.org/10.1590/S0100-40422002000100017

[8] R. M. Torresi, S. I. C. de Torresi, E. R. Gonzalez, *J. Electroanal. Chem.* **461** (1999) 161-166. http://dx.doi.org/10.1016/S0022-0728(98)00069-2

[9] E. R. Gonzalez, *J. Electrochem. Soc.* **143** (1996) L113-L115. http://dx.doi.org/10.1149/1.1836893

[10] S.N. Rasband, *Chaotic dynamics* of *nonlinear systems*, Wiley, New York, 1990.

[11] K. Malek, F. Gobal, *Synth. Met.* **113** (2000) 167-171. http://dx.doi.org/10.1016/S0379-6779(00)00194-6

[12] A. Bandura, *American Psychologist* **37** (1982) 747-755. http://psycnet.apa.org/doi/10.1037/0003-066X.37.7.747

[13] A. Hamnett, *Catal. Today* **38** (1997) 445-457. http://dx.doi.org/10.1016/S0920-5861(97)00054-0

[14] R. Manoharan, J. Prabhuram, *J. Power Sources* **96** (2001) 220-225. http://dx.doi.org/10.1016/S0378-7753(00)00683-2

[15] J. Lee, C. Eickes, M. Eiswirth, G. Ertl, *Electrochim. Acta* **47** (2002) 2297-2301. http://dx.doi.org/10.1016/S0013-4686(02)00075-0

[16] B.E. Conway, T.C. Liu, *Langmuir* **6** (1990) 268-276. http://dx.doi.org/10.1021/la00091a044

[17] S. Chen, M. Schell, *J. Electroanal. Chem.* **478** (1999) 108-117. http://dx.doi.org/10.1016/S0022-0728(99)00421-0

[18] R. Parsons, T. VanderNoot, *J. Eelctroanal. Chem.* **257** (1988) 9-45. http://dx.doi.org/10.1016/0022-0728(88)87028-1

[19] T. Iwasita, W. Vielstich, E. Santos, *J. Electroanal. Chem.* **229** (1987) 367-376. http://dx.doi.org/10.1016/0022-0728(87)85153-7

[20] B. Beden, F. Hahn, S. Juanto, C. Lamy, J. M. Leger, *J. Electroanal. Chem.* **225** (1987) 215-225. http://dx.doi.org/10.1016/0022-0728(87)80015-3

[21] S. Wasmus, A. Kuver, *J. Electroanal. Chem.* **461** (1999) 14-31. http://dx.doi.org/10.1016/S0022-0728(98)00197-1

[22] H. A. Gasteiger, N. Markovic, P.N. Ross, E. J. Cairns, *J. Phys. Chem.* **97** (1993) 12020-12029. http://pubs.acs.org/doi/abs/10.1021/j100148a030

[23] V. J. Kwasniewski, L. D. Schmidt, *Surf. Sci.* **274** (1992) 329-340. http://dx.doi.org/10.1016/0039-6028(92)90838-W

[24] T. Kobayashi, P. K. Babu, L. Gancs, J. H. Chung, E. Oldfield, A. Wieckowski, *J. Am. Chem. Soc.* **127** (2005) 14164-14165. http://dx.doi.org/10.1021/ja0550475
[25] P. K. Babu, J. H. Chung, S. T. Kuk, T. Kobayashi, E. Oldfield, A. Wieckowski, *J. Phys. Chem.* **109** (2005) 2474-2477. http://dx.doi.org/10.1021/jp040729k
[26] R. C. Hilborn, *Chaos and Nonlinear Dynamics*, Oxford University Press, Oxford, 2000.
[27] M. Schell, X. Cai, *Electrochim. Acta* **38** (1993) 519-527. http://dx.doi.org/10.1016/0013-4686(93)85007-L

Permissions

All chapters in this book were first published in JESE, by IAPC Publishing; hereby published with permission under the Creative Commons Attribution License or equivalent. Every chapter published in this book has been scrutinized by our experts. Their significance has been extensively debated. The topics covered herein carry significant findings which will fuel the growth of the discipline. They may even be implemented as practical applications or may be referred to as a beginning point for another development.

The contributors of this book come from diverse backgrounds, making this book a truly international effort. This book will bring forth new frontiers with its revolutionizing research information and detailed analysis of the nascent developments around the world.

We would like to thank all the contributing authors for lending their expertise to make the book truly unique. They have played a crucial role in the development of this book. Without their invaluable contributions this book wouldn't have been possible. They have made vital efforts to compile up to date information on the varied aspects of this subject to make this book a valuable addition to the collection of many professionals and students.

This book was conceptualized with the vision of imparting up-to-date information and advanced data in this field. To ensure the same, a matchless editorial board was set up. Every individual on the board went through rigorous rounds of assessment to prove their worth. After which they invested a large part of their time researching and compiling the most relevant data for our readers.

The editorial board has been involved in producing this book since its inception. They have spent rigorous hours researching and exploring the diverse topics which have resulted in the successful publishing of this book. They have passed on their knowledge of decades through this book. To expedite this challenging task, the publisher supported the team at every step. A small team of assistant editors was also appointed to further simplify the editing procedure and attain best results for the readers.

Apart from the editorial board, the designing team has also invested a significant amount of their time in understanding the subject and creating the most relevant covers. They scrutinized every image to scout for the most suitable representation of the subject and create an appropriate cover for the book.

The publishing team has been an ardent support to the editorial, designing and production team. Their endless efforts to recruit the best for this project, has resulted in the accomplishment of this book. They are a veteran in the field of academics and their pool of knowledge is as vast as their experience in printing. Their expertise and guidance has proved useful at every step. Their uncompromising quality standards have made this book an exceptional effort. Their encouragement from time to time has been an inspiration for everyone.

The publisher and the editorial board hope that this book will prove to be a valuable piece of knowledge for researchers, students, practitioners and scholars across the globe.

List of Contributors

Jyoti T. Bagalkoti, Vijay P. Pattar and Sharanappa T. Nandibewoor
P.G. Department of Studies in Chemistry, Karnatak University, Dharwad 580 003, India

Balaji Rengarajan, Ramya Krishnan and Kaveripatnam Samban Dhathathreyan
Centre for Fuel Cell Technology, International Advanced research Centre for Powder Metallurgy and New Materials(ARCI), II Floor, IIT-M Research Park, No.6 Kanagam road, Taramani, Chennai - 600 113, India

Seetharaman Swaminathan
Centre for Fuel Cell Technology, International Advanced research Centre for Powder Metallurgy and New Materials(ARCI), II Floor, IIT-M Research Park, No.6 Kanagam road, Taramani, Chennai - 600 113, India
Department of Chemical Engineering, A.C. College of Technology, Anna University, Chennai 600 025, India

Manickam Velan
Department of Chemical Engineering, A.C. College of Technology, Anna University, Chennai 600 025, India

Dasari Rekha, Puthalapattu Reddy Prasad and Neelam Yughandhar Sreedhar
Department of Chemistry, Sri Venkateswara University, Tirupati-517502, A.P, India

Dimitra Vernardou
Center of Materials Technology and Photonics, School of Engineering, Technological Educational Institute of Crete, 710 04 Heraklion, Crete, Greece

Antonia Bei
Department of Mechanical Engineering, School of Engineering, Technological Educational Institute of Crete, 710 04 Heraklion, Crete, Greece

Dimitris Louloudakis
Center of Materials Technology and Photonics, School of Engineering, Technological Educational Institute of Crete, 710 04 Heraklion, Crete, Greece
Department of Physics, University of Crete 711 00 Heraklion, Crete, Greece

Nikolaos Katsarakis
Center of Materials Technology and Photonics, School of Engineering, Technological Educational Institute of Crete, 710 04 Heraklion, Crete, Greece
Department of Electrical Engineering, School of Engineering, Technological Educational Institute of Crete, 710 04 Heraklion, Crete, Greece
Institute of Electronic Structure & Laser, Foundation for Research & Technology- Hellas, P.O. Box 1527, Vassilika Vouton, 711 10 Heraklion, Crete, Greece

Emmanouil Koudoumas
Center of Materials Technology and Photonics, School of Engineering, Technological Educational Institute of Crete, 710 04 Heraklion, Crete, Greece
Institute of Electronic Structure & Laser, Foundation for Research & Technology- Hellas, P.O. Box 1527, Vassilika Vouton, 711 10 Heraklion, Crete, Greece

Mouhammed Khateeb
Department of Basic Sciences, Faculty of Mechanical and Electrical Engineering, University of Al-Baath, Homs, Syria

Basheer Elias and Hazar Alksair
Department of Chemistry, Faculty of Sciences, University of Al- Baath Homs, Syria

Denys Gromadskyi, Volodymyr Chervoniuk and Sviatoslav Kirillov
Joint Department of Electrochemical Energy Systems, 38A Vernadsky Ave, 03680 Kyiv, Ukraine

Kasaram Roja, Puthalapattu Reddy Prasad and Neelam Yugandhar Sreedhar
Electroanalytical Lab, Department of Chemistry, Sri Venkateswara University, Tirupati-517 502, A. P., India

Punyasamudram Sandhya
Department of Chemistry, Sri Padmavati Mahila Visvavidyalayam, Tirupati-517 502, A. P., India

Julijana Velevska, Nace Stojanov and Margareta Pecovska-Gjorgjevich
Institute of Physics, Faculty of Natural Sciences and Mathematics, University Sts Cyril and Methodius, Arhimedova 3, 1000 Skopje, Republic of Macedonia

Metodija Najdoski
Institute of Chemistry, Faculty of Natural Sciences and Mathematics, University Sts Cyril and Methodius, Arhimedova 5, 1000 Skopje, Republic of Macedonia

Pavithra M. Krishnegowda, Venkatesha T. Venkatarangaiah, Anantha N. Subba Rao and Shubha H. Nataraj
Department of Chemistry, Kuvempu University, Shankaraghatta-577451, Shimoga, Karnataka, India

Punith Kumar M. Krishnegowda
Department of Materials Engineering, Indian Institute of Science, Bangalore-560012, Karnataka, India

Shreekant M. Patil, Vijay P. Pattar and Sharanappa T. Nandibewoor
P. G. Department of Studies in Chemistry, Karnatak University, Dharwad 580003, India

Tiago Brandão Costa
Programa de Pós-Graduação em Engenharia Metalúrgica (PPGEM), Escola de Engenharia Industrial Metalúrgica de Volta Redonda (EEIMVR), Universidade Federal Fluminense (UFF), 27255-125 Volta Redonda, RJ, Brazil

Tania Maria Cavalcanti Nogueira
Programa de Pós-Graduação em Engenharia Metalúrgica (PPGEM), Escola de Engenharia Industrial Metalúrgica de Volta Redonda (EEIMVR), Universidade Federal Fluminense (UFF), 27255-125 Volta Redonda, RJ, Brazil
Departamento de Metalurgia, EEIMVR, UFF, 27255-125 Volta Redonda, RJ, Brazil

Ladário da Silva
Programa de Pós-Graduação em Engenharia Metalúrgica (PPGEM), Escola de Engenharia Industrial Metalúrgica de Volta Redonda (EEIMVR), Universidade Federal Fluminense (UFF), 27255-125 Volta Redonda, RJ, Brazil
Departamento de Física, Instituto de Ciências Exatas (ICEx), UFF, 27213-145 Volta Redonda, RJ, Brazil

Salwa Fares Rassi
Department of Chemistry, Faculty of Science, Al-Baath University, Homs, Syria

Francisco Prieto-García, Judith Callejas-Hernández and Yolanda Marmolejo-Santillán
Área Académica Química, Universidad Autónoma del Estado de Hidalgo. Carret. Pachuca-Tulancingo, km. 4.5, C. P. 42076. Mineral de la Reforma. Hidalgo. México

Judith Prieto-Méndez
Área Académica de Agronomía, Universidad Autónoma del Estado de Hidalgo. Carret. Pachuca-Tulancingo, km. 4.5, C. P. 42076. Mineral de la Reforma. Hidalgo. México

Abideen Idowu Adeogun
Department of Chemistry, Federal University of Agriculture, Abeokuta, Nigeria
Electrochemical Pollution Control Division, CSIR - Central Electrochemical Research Institute. Karaikudi 630006, India

Ramesh Babu Balakrishnan
Electrochemical Pollution Control Division, CSIR - Central Electrochemical Research Institute. Karaikudi 630006, India

Said M. Sayyah and Fatma Mohamed
Polymer Research Laboratory, Chemistry Department, Faculty of Science, Beni-Suef University 62514, Beni-Suef City, Egypt

Sayed S. Abd-Elrehim
Chemistry Department, Faculty of Science, Ain Shams University, 11566 Abbassia, Cairo, Egypt

Rehab E. Azooz
Chemistry Department, Faculty of Science, Ain Shams University, 11566 Abbassia, Cairo, Egypt
Chemistry Department, Faculty of Science, Jazan University, 2097 Jazan, Saudi Arabia

Rehab E. Azooz
Chemistry Department, Faculty of Science, Jazan University, 2097 Jazan, Saudi Arabia

Abdelaziz Ait Sidi Mou, Aicha Ouarzane and Mama El Rhazi
Laboratory of Physical Chemistry and Bioorganic Chemistry Mohammedia (FSTM), Morocco

Borislav N. Malinovic and Tijana Djuricic
University of Banja Luka, Faculty of Technology, Stepe Stepanovića 73, 78000 Banja Luka, Bosnia and Herzegovina

Miomir G. Pavlovic
University of Belgrade, ICTM-CEH, Njegoševa 12, Belgrade, Serbia

Hossein Heli
Nanomedicine and Nanobiology Research Centre, Shiraz University of Medical Sciences, Shiraz, Iran

Fereydoon Gobal
Department of Chemistry, Sharif University of Technology, Tehran, Iran

Index